Praise for *Call of the Reed Warbler*

"Part lyrical nature writing, part storytelling, part solid scientific evidence, part scholarly research, part memoir, [this] book is an elegant manifesto, an urgent call to stop trashing the Earth and start healing it."

— *THE GUARDIAN*

"Charles Massy has written a definitive masterpiece that takes its place along with the writings of Aldo Leopold, Wendell Berry, Masanobu Fukuoka, Humberto Maturana, and Michael Pollan. No work has more brilliantly defined regenerative agriculture and the breadth of its restorative impact upon human health, biodiversity, climate, and ecological intelligence. There is profound insight here, realized by thirty-five years of farming on the ancient, fragile soils of the Australian continent, discernment expressed with exquisite clarity, seasoned wisdom, and some breathtaking prose of poetic elegance. I believe it takes its place as the single most important book on agriculture today, one that will become a classic text."

— PAUL HAWKEN, author of *Blessed Unrest*; editor of *Drawdown*

"I first met Charles Massy in 2015 when he visited the ranch of the Africa Centre for Holistic Management in Zimbabwe. Building on the work of many people, Massy has now written a compelling and comprehensive book on the importance of management being holistic — and how that will ultimately lead to a regenerative agriculture capable of restoring even the most degraded ecosystems and marginalized land in any climate and at any scale. He has done this with wonderful stories that take us on a journey of ecological literacy, supported by evocative insights into landscapes, science, and practical farming and living. *Call of the Reed Warbler* is a massive accomplishment and contribution to our collective work of building a new agriculture, a new Earth, and renewed human society and health."

— ALLAN SAVORY, president of the Savory Institute

"This book will change the way you think about food, farming, and the place of humans on the planet. Introducing us to leaders of the regenerative agriculture movement, Massy offers real hope that we may yet fashion a society that gives more than it takes."

— LIZ CARLISLE, author of *Lentil Underground*; lecturer,
School of Earth, Energy, and Environmental Sciences, Stanford University

"Conceptually rich and filled with examples of diverse innovators, *Call of the Reed Warbler* is the most comprehensive and engaging book I've read on regenerative agriculture. Charlie Massy contends humans have morphed from an 'Organic mind' into a 'Mechanical mind,' which is now evolving into an 'Emergent mind' — a change in consciousness that embraces self-organizing processes. He shows how the minds of the innovators in his book were opened to three key processes: First, they began to understand how landscapes function, how ecological system work, and how they are indivisibly connected. Second, they got out of the way to let nature repair, self-organize, and regenerate these functions. Third, they had the humility to 'listen to their land,' change, and continue to learn with that same openness. Massy concludes we can heal Earth, but only by transforming ourselves and our connections with the landscapes and communities in which we live. This book is a thoughtful step in that direction."

— FRED PROVENZA, professor emeritus, Department of Wildland Resources, Utah State University; author of *Nourishment*

"Charles Massy is a leader in the regenerative agriculture movement in Australia with a message of hope for everyone. Using his arid homeland as a touchstone, Massy thoughtfully counterbalances the damage done by industrial agriculture to our land and our prospects with evocative examples from around the world of a hopeful way forward. His beliefs are grounded in practical experience, his vision clear, and his words inspiring. *Call of the Reed Warbler* is a must-read!"

— COURTNEY WHITE, author of *Grass, Soil, Hope*

"*Call of the Reed Warbler* not only heralds the sound of an ecosystem functioning but also of a world awakening to regenerative agriculture. Charlie Massy is Australia's equivalent to Thoreau and Leopold and a practical regenerative farmer to boot. I can't think of anyone better equipped to pen a book like this, and to do so with such scholarship, integrity, and rollicking prose is a credit to Charlie and those whose journey he's portrayed. Easily my 'Book of the Year.'"

— DARREN J. DOHERTY, founder, Regrarians Limited

CALL
OF THE
REED
WARBLER

CALL
OF THE
REED
WARBLER

A NEW AGRICULTURE
A NEW EARTH

CHARLES
MASSY

Chelsea Green Publishing

White River Junction, Vermont | London, UK

Originally published in 2017 as *Call of the Reed Warbler: A New Agriculture, A New Earth*
by University of Queensland Press, PO Box 6042, St Lucia, Queensland 4067 Australia.

This edition published by Chelsea Green Publishing, 2018.

Project Manager: Alexander Bullett
Project Editor: Brianne Goodspeed
Proofreader: Angela Boyle
Indexer: Ruth Satterlee
Designer: Melissa Jacobson
Cover photo by Trisha Dixon

Printed in the United States of America.
First printing August, 2018.
10 9 8 7 6 5 4 3 2 1   18 19 20 21 22

Our Commitment to Green Publishing
Chelsea Green sees publishing as a tool for cultural change and ecological stewardship. We strive to align our book manufacturing practices with our editorial mission and to reduce the impact of our business enterprise in the environment. We print our books and catalogs on chlorine-free recycled paper, using vegetable-based inks whenever possible. This book may cost slightly more because it was printed on paper that contains recycled fiber, and we hope you'll agree that it's worth it. Chelsea Green is a member of the Green Press Initiative (www.greenpressinitiative.org), a nonprofit coalition of publishers, manufacturers, and authors working to protect the world's endangered forests and conserve natural resources. *Call of the Reed Warbler* was printed on paper supplied by Thomson-Shore that contains 100% postconsumer recycled fiber.

Library of Congress Cataloging-in-Publication Data
Names: Massy, Charles, author.
Title: Call of the reed warbler : a new agriculture, a new earth / Charles Massy.
Description: White River Junction, Vermont : Chelsea Green Publishing, 2018.
| Includes bibliographical references and index.
Identifiers: LCCN 2018019459| ISBN 9781603588133 (pbk.) | ISBN 9781603588140 (ebook)
Subjects: LCSH: Sustainable agriculture--Australia. | Food supply--Australia.
Classification: LCC S478.A1 M37 2018 | DDC 338.10994--dc23
LC record available at https://lccn.loc.gov/2018019459

Chelsea Green Publishing
85 North Main Street, Suite 120
White River Junction, VT 05001
(802) 295-6300
www.chelseagreen.com

CONTENTS

FOREWORD

The watchword of this remarkable book is "landscape function." Perhaps it sounds inscrutable, at least at first. Yet the phrase is at the heart of understanding and repairing the strained human relationship with our planet, especially when it comes to feeding ourselves. Charles Massy's analysis is detailed, meticulous and, quite literally, microscopic, and, at the same time, large scale, holistic and ecosystems-based. And it's precisely what's needed at this moment.

While some may initially think Massy's work is too specific to Australia to shine light on the agricultural and food system challenges we face here in the United States, this book is uniquely valuable precisely because of its specificity. The people and farms described show how recreating vibrant living landscapes is possible in even some of the world's most inhospitable terrain. Moreover, large portions of the globe, including much of the western half of the United States, have a similar climate. The overarching lessons and messages of this book — *the essential focus on landscape function* — are wholly applicable, regardless of the climate in which one resides.

In teaching us to pay attention to and understand how a landscape functions, Charles Massy describes ancient natural systems and humans' age-old understanding of them. It's only in recent centuries that our connection with and comprehension of these systems have been discarded. Thankfully, as the examples of the agricultural pioneers in this book illustrate, we are entering a reawakening phase. Modern science seems to be striving to catch up where it has fallen behind in understanding overlooked realms, such as the vast subterranean life in our soils.

A virtually unnoticed issue at the dawn of the new millennium, the way America eats and farms and how those things relate to each other and to our health has become a hot topic. School gardens and cooking classes are now widespread. Farmers markets, community-supported agriculture (CSAs), and farm-to-table restaurants are sprouting up everywhere. Even a plethora of new apps and other technologies have been developed to help farmers connect with customers and assist consumers in identifying food ingredients and supply chains. Most hopeful of all, a burgeoning group of men and women in their twenties and thirties are apprenticing on farms and ranches while finding creative ways to access capital and land for grazing and cultivation. We have overcome the problem of ignoring the problem.

At the same time, however, a worrying tendency to vastly oversimplify the intricacies of food and agriculture has emerged. This is especially true with respect to farm animals and the foods derived from them. Where Americans had always been advised to include ample milk, eggs and meat in their diets, they are increasingly being told to shy away from such foods. Nominally, this is because we are consuming more than we have previously and have now reached an unhealthy over-saturation point.

This is at odds, however, with official federal data about what Americans are actually eating. Obesity, diabetes and related chronic diseases have steadily risen over the past three decades while our consumption of animal fats, eggs, whole milk and red meats has fallen. What has risen, however, is the amount of processed foods (mostly carbohydrates) and refined vegetable oils we eat and the volume of sugary beverages we drink.

Nutrition writers like Gary Taubes have irrefutably documented the intimate connection of these trends. As Americans have reduced nutrient dense, real foods like meat, whole milk, and eggs, they are battling a chronic sensation of feeling unsatiated, even after eating. Sugar has been shown to be addictive, and it stimulates the appetite for still more sugar while being strongly linked with fatty liver disease and related negative metabolic effects. Meanwhile, the official US Department of Agriculture food pyramid (now called MyPlate), which advises Americans on what to eat, urges us to cut back our calories (suggesting our lack of willpower as the primary culprit for ever-expanding waistlines) and to choose "low-fat and fat-free dairy" products. Such messages reinforce the impression that animal fats are what ails us.

Lately, advice to cut back on animal-based foods has included corresponding environmental admonitions from various quarters. "Eat less meat" is advice that now saturates the messages from public health, medical and environmental organizations. The attorney heading up the food and agriculture program for the nonprofit the Natural Resources Defense Council told me in 2016 that based on concerns about water and air quality, resource usage, and, especially, climate change, the group's *primary* public advocacy message was "eat less meat."

"But how," I pressed repeatedly in a series of emails and conversations, "does that help us build a more ecologically resilient food system?" He never supplied a cogent answer.

In fact I have yet to hear anyone explain the means by which the "eat less meat" mantra actually works on the ground. How does it enhance *landscape function*, if you will, and bring about the truly regenerative, resilient agriculture that humanity must forge to continue life on this planet over the long term?

There is no good answer to this question because the opposite is actually true: The *absence* of animals on our landscapes is a key part of what's wrong with modern agriculture. Animals, plants and fungi are always woven together in nature in a complex and sumptuous tapestry.

I share criticisms of industrial systems, which function like factories, for the ways in which they raise farm animals, treating them as widgets. They remove animals from the land, concentrate them in inhumane and inherently polluting systems, and separate them from both their incoming food sources and their outgoing wastes, all of which should be part of a natural continuum. Industrial animal operations endlessly ignore and disrupt the natural connectedness of all living things.

When animals are present on landscapes they are usually managed in ways that fail to optimise potential ecosystem benefits.

But the failings of modern agriculture extend well beyond its treatment of animals. Humans cultivate plants in vast, uniform crop fields found nowhere in nature. Even most of the US organic production is taking place at large, non-diverse operations that bear no resemblance to nature's tangled webs. Our methods for generating food are fundamentally broken.

What, then, must we do? We must, as Charles Massy so beautifully lays out in this book, start by learning to understand how our lands and our waters are actually meant to function. Then we must creatively design agriculture — always tailored to specific climates, topographies and human capacities — that allows us to generate food while working *with* rather than *against* the ways these lands and waters are intended to work.

Farm animals are not the enemies in this process. They are actually indispensible allies. Animal impact is essential to restoring optimal landscape function. Or, as my friend, scientist Russ Conser, has brilliantly quipped (I now have a T-shirt that says this): "It's not the COW, it's the HOW!"

It is my deepest hope that this book will be widely read, understood and applied. We have not a moment to lose.

NICOLETTE HAHN NIMAN
May 8, 2018
Bolinas, California University

The history of social movements is a history of people operating in the cracks of superstructures. Of using the energies generated at the margins of systems and organizations. Of exercising considerable imagination, critical thinking, subversion and undutiful behaviour to destabilize and de-construct the authority of the inevitable... Taking back control and joining with others in collective action to achieve change is at the root of concepts like participation and democracy. It finds its impetus in human agency and can transform people's lives. As well as transforming views about oneself.

J. L. THOMPSON, 'Really Useful Knowledge'[1]

Since our break with nature came with agriculture, it seems fitting that the healing of culture begin with agriculture, fitting that agriculture take the lead.

WES JACKSON, *Becoming Native to This Place*[2]

INTRODUCTION

From the Ground Up

The dialectical or ecological approach asserts that creating the world is involved in our every act.

WES JACKSON, *Becoming Native to This Place*[1]

I t is late August, 4 am. I have broken the ice and begun writing this book. An appropriate metaphor, ice, as a hard frost embosses the earth, trees, the bleached-straw grass. As I walk to my farm-shed-cum-office, the Milky Way curves above me, each star pulsing sharply in the clear air, the Southern Cross slowly turning on its elbow. I can see the dark shape of what some Aboriginal nations call the 'emu in the sky', the emu's head nestling below the Cross. As my feet crunch the grass, a male magpie somewhere in the canopy of a candlebark gum maintains a gentle, melodious night-warbling.

Holding a pen, I ponder the clarity of this night-world and the journey I wish to relate. For this is partly a story about country. About my own country: country on which my family have lived for five generations and I for most of my life; country on which I grew up, running barefoot the long day, gambolling with poddy lambs, tadpoling and exploring thick bush. It is country on which I discovered the natural world and a lifelong interest in birds, mammals, reptiles and other creatures, along with their vegetative and subterranean abodes, on which I trapped and banded birds in study, collected butterflies and spot-lit owls and gliding marsupials, and on which I learnt to track wallabies and kangaroos, and to flush secretive nightjars from peppermint gum hollows.

But also country on which I learnt to hunt with gun and rifle; to fish for trout in nearby streams (first with worms, then fly); to milk a cow; to cut and split wood; to kill and butcher a sheep or steer; to muster livestock (sheep, cattle, goats); to ride and jump a horse; to drive farm trucks, tractors and motorbikes; to start diesel engines and pumps; and to use a shearing handpiece (badly).

And yet, though intimately my country, I came to realise that for a long time I didn't fully understand it. Consequently, at times, I caused immense damage to this country – in some paddocks, perhaps at least a few thousand years worth. I now know that if we want to profitably manage, nurture and regenerate country, then we need to fathom where it came from and how, what it is made of, how it works and functions, how it was managed before us, what organisms and vegetation reside on it, and how they in turn function and play a role. However, while the narrative that follows reveals snippets of my personal journey of ecological enlightenment, its core is the remarkable awakening of a new regenerative agriculture in Australia and elsewhere.

This book is about a new agriculture and the implications of landscape management: about healthy and unhealthy food and how modern industrial agriculture and the connected food system is not just poisoning us but is also confoundingly making us obese while starving us at the same time; about the potentialities of rural and urban societies and human well-being; and about the actions of seemingly ordinary farmers extraordinarily regenerating their landscapes. It is about a group of farmers and others providing answers to Paul Hawken's evocative question in his book *Blessed Unrest*: 'What would a world feel like that created solutions from the ground up?'[2]

And so, in the end, this book is about the future survival of Earth and humanity. This is a story for our story-loving species at this most critical of all moments in its entire time on Earth.

Our species is labelled *Homo sapiens sapiens* – allegedly for being doubly wise. To the contrary, we have for many centuries behaved in a self-destructive manner. Initially this behaviour was unwitting, but today it is not.

There is abundant evidence now that modern industrial agriculture and humanity's ongoing burning of stored fossil fuels is destroying Earth's life-sustaining systems, poisoning the foods we live on and divorcing us from a natural world we coevolved with. While consuming more resources than Earth's systems can replenish, we are hurtling towards multiple calamities. We must urgently and dramatically address the root causes of these impending disasters. This book is full of stories that show us how.

A widespread consensus is building among scientists that the Earth has moved out of the favourable Holocene period[3] and into a new, human-shaped geological epoch: the Anthropocene. Thus, for the first time, one species – humanity – is influencing (and may well determine) the future health and survival of life systems on this planet.[4]

What has precipitated the Anthropocene is that humanity has shifted from an organic, ever-renewing economy to one that is overwhelmingly extractive. This is because, as the late, brilliant Jesuit ecological thinker Thomas Berry pointed out, we now continue to 'see ourselves as a transcendent mode of being. We don't really belong here anymore.'[5]

Earth scientists such as Johan Rockström and Will Steffen have proposed a framework of nine 'planetary boundaries' within which humanity can safely operate. Already three if not four of these boundaries appear to have been crossed: climate change, rate of biodiversity loss and the distortion of key biogeochemical markers (namely the interconnected global nitrogen and phosphorus cycles). Of the remaining boundaries, we are rapidly approaching (or about to cross) a further three: ocean acidification, global fresh water use and changes to land cover and its systems by conversion to cropland and other uses.[6] As historian John McNeill put it, 'humankind has begun to play dice with the planet, without knowing all the rules of the game'.[7]

The expansion of industrial agriculture since 1950 is seen as a major contributor to this predicament. Thus agriculture is clearly vital in any attempts at addressing the imminent challenge to planetary survival. If we are to preserve our living systems, then new and transformative responses will be required. Much else is at stake – not least human physical and mental health. The current regenerative agriculture practices and thinking described in this book are all about life-saving transformation, and transformation not just in practice but also within the minds and hearts of individuals: farmers, consumers and decision-makers.[8]

Agriculture, in occupying thirty-eight per cent of the Earth's terrestrial surface, is both the largest user of land on the planet and humankind's largest engineered ecosystem. Because it is based on plants, which take carbon out of the atmosphere to make and store sugar through photosynthesis, and because these plants have roots growing in the ground, a healthy agriculture has the potential to bury huge amounts of carbon for long periods. Excessive carbon dioxide in the atmosphere (largely due to the release of long-stored carbon through burning fossil fuels) is, we now know, one of the key causes of the greenhouse effect. Absorption of excess carbon dioxide also makes the world's oceans more acidic.

Moreover, when a healthy agriculture puts more long-lasting carbon into the soil while minimising the loss of such carbon, this in turn has a major impact on the water cycle and its crucial role in thermoregulation (i.e. climate control) of our planet. But the problem is that traditional industrial agriculture – through practices such as burning vegetation for land clearing, using fossil fuels (in fertilisers and chemicals, and to power farm machinery), overgrazing, ploughing and fallowing – emits, rather than stores, carbon.

But it need not be this way. An ecologically and socially enhancing agriculture – what I call 'regenerative agriculture' – can reverse this harmful carbon-emitting signature of industrial agriculture. It can do this via various methods, but all are based around revegetation and inculcating healthy, living soils (that is, soil containing plants, insects, bacteria, fungi and other organisms). The practice of regenerative agriculture across vast swathes of the world's farmlands, grasslands, marginal savannahs and arid regions can provide a major solution to climate change as well as confronting the other key Anthropocene planetary boundary challenges.

However, the trashing of landscapes and life-supporting systems is not the only negative impact of modern industrial agriculture. Another is the way it produces food and then how it processes, distributes, markets and sells it. At the same time as we are degrading the air we breathe, we are also denaturing the food we eat and water we drink and lacing them with a witch's brew of deadly poisons. So the confronting truth is that much of the modern threats to human health, including the near exponential increase in modern chronic diseases, is due to our industrial systems and especially those embedded in industrial agriculture.

What is alarming about this modern health disaster is that four of its major causative or exacerbating factors are not on the public or political radar. The first two relate to the fact that our bodies largely coevolved with our environment when we were hunter-gatherers. Consequently, we are programmed to lay down fat in our bodies so as to survive lean times. Modern industrial diets and lifestyles mean we maladaptively keep putting on fat (especially chronic, disease-causing internal fat). The second factor is that we also became adapted to eat a widely varied hunter-gatherer diet (and one packed with hundreds of different nutrients). In modern industrial society, we are only fed a minuscule number of these crucial nutrients. When the decline in nutritional value of modern-day foods is added to this, the health consequences are awful.

The third causative factor in our modern health crisis is that the world's most widely used agrochemical – glyphosate – is penetrating modern foods. It is now being implicated as a primary causal factor in many of today's major diseases. Finally, there is the effect of epigenetics, the newly apprehended but little-known ways in which genetic mechanisms are switched on or off by the surrounding environment. All four factors are essentially invisible until they create damage, and when they do, the effects of the first three are amplified by the fourth, especially in future generations.

What makes this quartet of factors so deadly is that we are genetically hard-wired to live off our natural environment. This includes eating food that is free of man-made chemicals. However, while we can't change this genetic wiring, we can change our landscapes, and thus the food and water that they supply. This is the remarkable story I wish to relate in this book because, by shifting away from modern, chemically based industrial agriculture to one that regenerates Mother Earth and its systems, we address the Anthropocene boundary crossings at the same time.

The other day, I was driving to town with my son-in-law Andrew and grandson Hamish to watch the latter play soccer. On the way, we passed a farmer on his tractor, boom-spraying a paddock with glyphosate. Hamish, all of nine years old, turned to me with a puzzled look on his face. 'Grandpa,' he asked, 'why do people have to kill things to grow things?' I was speechless for a moment. It was a profound question, but a question with a simple answer: 'You don't have to kill things to grow things.'

My journey in coming to understand this and the importance of regenerative agriculture really began when I was twenty-two. My father, forty-six years older than me, suffered a sudden and massive heart attack. I immediately left university and came home to take over management of the farm. While I finished my degree part-time, my education on the farm began from day one.

Growing up on a farm does not equip one for management. It takes years to become hardened to physical work, and an immature youth – at times in conflict with his father – is not ideal material for absorbing important lessons. Nevertheless, I had no choice but to begin learning. In my first ten days, I went out one morning to muster a paddock. To my horror, I found twenty-five sheep dead, and many others unwell. It was mid-autumn, a dry March, and thus a time of short grass and heavy dews. The deadly intestinal sheep-worm known as barber's pole thrives in such an environment, and it can breed overnight to prodigious numbers. Too inexperienced to read the situation and take preventative action, I walked into a barber's pole 'smash'. Later I would come to realise that such an occurrence was much more than a chemical health-management issue – it was also about landscape and ecological management – but had I been told this at the time I would have thought such views were the ravings of a lunatic.

After this baptism of fire, I assiduously set about learning management skills. I sought my father's counsel; I co-opted Department of Agriculture officers; I read the scientific and departmental literature; and I perused my father's books on managing lucerne, 'improved' pastures and ruminant (cud-chewing) animals. I also sought advice from those regarded as the district's best and most progressive farmers. I spoke to leading stock-and-station agents, and I enlisted a sheep classer to help me develop a merino sheep stud. Within a decade, I had become what I thought was a competent livestock and pasture manager for the Monaro region.

In short, I imbibed the voices of knowledge and experience surrounding me, copying their practices. Only aware of one path, I assiduously set my feet upon it. Therefore, in those first few years, I was being inducted into the dominant Western industrial-agriculture approach.

I find it sobering now that, despite my 'education' over the next two decades and the fact I gained my living from the Earth and its natural systems, I actually understood very little. Irrespective of my university training in plant and animal physiology and in soils, plus my own interest and training in the natural world and holistic human ecology, I was somehow oblivious to the existence of an entirely different management approach and its accompanying body of knowledge. In time, this 'blindness' meant the escalation of a debilitating debt, before, finally, I was able to open my mind to another way of seeing and thinking.

This alternative view held that soils were not inanimate chemical boxes, that our farm was instead a complex living entity of dynamic cycles, energy flows and networks of self-organising functions and coevolved nebulous systems beyond imagining. Later still, I would discover that such a parallel universe paradoxically comprised both the most ancient Indigenous and yet also newest scientific knowledge, and that it related profoundly to human health, to farm and animal health, and to planetary health. Moreover, this approach could be just as (if not more) profitable in both an economic and ecological sense. Most certainly, it could feed the world without destroying the environment at the same time.

The structure of this book reflects the personal journey I went on in becoming a regenerative farmer, as well as the open-ended learning approach exhibited by many regenerative farmers.

Part I, 'Into the Anthropocene', paints the background story to the unique subcontinent in which we modern Australians attempt to farm and live. It reveals how one human group – Indigenous Australians – evolved a long-lasting approach to land use but then how a different human mind – the Mechanical – usurped this long-term approach and helped to precipitate us into the Anthropocene. Part I also traces the rise of industrial agriculture and a nascent ecologically based agriculture.

Following this is the extensive Part II, 'Regenerating the Five Landscape Functions'. To move to a position where we can regenerate Earth, its systems and human health, we need to gain a working understanding of how landscapes and natural systems function (i.e. to gain some ecological literacy). So Part II takes us through how we can enhance the five key landscape functions: the solar-energy cycle (primarily aimed at fixing as many plant sugars as possible, via photosynthesis); the water cycle (primarily by improving the soil and vegetation ecology so as to store and recycle as much water as possible); the soil-mineral cycle (through having healthy living soils that contain and recycle a rich lode of diverse minerals and chemicals); dynamic ecosystems (through encouraging maximum biodiversity and health of integrated ecosystems at all levels); and the human–social component (through humans working in harmony with – not against – nature's functions so as to enable landscape regeneration). Each key function is described and illustrated through the remarkable stories of leading regenerative farmers who have tuned in to these systems to great effect. The journey through Part II also reveals how all five landscape functions are indivisibly connected: if one is degraded, then all suffer – as does the landscape.

The final section of the book – Part III, 'Transforming Ourselves – Transforming Earth' – looks at the capacity for regenerating Mother Earth, our societies, and animal and human health once we allow ourselves to gain a new understanding and act in accordance with it. We must first transform ourselves before we can regenerate Earth and human health. Finally, the last chapter of the book, 'Towards an Emergent Future', ties the many threads together.

As we accelerate further into the Anthropocene epoch, we are entering unknown and frightening territory. Though spoken in 1967 and in a different context, Martin Luther King's words are strikingly pertinent today. For, in a famous (indeed, prophetic) sermon in Riverside Church, New York City, on 4 April 1967 (exactly a year before he was assassinated), he emphasised the urgency of taking action today or running the risk of being too late.[9]

For me these are not nihilistic words but a galvanisation of hope. This is because while we are a human civilisation with its ruling metaphors and 'Mind' in deep trouble, it is from such utmost crisis that transformative ideas and actions emerge.

My examination in this book not only of my own journey but also of the remarkable transforming and transformed farmers featured in these pages led to an inevitable conclusion – and became the reason for writing this book. I realised that those transformative farmers were, and are, at the forefront of an underground agricultural insurgency.

In their Earth-sympathetic thinking, and in their connections to like-minded urban sisters and brothers (who are equally passionate about healthy food, human and societal health, and about the Earth and its natural systems), they are part of a powerful vanguard that is rapidly gathering momentum both in Australia and across the world. This movement is returning humans and societies to the state of health that our evolutionary history has designed us for and can turn around our destruction of Mother Earth and human societies as we enter this potentially cataclysmic Anthropocene era. Collectively, these farmers provide a template for regenerating Earth.

PART I

INTO THE ANTHROPOCENE

CHAPTER 1

A Gondwanan Ark

> *Populations are beyond good and evil. They grow in response to the availability of space, food, and water. When too numerous, organisms either perish or transcend themselves.*
>
> LYNN MARGULIS and DORION SAGAN, *Microcosmos*[1]

Australia's Unique Ecosystems

Australia is a 'one-off' when we step back and examine world biogeography and its intersection with human culture. It is the only inhabited continent 'entire unto itself'; it contains some of the oldest fossils and geological specimens found anywhere on Earth; and second to Africa, it has been permanently inhabited by modern humans for longer than any other continent.

However, while our Australian landscapes and their ecologies have been evolving for up to 3.9 billion years, a key factor in the distinctive shaping of the Australian continent was that of continental drift: the slow movement of giant crustal plates across the surface of the globe (behaviour that has probably been going on for around 1200 million years).

More than 500 million years ago, Australia was part of a supercontinent called Gondwana, which contained most of the now separate land masses of the southern hemisphere (including Antarctica), as well as India and the Arabian Peninsula. Over the next 300 and more million years, this supercontinent traversed the oceans and hemispheres, crashing and conjoining onto a colossal global land mass, and then breaking apart to reform as Gondwana again around 180 to 160 million years ago.

Crucial for our story is that Gondwana itself then started splitting apart – a process that took another 100 million years or so. As a result, in the last thirty million years, a now distinct Australian continent finally drifted northwards from Antarctica, and from there began to develop its own flora and fauna out of its unique biogeochemistry.

It is this uniqueness that should throw up a bold warning signal to any human cultural group approaching: 'Beware! Handle with extreme care!'

Unless we understand how this ancient Gondwanan continent is so different from the others, we cannot hope to live and farm here with any longevity. Therefore, in this and the next chapter I will set the scene for the remainder of the book with a quick gallop across time and landscape (both biophysical and mental).

As the environmentalist Tim Flannery has pointed out, three great factors have shaped – and continue to shape – Australia's unique biota and ecological systems. These factors are continental drift, regional geology and climate.[2] Overlaid on all these factors is the great element of time: the fact that the Australian continental massif is very, very old, and that the first three factors impacted on this ancient land over extremely long periods.

Given this backdrop, the crucial period for our story starts some sixty-five million years ago, with the sudden mass extinction of the non-avian dinosaurs, ancient sea life such as corals, most mammals and other creatures. This was due to a massive bolide (an extraterrestrial object of significant size, such as an asteroid or a comet) colliding with Earth on the north coast of the Yucatán Peninsula in Mexico to form the giant Chicxulub crater.

Ironically for Australia, as historian-palaeontologist Richard Smith observed, this extinction event 'cleared the global stage', leading 'to the appearance of the Australia we know today'.[3] That is, this massive biogeochemical transition ushered in a new biological age and shaped the essentials of our present Earth. What is crucial to Australia's unique history is that the Chicxulub disaster predated the separation of the Australian continent from Antarctica by about twenty to thirty million years. So when Australia went its separate way, as Smith said, it 'sailed off alone, an island ark adrift on a sea of change', carrying a cargo of flora and fauna that had become extinct in other continents.

The result was a wide suite of species and families adapted to, and unique to, the Australian continent. This is seen in the rise of mammals (including marsupials), the burgeoning of insect diversity, the full emergence of flowering plants (and, of particular significance to this story, shrubs and grasses), and the spread of snakes, lizards and birds (especially the perching or songbirds).

During these ark times, in the words of pioneering palaeontologist and environmental writer Mary White, Australia completed a 160-million-year triple phase of 'rifting, drifting and drying: akin to the gestation and birth, maturing and ageing of a living organism'.[4] However, this sequence of challenge and change (underpinned by Darwinian selection processes) is why so many unique life forms evolved to occupy vast numbers of eco-spaces in Australia.

Over time, the subsequent drying of Australia saw more than 700 species of eucalypt evolve (and a similar number of acacia or wattle species), with open gallery forests supplanting rainforests. As aridity increased, these forests yielded to wider, more open spaces, plains and steppes. Accompanying this was the emergence of 'sclerophylly' or 'scleromorphy' across a range of plant classes, including eucalypts and acacias. These terms describe a wide range of traits, including a toughening, thickening and shrinking of the leaves to equip them better for our harsh, dry environment.

As continental drying increased, the open spaces allowed wind to carry the pollen of various grass species over broad distances. Inevitably, hopping and grazing marsupials (uniquely adapted to traversing these environments) evolved in these spaces. The post–Gondwanan ark effect (combining drifting and drying) means that eighty per cent of all Australian flowering plant, mammal, reptile and frog species are endemic to Australia – not to mention most of its freshwater and marine fishes and half its birds.

Another key distinction is that Australia's soils are far less nutrient-rich than, say, northern Europe's and North America's. These northern-hemisphere soils are extremely young by comparison with Australia's, so they have what Australia doesn't: washed or wind-blown soil deposits from glaciation. Such activity, in grinding down rocks off young mountain chains and landscapes, creates nutrient-rich soils. Australia's ancient rocks, by contrast, are far more stable, meaning the continent's soils, in forming much more slowly and over a long time frame, were constantly leached of nutrients. Adding to this is the fact that Australia, by comparison with many other regions, had very little volcanic activity, which also serves to renew the soil's nutrients.

Australia's soils are not only very thin but also generally subjected to high temperatures, which, in combination with the subcontinent's dry climate, inhibits active soil biology. As we shall see in Part II, this latter factor plays a crucial role in building soil. All up, therefore, by comparison with the rest of the world, Australian soils are leached of both macro- and micronutrients – the mega-catalysts for life. They are low in phosphorus, nitrogen and soil organic matter (that from plants and other organisms at various levels of decomposition in the soil), all of which are crucial to ongoing building of rich and healthy soils. These deficiencies contribute to the fact that Australian soils are also generally poor in structure and tend to be restrictive of water drainage and plant growth.

Combine all this with the fact that Australian soil receives comparatively few nutrients from ocean-derived aerosols (sea spray blown into the atmosphere), and also little dust from other continents, and the result is substantial nutrient poverty. Significantly, this relative nutrient deficit occurs in many of the continent's key farming regions.

Linked to all these features is the fact that Australia is composed of low-productivity ecosystems. This is hugely relevant to the continent's unique biotic evolutionary adaptations (like scleromorphy) but also to its bushfire proneness.

In short, by having some of the nutrient-poorest, worst-structured, most fragile and driest soils on Earth, Australia ended up with a glorious and voluminously wide adaptation of intriguingly new plant and animal forms. However, this also meant the continent was a disaster waiting to happen if the wrong land-use technologies and worldviews were applied to it.

Then there is the critical issue of salinity. Australia's unique age and climate, in combination with the complex fragility of its soils and landscapes, deposited highly mobile salt in its soils and groundwater. This natural salinity left a ticking time-bomb for any foreign settler not understanding this ancient continent. This is because inappropriate land use releases or accesses these salt deposits, causing secondary salinity. This occurred after European settlement in 1788, and today is a massive problem of the order of millions of acres – and rapidly increasing.

In its final phase of northerly drifting as an island ark, Australia found itself in the dry, wind-blown mid-latitudes. Instead of regular, generally dependable and predictable seasonal shifts in temperature and precipitation (as in the Europeans' home countries), Australia's climate – and in particular the critical factor of moisture – is determined by two major and irregular ocean temperature-driven systems. These are the El Niño–Southern Oscillation (ENSO) cycle and the Indian Ocean Dipole (IOD) system.

While, after Antarctica, Australia is the driest continent, what exacerbates this is the great variability and irregularity of the rain that does fall. Moreover, periodic drought events are a key factor in precipitating bushfires in Australia. This variable, unpredictable, wet–dry cycle of Australia's continental climate means a consequential inhibition of biological productivity, and is a key reason why over a third of Australia is classified as desert, and seventy to eighty per cent is classed as arid (less than 250 millimetres of rainfall a year) or dry (less than 750 millimetres). This aridity is the fundamental reason why a similar-sized land mass in Europe (with its annually predictable climate combined with rich soils and greater humidity) can sustain nearly thirty-three times more people than Australia. And, even then, Australia's attempts to sustain its mere twenty-plus million people and export food for forty million more is causing an ancient, coevolved and interrelated group of ecosystems to come apart at the seams.

A further climatic factor is the extensive nature of Australia's dry centre and almost entire semi-arid continental environment. Heated, dry air coming off an arid interior that is swept by prevailing westerly winds means not only a generally less humid air compared with parts of northern Europe but also extremely high rates of water evaporation (where soils and plants dry out fast following rain). The latter factor is still vastly underestimated by Australian agriculturalists, gardeners, policymakers and most non-Indigenous inhabitants.

The extraordinary coevolution of organisms in Australia's ecosystems and the resultant complexity of different ecosystems across the continent has resulted in what can be described as the 'writing' of the Australian landscape: its biogeochemical uniqueness, patterns, nuances, amazing beauty and downright startling innovations and differences. Nowhere else on Earth has 'nature-writing' like this. And therein lies the potential for tragedy, for such writing is indecipherable to new arrivals. For it takes time and patience to gain Australian ecological literacy: to learn how to read the landscape. In other words, Australia is what it is, not a splodge of plasticine waiting to be remoulded. Few land managers understand this still, with but a tiny minority appreciating that humans must adapt to this ancient land and its climate and biota, not the other way around.

More than fifty millennia apart, two distinctly different human groups settled this continent – but with startlingly different results. The first group were able to survive and build a sustainable human culture until the arrival of the second group. Their example is largely unheeded to this day, but it has some lessons to teach. Consequently, I now turn to what these lessons might suggest.

Two Thousand Generations on Country

The great irony for the Australian continent – the flattest, driest and second-hottest (after Africa) on Earth – is that humans first settled here because of ice.

One momentous day – probably around 65,000 years ago or earlier – at a time of the lowest sea levels in the midst of a great ice age, a small group of people were poised on the mud flats of southern Sundaland (the greater south-east Asian land). These first Australians would have known land lay beyond, though none could have imagined it comprised the huge, uninhabited continent of Greater Australia, or Sahul, encompassing New Guinea, mainland Australia and Tasmania all in one. Perhaps it was 'land clouds' formed by cooling convection over higher features on Sahul that was the clue. Or even migrating birds. Or most likely it was distant smoke from lightning-struck bushfires, or the red glow of such fires at night – for Australia has long been a blazing continent.

We don't know what drove this small group of people to make the dangerous crossing, but the result comprised an experiment of epic proportions, which in time yielded one of the greatest ever sustainable partnerships between humankind and the ecosystems they occupied.

It is hard to imagine what this first settlement was like. First, for eighty per cent of the time that humans have occupied Australia, it existed as Sahul, and was up to thirty-four per cent bigger than today's island continent. But, on landing, these 'super nomads' would have fanned out along the coast, and then along rivers to penetrate the inland. It is possible that they settled a fair amount of Greater Australia within only a few thousand years. The Monaro region where I live was occupied at least as early as 25,000 years ago. Second, these first people would not have entered a benign land. Being foragers and gatherers as well as hunters, they would have had to contend with an array of new poisonous plants and fearsome but unique southern megafauna predators, not to mention the world's greatest diversity of poisonous snakes, spiders and sea creatures.

Clearly these early Australians were skilled survivors, because they escaped annihilation in uninhabited Sahul and quickly spread and expanded in population. Millennia of trial-and-error learning were the key.

The extraordinary occupation of many niches in Australia by at least 35,000 years ago was accompanied not just by the development of highly complex social mechanisms but also by a unique and interrelated cultural development. The latter combined practical knowledge of the occupiers' new environment with complex cosmological, religious and kinship systems, and it resulted in sophisticated and sustainable management of their multiple ecosystems.

This integration of culture and sustainable practices occurred because the peoples of the region had been strongly shaped by the most challenging of violent climatic oscillations over tens of millennia (including a ten-thousand-year, drought-like ice age). Such challenges engendered an unimaginably long-term perspective and left an indelible imprint on collective memory, and also on the biological evolution of both the peoples themselves and their surrounding biota.

So, what cultural mechanisms did these super nomads develop that allowed their long-term, sustainable survival – a survival that largely preserved and diversified their sustaining ecosystems?

Not just early Europeans but probably the vast majority of present-day, non-Indigenous Australians were, and still are, largely ignorant of the fact that much of Australia by 1788 was very much a 'cultural landscape': a managed and modified human environment.

Along with almost all early European observers, Captain James Cook assumed that Australian Aboriginal people merely wandered across the landscape in an ad hoc manner, passively relating to their surroundings as they harvested and hunted what they could find so as to survive. Ironically, this was despite him also observing that Australia was a 'continent of smoke': smoke that, as we shall see, was evidence of an extraordinarily sophisticated and continent-wide land-management system.[5]

The people doing the managing, across most ecological niches of the landscape, belonged to 250 or more Aboriginal nations. This meant that each nation (with its own distinct language) was indivisibly 'married' to its own particular area of land or 'country' in a truly organic sense. Such occupation in effect defined 250 or so eco-regions – for it was an ecologically based spatial and human arrangement. Groundbreaking historian Bill Gammage concluded, following a long-term study of Australia's Indigenous land management, 'What they did was remarkable. No other world civilisation achieved it. They were ensuring that all species had a habitat. People could use those resources, plants and animals, and sustain their society, keep its resources abundant and unchanging without risk to their future.'[6]

This is not to say that Australia's Indigenous nations were all 'noble savages' who never wrought ecological havoc or never had dysfunctional social lives and practices at times. For the truth is that Indigenous Australians, like all humans, would have made ecological mistakes or wouldn't have adjusted to shifting circumstances. At times, therefore, they would have overused some resources. For example, Aboriginal people are strongly implicated in the last great wave of megafauna extinctions and possibly also in the extinctions of the Tasmanian devil and thylacine on the mainland (especially after the dingo was introduced). As leading regenerative agriculture thinker Bruce Maynard states in an attempt to balance this ongoing and often contentious discussion, 'I think the more exciting and substantive story is how the Indigenous populations walked the tightrope of conditions they were presented with "balancing" as they went ... What remarkable feats of mental and physical gymnastics!'[7]

The key feature about ancient Indigenous Australia is that the complex cultural management of landscape was dependent upon understanding the weather and long-term climate cycles along with the natural environment and its biota. This knowledge was then combined with a sophisticated interrelation of religious and cultural practices.

These practices revolved around the 'Dreaming', which, in the Arrernte (Aranda) language of central Australia, means the 'Eternal' or 'Law'. Central to the Dreaming is belief in a mythical, creative, timeless and cyclic past when ancestral beings existed; when they performed heroic deeds, and when they created, moulded or enhanced key features of the landscape, including creating plants and animals; and when they laid down social customs that are still followed today by some Aboriginal nations.[8]

In describing the Dreaming as 'a complex network of faith, knowledge and ritual that dominates all spiritual and practical aspects of Aboriginal life', anthropologist Josephine Flood explained that it thus 'lays down the structures of society, rules for social behaviour and ceremonies to maintain and increase the land's fertility'. Crucially, Flood points out that the 'Dreaming comes from the land; it is a powerful living force that must be nurtured and maintained'.[9] Totems play a powerful role in this. Rules and social behaviours around, for example, protecting one's totem animal or plant are part of The Law (called *Tjukurpa* by Western Desert Pitjantjatjara). The Law is that 'body of religious and cultural knowledge that is used to inform and direct Aboriginal society'. In being derived from the Dreaming, such law is believed to be unchangeable.[10]

The Dreaming is grounded, Bill Gammage said, 'on ecological realities, whatever its social applications'. Crucially, as Gammage concluded, 'All Australia obeyed the Dreaming. By world standards this is a vast area for a single belief system to hold sway, and is itself cause for thinking Australia a single estate, albeit with many managers.'[11]

Integral to the arrangements of the Dreaming is a range of social structures and kinship connections. Extensive kinship networks become vital during times of climatic stress, or where resources are located in another clan's territory. By exercising kinship, groups are enabled, without conflict, to access those resources. This complex mix of religious, social and spiritual beliefs is therefore focused on long-term sustainable management and harvesting of the landscape so as to support diverse groups.

Linked to this belief system are Dreaming tracks, or songlines, which are the routes taken by ancestral beings in the Dreaming. As the ancestors formed the land and laid down The Law, they 'sang-up' the country into life. People gained direct access to the Dreaming and were able to navigate the landscape by performing the right songs and ceremonies at 'story' places along a ritual Dreaming track. Crucially, songlines are far more than navigational aids. In anthropologist Lynne Kelly's words, such singing tracks weaving across the landscape 'enable every significant place to be known' and connected in a fixed order. The result 'is a sung chart of the ancestral being's creative journey or origin story'.[12] Songline stories also give details of animal, plant, person, ancestor, ritual, dance and more. In short, such landscape paths allow, for an oral culture, the memorisation of a huge amount of information that comprises 'encoded knowledge'.[13]

Australia is criss-crossed by songlines, some of them 2000 kilometres or more long. Such songline connections also allowed different bands, clans and tribes access to areas, precious waterholes and resources outside their home range. This is because songlines also acted, in Bruce Maynard's words, as ecological 'spines' – tracing 'our greatest ephemeral rivers along with the man-made wells that traversed the country' and thus allowing 'the movement of people to utilise large areas'.[14]

Evolving out of, and indivisible from, this extraordinary framework of social and religious beliefs was one of the most sophisticated, widespread and complex land-use systems devised anywhere on Earth. The chief land-management practice was 'fire-stick farming', whereby Indigenous Australians would adapt the landscape to their needs by regularly using fire-sticks to burn vegetation. Therefore, while not looking like a farmed and 'civilised' landscape to the early Europeans, much of Australia was nevertheless culturally 'farmed' and 'harvested'. The difference was that, unlike virtually all agriculture since domestication millennia ago, the Australian cultural approach was sustainable in the very long term.

It therefore appears that much of Australia, as Bill Gammage described, 'was patterned to suit the animals, plants and therefore people of Australia'. Such a widespread and systematic approach thus comprised, said Gammage, 'a sophisticated tool to create the right resources in the right places'. Moreover, he explained, this was designed to make Indigenous peoples' resources 'abundant, convenient and predictable'.[15]

Such a process was clearly regenerative, and it has now become clear that Aboriginal fire-stick farming played a significant role in maintaining the biodiversity of Australia – whose vegetation is largely fire-adapted. To farm successfully in this way, as anthropologist Deborah Bird Rose pointed out, involved a 'detailed knowledge of soils, landforms, surface and under-ground water, and types of vegetation, as well as time of year, time of day, and type of wind'.[16] From working with local Ngarigo Aboriginal senior law-man and fire expert, my friend Rod Mason, I can personally attest to this highly sophisticated and skilful knowledge base.

Aboriginal burning created vast grasslands out of mixed forests and shrublands (hence the frequent 'parkland' descriptions of early explorers and settlers). But Aboriginal mosaic-burning regimes also maintained heathland, woodland, forest, scrub, grassland and sedgeland.

The net result of the sophisticated use and non-use of fire as a tool was that Aboriginal people were active land managers, whose management strategies were able to sustain a complex mosaic of ecosystems continent-wide in Australia: a form of proto-agriculture.

There are also many other examples of practical, even advanced, forms of proto-agriculture in operation in Australia, both in 1788 and, based on archaeological evidence, well before that. As Aboriginal man and historian Bruce Pascoe has revealed, this includes active long-term assistance of the production and husbandry of food (including basic cultivation and the planting of seeds of edible food species), the sustainable management and protection of vegetation that yielded edible resources (like wattle gum), and evidence of fish and eel trapping and husbandry across the continent. Food storing (both dried meat and vegetable foods such as desert raisins and wild figs) was also common, as was the storage and harvesting of grass seed, which was then ground into flour for both unleavened and leavened bread.[17]

Therefore, in 1788, much of the continent comprised a cultural landscape shaped by millennia of sophisticated land management. Importantly, the focus of this integrated, active cultural practice was ecological. It was aimed at the creation and maintenance of diverse, healthy and regenerative eco-systems continent-wide that would last indefinitely.

This holistic integration of different beliefs dovetails with the Indige-nous use of the term 'country'. Deborah Bird Rose evocatively defined country as 'a nourishing terrain'. She described it as 'a place that gives and receives life. Not just imagined or represented, it is lived in and lived with.'

She added, 'Country is a living entity with a yesterday, today and tomorrow, with a consciousness, and a will toward life. Because of this richness, Country is home, and peace; nourishment for body, mind, and spirit; heart's ease.'[18]

Anthropologist David Turner observed that Aboriginal Australia is made up of a series of 'promised lands', each with its own 'chosen people'.[19] Every Indigenous person has a *country*, where one's ancestral land has both spiritual and universal components as well as the practical and local. The concept of country is dynamic and multi-layered, as it refers to both the physical and the cultural: to the land above and below ground, soil, minerals, air, water, every living organism, and to stories of the Dreaming.[20]

Bob Randall, Yankunytjatjara elder and traditional 'owner' of Uluru, insightfully expressed his view of the concept of country. He recently stated, 'We lived on the land as people of the land. To us, it was a natural way of being – being part of all that there is ... Life is the binding and connecting way, the oneness is ... if you're alive, you connect to everything else that is alive. But that oneness included everything that was around us and you were raised with that from a child ... See, my people see land ownership as being totally different to the English way of ownership, because our way used to be "The land owns us", and it still is that to us. The land grows all of us up ... We say "the Granny Law" ... [it] has given me my responsibility now that I have grown up, to care for my country, care for my mother, care for everything that is around me. The oneness, the completeness of that oneness. To be responsible in both caring, in every single way, what we call the "Kunnini", caring with unconditional love with the responsibility.'[21]

As Bill Gammage described, 'Aboriginal landscape awareness is rightly seen as drenched in religious sensibility, but equally the Dreaming is saturated with environmental consciousness. Theology and ecology are fused.'[22]

In other words, according to Gammage, 'Songlines distributed land spiritually; "Country" distributed it geographically', and 'land, water and their sites and knowledge' were 'in the care of a family under its head'.[23] There is thus a network of countries that form an interlinked, natural continuum, where each landowner (as Bob Randall so powerfully evoked) is obligated to care for their own country.

Given that Indigenous people depended on land for long-term survival, were obligated to care for it, and had deep ecological and spiritual

knowledge of their country, unsurprisingly they developed intimate associations with the landscape and were deeply embedded within it.

This world view (what is called the 'Organic mind') is a universe away
from the ideology that post-Enlightenment, 'rational', mechanically
minded Europeans brought after 1788. The difference was that Aboriginal
relationships to land, embedded in the concept of country, linked people to
ecosystems, rather than giving them dominion over them. As we shall see
directly, it was dominion thinking that underpinned the 'Mechanical
mind' when it encountered Australia.

A few years back, I thought I was making progress in coming to understand
the depth, longevity and at least some of the nuances involved in an Indigenous conception of country. But an incident at home alerted me to the
fact I had a long way to go.

Around fifteen years ago, I was privileged to spend two days on a bush
block east of Nimmitabel (one hour's drive from us towards the coast) with
local Aboriginal Ngarigo elder Rod Mason. Since that time, Rod has visited our farm on a number of occasions, as this is his ancestral country and
he knows he is welcome here to visit and seek its resources. I have also visited his piece of country down the coast: savouring his freshly caught fish
flavoured with local bush native pepper and lemon. Over the years, I have
come to befriend this gentle, wise yet strong senior law-man. More recently,
we have also begun hosting cool patch-burning workshops, which Rod
runs, where he quietly trickles out a deep and ancient knowledge, underlaid by his subtle sense of humour.

But on this particular day, Rod took the scales from my eyes. At the
front of our garden (a large, rambling hectare or so of mixed vegetation,
fruit trees and vegie-garden beds) is a tree that I remember growing up
with. It is a kurrajong (*Brachychiton populneus*), of ancient Australian lineage. However, kurrajongs are simply not found anywhere on the high,
cold parts of the Monaro. I don't know of any others until you drop down
into warmer country near Canberra in the north or lower Snowy River in
the south. Moreover, my father told me that, before he arrived to 'take
possession' of our farm in the 1920s, there had originally been seven kurrajongs in the garden. I could only remember two survivors, and one died
twenty or so years back.

So this particular morning when Rod Mason arrived with a few ecologists for a survey of our bush, for some reason I was prompted to ask Rod what he thought of the 'out of place' kurrajong over on the garden's edge. Subliminally, I was aware that kurrajongs were a key resource tree in Aboriginal culture, providing shelter and bark fibre for strong woven bags, nets and line; wood-boles for coolamons; seeds for a coffee-like drink; and leaves and such for different medicines – but also, in Rod's words, as 'habitat for animals and other creatures to visit'.

Rod walked over to the tree and began closely examining it. But then I could see him getting quite emotional – even hugging it at one point. He told me that yes, such trees don't normally grow here, and then he pointed out several long, vertical strips where Aboriginal women had stripped bark to make fibrous material, and he pointed out a hollow on the trunk where a once-protruding branch bole had been knocked out to be fashioned into a coolamon.

Moreover, even in warmer country, kurrajongs are slow-growing, and as we examined this tree's height, giant trunk and a root-mound some one and a half metres higher than the surrounding ground, it was obvious this tree was at least 400 years old, and probably way beyond that.

Rod then told me his people's story. The oral history of the Ngarigo reveals that they originally came down from the western desert country following the 'ice Dreaming' songline. This reached to the west of Lake Mungo in western New South Wales and then linked to another songline to cross the mountains onto our country. Once settled, the Ngarigo would retreat to the coast for the winter. It is not clear when they arrived on the Monaro: perhaps 12,000 to 15,000 or so years ago, after the great ten-thousand-year drought of the Glacial Maximum period; perhaps earlier. Archaeological evidence reveals it may even have been during the great drought itself, some 23,000 or more years ago.

When the old women walked their songline track, said Rod, they always carried seeds of their favourite food and resource plants. These would be sown and carefully cultivated at their chosen, and spiritually significant, camping places. Clearly, our front garden was one such place specially chosen to plant kurrajongs. Rod said that originally there probably would have been a grove of fourteen or so trees, and that this was once an important ceremony place. He added with a serious mien, 'There's babies born here, but also "sorry business" was here.'

It was therefore understandable that, standing in the presence of a tree that represented the deep cultural actions of a dispossessed people and ancestors with a long, long view, Rod was powerfully moved. And as an outsider, I could only contemplate this touchstone of connection mixed with a profound sense of loss. Clearly, this old tree represented an integral, millennia-long connection to country – both biophysically and spiritually: an ancient connection deeper than we can imagine, and linking us indivisibly with the natural world.

Tragically, this ancient human–land connection and linkage to country was both unknown and trampled on after 1788.

A Clash of Hemispheres

Earlier in this chapter, we saw that the ancient land of Gondwanan Australia expressed its own nature-writing. This was then adjusted and shaped by Australia's Indigenous people over fifty or more millennia, resulting in a modified but still organic palimpsest. After 1788, however, this would all change, as a new, alien writing was soon to be brutally stamped on Australia's landscapes. The key question for our collective survival thus becomes: would the new writing totally erase the old?

When the British arrived to establish a colony in Australia in 1788, much had changed in the cultural evolution of the human species. Relatively recently, a domesticated agriculture had evolved. From the Middle East specifically, these agricultural techniques (refined over ten millennia or more) had reached Western Europe. By the late eighteenth century, this new cultural approach was itself undergoing another burst of revolutionary refinements.

But this latest agriculture had been adapted to a climatic and biogeochemical environment that was totally different from the island continent of Australia. Consequently, these techniques wreaked rapid and awful havoc. The 1788 settlement and subsequent migrations thus proved a 'clash of hemispheres': and not just north/south nor left brain/right brain, but also between mindsets. It was a clash between a modernising, dominating Mechanical mind and an adaptive, ancient Organic mind. This clash and its awful consequences had already been replicated on other colonised continents across the globe.

It is not possible within the scope of this book to catalogue the nationwide destruction of both Australia's Indigenous people and their cared-for countries that followed white invasion and settlement. This is extensively covered elsewhere by a range of authors, though still not adequately acknowledged at levels of political and national consciousness. We also fail to fully appreciate that the land and its Indigenous people were so inextricably bound that the destruction of one inevitably entailed the destruction of the other.

As the anthropologist W. E. H. Stanner stated in his famous 1968 Boyer Lectures *After the Dreaming*, the 'great wrecker had been the pastoral industry'.[24] That is, the influx of European farming methods had a catastrophic impact on both Indigenous societies and the land. This was bluntly encapsulated in 1930 by historian W. K. Hancock when he stated that 'in truth, a hunting and pastoral economy cannot co-exist within the same bounds'.[25]

It was because of the extraordinary social adaptation to Australia's distinctive environment that Aboriginal societies so quickly collapsed when their system was disrupted. Or why 'they responded so quietly to being worsted', as Stanner put it.[26] Quite simply, it was their 'otherness', which meant they were not an organised, martial people but rather were disparate in their ecologically based countries.

And the undoing was awful and rapid. For the Aboriginal and Torres Strait Islander people encountered a politically organised, aggressive and conquering people – and, crucially, one imbued with a post-Enlightenment Mechanical mind. The Ngurunderi people likened them to ants who swarmed over the land.

Though still debated, it is probable the Indigenous population in 1788 was somewhere between 750,000 and one million or so. Within only 140 years (six generations at most), Australia's Indigenous population had collapsed to but six to eight per cent of 1788 levels, or to a mere 60,000. Paraphrasing Jared Diamond, it was a clear case of guns, germs and steel that were responsible: though given Australia's unique social–ecological structures, to this can be added 'introduced livestock'. Thus, to horrific violence and genocide was added the equally deadly killers of disease to a population with no immunity to modern diseases.[27] Moreover, the elderly

– with generations of orally transmitted survival knowledge – would have been among the most susceptible to the new diseases.

Inevitably, Australia's diverse countries and their biotic communities immediately began to suffer the same fate as their Indigenous inhabitants. This was because the carers, nurturers and skilful managers of vast complexes of country either died or were expelled, and were replaced by exploitative Western land managers with their host of novel plants, animals and technologies.

With the Indigenous 'hand on the tiller' sequentially taken off, a radiating wave of simplification of diverse ecosystems spread and then cascaded across the continent. And this without the devastating impact of European agricultural modification of the land, which soon followed. This involved vast landscape degradation through overgrazing by foreign domesticated animals (primarily sheep and cattle, and later escaped 'ferals' such as horses, goats, donkeys, camels and water buffalo), unprecedented over-clearing and annihilation of eucalypt forests and the Big Scrub (a huge rainforest in eastern Australia, containing unique species such as the red cedar), and the wanton, senseless destruction of once-numerous birds, mammals and other life forms. Armed with ever-evolving modern weaponry, the settlers became more efficient at slaughtering both humans and animals as the nineteenth century wore on. For many unique plant and animal species, the *coup de grâce* was delivered by the spread of northern-hemisphere 'weedy' plants and the introduction of rabbits, foxes and cats.

What needs brief examination, however, is what lay behind the rapacious settlement of ancient Sahul post-1788. That is, what baggage of mental beliefs could precipitate such a brutal and ongoing conquest of over 200 different Indigenous nations and their respective cultural-ecological countries? I address this in the next chapter.

Compounding the tragedy for Australia is that it was conquered and settled by white invaders at a unique moment in human and world history. Modern agriculture had conjoined with industrialisation to produce a 'compressed double revolution'.[28] Environmental historian Cameron Muir perceptively touched on deeper issues when he observed:

> There is a dimension of 'war' about the way settler Australians have approached their land – understanding it as

'mongrel country', rather than a functioning ecosystem poorly adapted to the expectations of Western agriculture ... The same society that executed massacres caused ecological degradation on the nineteenth-century grasslands.[29]

Besides wide reading and various talks with old-timers, my direct connection to the perplexing behaviour of not just nineteenth-century settlers and colonists but also those in the twentieth century has been via talking to my father, and through him to the grandfather I never met.

My father told me that my grandfather's idea of a 'great day' on our property in the late 1920s – shortly after he purchased it – was to set out in the morning with a sandwich lunch to first ringbark a few trees (i.e. kill them off by cutting away the inner living bark in a ring) and then, around lunchtime, to set fire to the carcases of trees and logs already dead while he boiled a billy. In the late afternoon (depending on the wind and weather), he would then take his gundog and gun and venture forth into favoured lee sides of hills, where he could flush quail and other ground birds as they came out to swivel, preen and bathe in their delicate dust-hollows.

Today, evidence of this 'pioneering' activity can be seen in many of our paddocks: great hollows of old, dug-out rabbit warrens in sandy granite country; the odd ghostly dead tree and a channel of weathered axe cuts around the girth of its remaining aged stump; and the sad remnants of once-vibrant grassy woodlands. These include isolated snow gums or ribbon gums, and perhaps a blackwood hickory, blackthorn, lomandra or silver wattle all alone and clinging for life in a rock crevice inaccessible to grazing livestock.

I especially remember one story my father told me of their days spent 'cleaning up' country in the 1930s on our farm. They were burning timber that had dried after ringbarking. There was one particular tree – a giant ribbon gum whose stump I can still see today – to which they set fire. After an hour or so of heat and smoke, a sugar glider appeared high up from out of a hollow. He then set sail and one of the men caught him in his hat after a sixty-metre glide. In quick succession, seven more followed, landing in exactly the same place, as excited men took turns to catch them in their hats. But I have often wondered what happened then. Did they let the dogs kill the beautiful marsupial gliders, who always nest in family groups? Or did they let them go, and if so to where? Because all local tree territories would have been taken. Somehow, this displaced family of sugar gliders seems a poignant symbol of all the loss and disruption that came after that first settlement in 1788.

However, I have come to realise that it is too easy, too facile, for me to blame those nineteenth- and early twentieth-century generations. They were the product of their times, and I question whether we would have behaved much differently. In short, and as we shall see directly, they were the result of three to four centuries of so-called 'Enlightenment' thought, whereby, in overthrowing the ancient Organic mind for the Mechanical, they had arrived at a place in their heads and hearts where they could give no second thought to the destructive way in which they treated nature, the land, and the humans they deemed 'primitive'.

One thing stands out clearly from the destructive and tragic events of the post-1788 invasion: the face, substance and functioning landscapes of Australia were radically changed. Indeed, in many places the ancient nature-writing was largely erased. Stamped in its place over vast swathes was a hugely altered land of simplified monoscapes and biota, of destruction and functional degradation.

Les Murray wrote:

> It will be centuries
> before many men are truly at home in this country[30]

Is it therefore too late to resurrect some of these landscapes, to create a regenerative palimpsest? The answer, as we will see, is an emphatic 'No!' But to understand why, it is imperative we comprehend how human worldviews and culture-ruling metaphors evolve, change and can be changed.

CHAPTER 2

Emergence of the Mechanical Mind

So destructive has the agricultural revolution been that, geologically speaking, it surely stands as the most significant and explosive event to appear on the face of the earth, changing the earth even faster than did the origin of life.

WES JACKSON, *New Roots for Agriculture*[1]

Nature as a Commodity

In mid-2014, a chilling story swept across Australia's national airwaves and print media. While attempting to prevent unlawful vegetation clearing on potential cropping land, fifty-one-year-old New South Wales government environment compliance field officer Glen Turner had been gunned down by a farmer.

The prior two to three decades had seen unprecedented, largely illegal destruction of huge swathes of valuable grasslands and woodlands in northern New South Wales. This was in the wake of the most recent stage of the modern agricultural revolution and the advent of direct-drilling (sowing seed and fertiliser with specialist machinery without cultivating the soil) and weed control by herbicides. In 2010–11 alone, the area cleared was 605,150 acres (245,000 hectares). It was pioneering days all over again, albeit with more modern tools.

The illegal clearing was in defiance of environmental laws that were seen as whittling away at farmer autonomy and freedom. Beliefs about the right to do what they wanted with freehold land had festered like corrosive acid in one family in particular near the town of Croppa Creek, and especially in its seventy-nine-year-old patriarch, Ian Turnbull. For years, he and his family had been engaged in bitter legal disputes with the New South Wales Department of Environment over the family's blatant pioneering approach to stripping most of the vegetation off 'their' land.

As part of this legal battle, Glen Turner (who was generally liked and respected by farmers) had become the fall guy in handling the legal cases and liaison situation. But on 29 July 2014, matters boiled over for Turnbull. Near dusk, officer Turner and a colleague were investigating further illegal clearing issues on a dirt road adjoining the boundary of one of Turnbull's properties. Suddenly Turnbull pulled up alongside the officers' vehicle. As journalist Richard Guilliatt recounted, 'the old farmer' deliberately and systematically 'fired his rifle from a distance of 10–15m ... a fifth shot hit Turner in the back, killing him. After the final shot, Turnbull told Turner's companion: "You can go now – I'll be home waiting for the police."'[2] Turnbull was subsequently arrested for murder and in mid-2016, still unrepentant, was convicted. He was to spend the rest of his life in jail and died there in March 2017.

This case reveals the mental constructs that have become deeply embedded in the subconscious of many post-1788 settler-descendants in this country. Only such powerful drivers could lead to the irrational explosion of hate that ended in this murder on a lonely and dusty road in northern New South Wales. They are part of the Mechanical metaphor of Western society, which says, among other things, that man is separate from, and so can dominate and is master over, nature.

The question is where did this Mechanical mind come from? For it was this thinking that lay behind the extraordinary desecration of our Gondwanan ark and the longest-lived culture in the world after 1788 (and which continues still). We will therefore go briefly back in time again because the emergence of a world view like the Mechanical mind (and more recent expressions such as economic rationalism and neo-liberalism) goes to the very heart of this book and its story of landscape and societal regeneration.

Prior to the beginning of agriculture, a world view called animism had reigned for 200,000 or more years in many human societies. This view constituted the Organic mind, as it did not see humans as being separate from their environment, or from an objective reality. Instead, reality was an interconnected, spirit-filled landscape in which non-human entities – plants, animals, inanimate objects (such as rocks, rivers, mountains) or phenomena (such as thunder, wind, shadows) – possessed a living soul or spiritual essence and had awareness and feelings. Animism's significance was that it contained strong ethical and spiritual implications for nurturing and sustaining the Earth: an associated value system.

The shift from this Organic mindset to our society's dominant Mechanical world view was triggered by the development of domesticated agriculture some 10,000 or so years ago. In time, domestication meant that plants, animals and other natural phenomena became manipulable property, as opposed to sacred beings or entities. Consequently, from the dawn of agriculture until the Renaissance of late fifteenth-century Europe, humans on the European and south-west Asian continents, in particular, began a slow process of progressively throwing off the long, coevolved Organic mind that had previously bound them to nature, Mother Earth and a spiritual world.

This new cultural practice of agriculture and its resultant abundance of food eventually led to population growth and intensified urban living, which culminated in the 'Urban Revolution' and the appearance of the first cities. Thence came the rise of large-scale political and social systems: all an increasingly far cry from our genetic evolutionary conditioning as hunter-gatherers invested in the Organic mind.

Humans now began to focus in on themselves and their societies (the beginnings of 'humanism'). Part of this involved people beginning to apprehend the power of the human mind to manipulate the Earth and its resources. Therefore, a massive shift in value systems, ethics and morals began to occur. Less and less was there an unquestioning recognition of the worth of all natural things, or of the concept of cyclical renewal, of what ecological historian Carolyn Merchant described as 'the binding of nature into a close-knit holistic unity'.[3] So human cultural evolution now radically shifted.

Leading up to the Renaissance and Reformation, there was a coalescence of powerful forces in Western culture that weakened the remnants of the Organic metaphor and opened the way to an inculcation of the Mechanical metaphor.

A key influence was Judeo-Christianity. This gave us, as Carolyn Merchant pointed out, 'The idea of man as nature's guardian and caretaker ... a managerial interpretation of the doctrine of dominion'. This then blended with classical and pastoral attitudes towards nature as being something that could be ploughed and cultivated, used as a commodity and manipulated as a resource, tamed and subdued for human benefit – particularly by males. This world view also saw females as passive and receptive: thinking incorporated into the new Mechanical world view. Such a mindset was easily and quickly adapted from the sixteenth century through to the eighteenth-century Enlightenment: that crucial phase of the scientific revolution and the evolution of the market economy.[4]

Thousands of books have been written about this epoch-changing phase that occurred from around 1500 to 1800 in Western Europe, which involved a major shift in the Western mind from an Organic to a Mechanical model. The Mechanical model meant humans perceived the world as a place where matter and nature were inert constituents of a new, machine-like world – one capable of manipulation. I will only pull out a few key points on this remarkable period, but the consequences were multiple.

The major phase of the metaphorical shift to the Mechanical, argued Carolyn Merchant, really began when 'the image of an organic cosmos with a living female earth at its centre gave way to a mechanistic world view in which nature was reconstructed as dead and passive, to be dominated and controlled by humans'.[5] The result, in modern times, is that Western society no longer regards nature as a nurturing mother, nor holds her in reverence, nor treats her with respect. As historian Henri Frankfort wrote in 1949, 'The world around us has become an "it" rather than a "thou".'[6]

This also paved the way for the rise of capitalism, which goes hand in hand with the Mechanical mind. The machine image under Descartes and fellow 'mechanists' in the seventeenth century, which invoked human (and especially male) power over lifeless, mechanistic nature, was a forerunner to modern capitalism as it became the foundation stone of materialist reductionism, empiricism and objectivism. In effect, Descartes had 'de-souled' the Earth.

The next step was the linkage of the ideas of Bacon, Descartes, Newton and peers to that of economic and political theory, thereby embedding a capitalist philosophy. In this way of thinking, nature had no value except where it was reduced to a means to human ends – just an instrument for use. This bridge and its accompanying moral hand-washing was made by John Locke and Adam Smith.

Locke's ideas on 'rationality' helped establish a platform for the value system of the European Enlightenment: that one could profitably appropriate the whole sphere of nature as 'reason's own individual property'. The acquisition of private property helped enable the idea that humanity could 'progress' from the 'state of nature' into ordered, civil society, where the natural world had no 'rights'.[7]

Adam Smith's contribution to the evolving master discourse was to incorporate new thinking on progress. This laid down a system of capitalist economic laws built on the advancement of science and technology, property and polity. In the process, morals and values regarding the Earth, nature, women, colonised lands and their indigenous people, and other 'creatures' began to be jettisoned. Thus it was Smith's thinking regarding the market system's slow and steady growth that opened the way to the destructive shift to a capitalist market economy and economic rationalism: the belief that continual growth is necessary and desirable.[8]

The transition to capitalism marked the moment when the traditional organic model of communal, interdependent society (one that emphasised the whole as well as the parts) was undermined and transformed by competitiveness and acquisitiveness. Inherent now was an intellectual arrogance towards nature, which was regarded as the raw material for wealth creation, with little to no ethical restrictions on this.

It was only upon analysing this entire story that it became clear to me why those early colonists and settlers who reached Australia were able to plunder, destroy and kill so unfeelingly, and why these attitudes have deepened with the rise of the 'culture of greed', economic rationalism and neo-liberalism.

What also became clear is that this monumental shift in the collective human subconscious has huge relevance to the personal mental constructs of modern Australian farmers and to their consequent land use (as was so powerfully demonstrated at the start of this chapter).

This mindset has awful complications, and has led to a blind belief in the tenets of economic rationalism, in limitless but necessary growth, and in the virtues of prioritising wealth. In the words of the contemporary philosopher John Ralston Saul, the West in particular (but now spreading globally) 'has become addicted to a particular set of illusions in order to avoid coming to terms with its own reality'.[9] This reality is that our unsustainable behaviour has precipitated us into the Anthropocene.

In the most fundamental of respects, therefore, by this twenty-first century, the Mechanical mind had gone 'rogue'. And it had gone rogue for the craziest of reasons and the most seductive of all illusions: consumption and greed.

The Rise of Industrial Agriculture

One evening at dusk, after I finished some sheep classing (advising clients and grading their sheep) near Cootamundra in the New South Wales sheep–grain belt, I headed south towards Wagga and a motel bed. Along the way, I was forced from the road by a monstrous tractor and spray rig. The way its headlights were close together and surrounded by orange and red warning lights, in conjunction with its folded spray arms, made it appear like some giant arachnid monster – but far more deadly.

And as I drove through the night, I pondered how we had come to this: a situation where some farmers can go into debt for giant spray rigs, tractors and combine-harvesters while their place of abode – the farmhouse – often remains basic, unloved, unpainted and just a dormitory for sleeping in at night. More than likely, there would also be piles of clothes in laundry baskets that reeked of glyphosate, and the ground behind the farm sheds would be strewn with hundreds of empty chemical drums.

The truth is that modern industrial agriculture is the inevitable spawn of the Mechanical mind. Prior to the late nineteenth century and the burgeoning of farm machinery and inorganic fertilisers, all agriculture was largely ecologically based or 'organic', using predominantly human and animal power in conjunction with natural material cycles (including the fertility cycle). But nineteenth- and twentieth-century science changed all that, particularly through advances in chemistry, biochemistry, engineering and then genetics.

This shift occurred because agriculture needed to be industrialised to support the industrialisation of the wider economy. Therefore, in less than two centuries, agriculture in developed nations has stopped being a system that was essentially self-contained, labour-intensive and biologically based, and thus was well organised for the capture of immediate natural solar energy. Today it is a fossil-fuelled, energy-intensive system that is hugely dependent on mechanical and chemical intervention, that disempowers humans and their communities through a labour-saving focus, and that is entirely market driven. The consequence has been huge environmental costs. Not the least of these is the destruction of long-coevolved, complex, self-organising ecosystems, the pollution of the environment with high-entropy waste products, and a poisoning of the consumer with the very food the industrial system has been so carefully designed to produce.

As the story of the rise of industrial agriculture is comprehensively covered elsewhere, I will only indicate key signposts along its journey. Agriculture has gone through a series of technological revolutions and refinements since its inception. By the late eighteenth century, these had culminated in what is known as the fourth or 'Early Modern' agricultural revolution. This hinged around superior crops, pastures and animal genetics, and an increased focus on animal rotations and green manures to maintain (even enhance) soil fertility. But under the pressures of rising populations, the emphasis then started to fall on fertilisers.

The real move towards fertilisers began in the German states in the nineteenth century. Curiously, a man interested in healthy soil humus (the dark organic material formed by decaying plants and animals) became known as the 'Father of German Scientific Agriculture'. It was Albrecht von Thaer (1752–1828) who, though initially practising medicine, set out to lay down the principles of a 'rational agriculture', with an early focus on agronomy (using plants for food, fuel and land reclamation).

However Thaer slipped up, for he worked in advance of modern knowledge about photosynthesis and the uptake by plants of elements such as carbon, nitrogen and oxygen from the air. So Thaer, in being partly right, argued that soil humus was the main plant nutrient and a major source of soil fertility. He called this the 'humus theory' – which later became the forerunner of modern organic agriculture. But Thaer had ignored the area of mineral balance and a wider range of elements.

Thaer's slip-up opened the way for two famous German chemists: Carl Sprengel (1787–1859) and his teacher Justus von Liebig (1803–1873); the latter an outrageously confident, innovative and self-promoting entrepreneur with an eye to the main chance. These two pioneered research into how plants gained nutrients (then called mineral salts) from both the soil and the air. In propagating their 'mineral theory of plant nutrition', they debunked Thaer's humus theory and his focus on plant carbon derived only from the soil. In effect, this was to swing the pendulum in agriculture from the organic side to the extreme opposite: that of high-input and non-organic industrial agriculture built around the new mineral nutrition theory.[10]

Liebig subsequently became famous for laying the foundations of modern industrial agriculture with his 'balance-sheet' theory. This stated that the harvesting of crops removed nutrients from the soil, so the onus for farmers was to put them back in. Complementing this was Sprengel's and Liebig's 'Law of the Minimum', which said 'a plant needs twelve substances to develop, and it will not grow healthily if any one of these is missing'.[11]

The result was the still prevailing 'NPK mentality' of restoring soil health by fertilising it predominantly with the three nutrients of nitrogen, phosphorus and potassium (N, P and K). And so the Mechanical mind triumphed with its reductionist view, which, in the biting words of the famous American soil scientist Charles Kellogg, reduced soil to 'a more or less static storage bin of plant nutrients'.[12]

Liebig also went on to develop an extract of malt, patent the production of concentrated milk powder for babies, and develop a meat extract derived from South American beef products. In the words of his biographer William Brock, he thus provided 'a model for the industrialization of the food processing industry upon which twentieth-century society is rooted'.[13] In other words, Liebig moved chemistry into the marketplace, positioning it for a significant role particularly in food production, nutrition and public health.

In 1843, the English agricultural scientist John Lawes formulated his 'patented manure', or the first modern fertiliser. Soon after, he and chemist Joseph Henry Gilbert built the first modern fertiliser factory. The use of grazing animals in a rotation system was now increasingly dispensed with in the new system known as 'high farming', which was characterised by a central focus on inorganic chemical inputs. A tide of mechanical-industrial thinking now swept away millennia of organic thinking. This brought into full flood the Mechanical mind when applied to agriculture. It would take until after 1930 for a concerted alternative or organic approach to industrial agriculture to be publicly activated and promulgated – as we shall see directly.

In the meantime, the rapid growth of cities during the industrial revolution meant there was a progressive abandonment of the recycling of nutrients between town and country: what Karl Marx perceptively called the 'metabolic rift'.[14] This non-cycling of city and town nutrients (that is, returning human waste products to the fields, as had occurred in the past) pushed farmers to think of alternative sources of soil enrichment – a demand quickly met by the burgeoning business of producing 'artificial manures'. And so the modern fertiliser industry took off.

But it hit a roadblock: an inability to produce enough nitrogen to meet exploding demand. However, again the chemists came to the rescue because in 1909 the German Fritz Haber first produced liquid ammonia from atmospheric nitrogen. He soon partnered with BASF chemist Carl Bosch to commercialise the production of what became the huge nitrogen fertiliser industry of today – called the Haber–Bosch process. Animal nitrogen recycling was now largely banished (as were the animals) from the mechanical world. Traditional crop rotations and fallowing were also increasingly abandoned in favour of a Liebig-like topping-up of the 'soil box' under continuous cultivation.

And so the world agrochemical industry went from strength to strength. This process was accelerated by the Second World War and concomitant technological advances. Two wartime chemical pathways in particular fully ushered in the testosterone-driven era of full-on industrial agriculture. The first came from munitions in partnership with the petrochemical industries. After the war, this led to ammonium-nitrate munitions stockpiles being converted into an abundant and cheap source of nitrogen, but made from non-renewable fossil fuels via the Haber–Bosch process. These fertilisers were developed and promulgated in tandem with other water-soluble, synthetic, purified fertilisers (such as superphosphate and its successors). The second chemical revolution came from the petrochemical industry again: the accelerated use of wartime synthetic pesticides for agriculture, led by the notorious insecticide DDT.

From the end of the war, this new chemical gospel then began to be rolled out – first in the developed world and then later in the developing world under the beguiling but ultimately disruptive and destructive program known as the Green Revolution. The result today is that industrial agriculture is dominant across the globe. It is driven by many of the world's largest and most powerful transnational organisations in the fields of chemicals, pharmaceuticals, genetic engineering, food manufacturing and global commodity trading, and it is universally backed by the most powerful nations of Western capitalism, while being equally embraced by leading non-democratic nations such as China and copied in their wake by developing nations. Many of the latter have rapidly ditched (or are in the process of ditching) an older peasant- or smallholder-based community agriculture that was dependent on human labour and traditional knowledge, craftsmanship, community and minimum industrial inputs, and especially the careful creation and recycling of humus. As Australian regenerative agriculture innovator P. A. Yeomans said in 1971, this completed the 'bastardization of agriculture'.[15]

Today, industrial agriculture is underpinned by seven main practices (along with a concomitant heavy dependence on fossil fuels): (1) intensive tillage; (2) monocultures; (3) application of synthetic fertilisers; (4) intensive irrigation; (5) chemical pest and weed control; (6) manipulation of plant and animal genomes; and (7) factory farming of animals.

Moreover, conventional industrial agriculture is built around two related goals: the maximisation of production and the maximisation of profit. This is why the approach is frequently termed 'productivism'. From a global perspective, problems have arisen because its practices tend to be implemented and integrated (a) without regard for their unintended, long-term consequences, and (b) without consideration of the ecological dynamics of agro-ecosystems. The frequent result has been an intermeshing of counterproductive practices that are destroying ecological systems and which are full of inherent paradoxes (such as escalating weed and insect resistance to chemicals, and the destruction of natural pest predators). These paradoxes appear to be largely ignored by practitioners and supporters of what is a powerfully dominant system.

The reason industrial agriculture is so powerful and dominant is because it is the logical spawn of modern capitalist society's world view and its enmeshment with the most powerful commercial and political entities on Earth. The interconnections of these elements, in conjunction with the Mechanical mind and an implicit belief in economic rationalism and neo-liberalism (including a blind faith in the benefits of free trade, for example), is due to what is called 'thick legitimacy'. This refers to the process whereby the agronomic and even agroecological and scientific legitimacy of industrial agriculture, in the words of social scientists Maywa de Wit and Alastair Iles, 'grow out of a web of legitimation processes in the scientific, policy, political, legal, practice and civic arena'.[16] The net result is to allow industrial agriculture and its food systems to pass 'credibility tests' (irrespective of the real truths) and to gain the status of being 'normal' and 'healthy'. This, therefore, cements the 'credibility' and normalcy of the productivist mentality: the belief that food output should be prioritised at the expense of other agricultural, natural and human values. In the process of this, the growth of alternative 'organic' and particularly regenerative farming systems are both stultified and even maligned in what amounts to a de-legitimisation process that erodes their credibility in the public's eye.

Therefore, the consequence of the rapid post-war roll-out of industrial agriculture is that just over seventy years of intensive modern farming has been disastrous for planet Earth. Such an approach to agriculture is not only unsustainable but also, as seen previously, a major contributor to tipping us into the Anthropocene through effects that are threatening the very life-support systems of the planet. Earth activist Vandana Shiva calls the impact of industrial agriculture 'Eco-Apartheid': an 'ongoing war against the Earth ... which has its roots in an economy which fails to respect ecological and ethical limits'.[17]

But, like it or not, we are inescapably in the Anthropocene. And that is where the story of this book becomes hugely relevant, because new and transformative responses will be required to replace or modify those we have now. The regenerative agriculture approaches described in this book are all about such transformation.

Moreover, such a transformative movement is enormously courageous because this approach is in the vanguard of attempts to transition our globe out of the life-threatening Anthropocene era and into one that is more self-sustaining: a transition that has no historical parallel.

The Genesis of an Alternative Agriculture

The act of cultivating land to raise crops and animals did not emerge solely in the Fertile Crescent in south-west Asia 10,000 or so years ago. It also arose independently at various times in China, India, New Guinea, South and North America, and probably also in parts of Africa – and, in some respects, in Australia also. Elements of this ancient, 'organic' agriculture are still practised around the world, where people, in the words of Wes Jackson (co-founder of The Land Institute), have looked 'to natural eco-systems precisely because they have featured recycling of essentially all materials and have run on contemporary sunlight'.[18]

Other people have more consciously sought to establish a sustainable agricultural system for our times, based on 'nature's wisdom'. This movement started in the nineteenth and twentieth centuries and now goes by various different names, but in its early phases it was called 'organic agriculture'. At its heart was a reawakening of Albrecht von Thaer's humus theory, but this time without the fatal flaws.

A number of leaders of the European Romantic movement in the nineteenth century – in protest against the Mechanical mind – espoused the best of the Organic mind, both generally and in specific relation to farming. One was German genius Johann Goethe (1749–1832), whose ideas were a major influence on Rudolf Steiner when he founded the first modern organic-farming approach, called biodynamics, in the 1920s (for this story, see Chapter 11). Meanwhile, as early as 1881, Charles Darwin weighed in with a book on earthworms, which promoted their crucial role in generating soil health.[19]

Accelerating from the early twentieth century to the end of the Second World War, in what can be called the first wave of ecological-agriculture thinking, writers and activists began articulating a new practice. This was triggered by an increasing awareness of the deterioration of farming and wider ecosystems, of soil health in particular and thus food quality, all of which were connected to overall declining human health in Western nations.[20]

This period saw the appearance of many of the principles and practices of the new organic-farming movement, particularly coming out of the United Kingdom, led by such leading lights as Sir Albert Howard and Lady Eve Balfour (the principles of Howard's knowledge having been derived from ancient Indian agricultural practices). But the USA also made an increasing contribution to the field from this time, notables including soil scientist William Albrecht, microbiologist Selman Waksman and regenerative agriculture pioneer Jerome Rodale.

Phase two of the modern and ongoing ecological-agriculture movement began in the late 1950s, as a counter-narrative to the post-war boom and emerging dominant industrial-agriculture practices (including its flagship of the Green Revolution). This wave, which was now calling itself 'sustainable agriculture', along with the earlier term of 'organic agriculture', coincided with the 1960s break-out in Western society of new thinking, feeling, music-making, experimenting and philosophising that is known variously as the New Age Generation, the Counter-Culture or the Aquarius and hippie movements. This saw the USA taking leadership of the modern organics and agroecology movements. Some of the high-profile American pioneers of this surge in new thinking about ecology, microbiology, soil health, food nutrients and holistic systems were Rachel Carson, Donella and Dennis Meadows, Barry Commoner, Lester Brown, Paul Ehrlich and Amory Lovins, along with the German E. F. Schumacher.

Paralleling this movement, there also occurred the rise of a modern agrarian tradition in the USA. Started by Aldo Leopold in the 1930s, this continued with the likes of Wendell Berry and later Wes Jackson, Fred Kirschenmann and others.[21] Internationally, many new thinkers and doers were popularising old and new eco-agricultural ideas and methods (for example, in Japan, Mokichi Okada and Masanobu Fukuoka).

Australia too had its own small band of pioneers, who helped foster the modern organic and ecological-agriculture movements here. Early post–Second World War Australia – as an agricultural exporter and because of its history – was still closely tied to Britain. So alternative-agriculture thinkers in Australia had access to emerging organic ideas via publishers such as Faber & Faber, who published Eve Balfour's *The Living Soil* in 1943. A group of organic farmers founded the world's first 'organic' society in Sydney as early as October 1944, and there were a number of farming and garden compost societies also.[22] In late 1958, Balfour came out from the United Kingdom to stay in Australia for a year.

Key Australian eco-pioneers included P. A. Yeomans, who evolved the Keyline Plan (see Chapters 7 and 16); popular author Elyne Mitchell, who wrote her prescient *Soil and Civilization* in 1946; and early broad-acre and organic-farming trailblazer Colonel H. F. White, who collaborated with nutritionist Professor Stanton Hicks to publish the landmark book *Life from the Soil* in 1953.

The third wave of organic farming began across the world in the 1980s, in response to the acceleration of industrial agriculture. This latter industrial phase saw the unleashing of new gene technologies and genetically modified food; a massive escalation in the use of chemicals; the sophisticated and intensified confinement of animals in feedlots; a vastly increased mechanical, standardised and monocultural approach; and the increasing dominance of giant multinational businesses in all aspects of agriculture. An outcome of this in the Western world was the astounding and alarming rise of multiple disease epidemics (including auto-immune diseases, cancers and obesity, plus issues such as allergies). And there were also dire knock-on effects for developing nations, including the disenfranchisement, disempowerment and pauperisation of peasant farmers.

It is one strand of this third-wave organic-farming response that the rest of this book is focused on: namely, regenerative agriculture – which contests the dominant industrial model. The term is derived from the verb 'regenerate', which has a number of connotations in the *Macquarie Dictionary*: 'to effect a complete moral reform in'; 'to re-create, reconstitute, or make over, especially in a better form or condition'; 'to generate or produce anew; bring into existence again'.

Regenerative agriculture therefore implies more than just sustaining something but rather an active rebuilding or regeneration of existing systems towards full health. It also implies an open-ended process of ongoing improvement and positive transformation. This can encompass the rebuilding or regeneration of soil itself, and of biodiversity more widely; the reduction of toxins and pollutants; the recharging of aquifers; the production of healthier food, clean water and air; the replacement of external inputs; and the enhancement of social capital and ecological knowledge.

The social–ecological element is a key aspect of regenerative agriculture. In addition to improved physical and mental human health, what this aspect also entails is the promotion of vital, coherent rural cultures and the encouragement of values of stewardship, self-reliance, humility and holism, particularly within the context of family farming. This inevitably entails resistance and insurgency against the ruling paradigm.

Robert Rodale, of the Rodale Institute, was the first to begin popularising the term 'regenerative' in relation to agriculture, in 1987, to 'distinguish a kind of farming that goes beyond simply "sustainable"'.[23] The Rodale Institute (founded by Robert's father, Jerome) also emphasises that farming should become 'once again, a knowledge-intensive enterprise, rather than a chemical- and capital-intensive or technologically driven one'.[24]

It is to this radical and transforming challenge that we will now turn. For this is a move that is not so much an 'about turn' – back to something embedded in the old Organic mind (though this is part of it) – but a truly radical and transformative turn that overthrows (and yet utilises the best components within) the Mechanical mind. Involved, therefore, is an evolutionary shift towards a new mind: one that 'looks to natural ecosystems' and 'nature's wisdom'.

PART II

REGENERATING THE FIVE LANDSCAPE FUNCTIONS

CHAPTER 3

An Indivisible, Dynamic Whole

The true problem of agriculture, and all other land-use, is to achieve both utility and beauty, and thus permanence. A farmer has the same obligation to help, within reason, to preserve the biotic integrity of his community as he has, within reason, to preserve the culture which rests on it. As a member of the community, he is the ultimate beneficiary of both.

ALDO LEOPOLD, 'The Land-Health
Concept and Conservation'[1]

O ver recent years, I have been refining a teaching model on ecological literacy; testing and developing it through talks to farming groups and through teaching university students. I call this the five landscape functions model, which is part-based on the work of Allan Savory.

From his ranch Dimbangombe near Victoria Falls in Zimbabwe, wildlife ecologist Allan Savory has sparked a revolutionary grazing management system that regenerates landscape functions and complex ecological systems. He calls it holistic planned grazing, and it is having remarkable impact across many millions of acres of the world's degraded landscapes, including on our own farm. The approach constitutes one of the greatest forward leaps in agriculture since domestication 10,000 years ago. This is because it combines ecological systems-thinking with modern management and scientific knowledge. As a result, it has turned domesticated pastoral agriculture into a new system of regeneration that can combat climate change, land degradation and desertification while delivering nutrient-dense, health-giving food.

However, holistic planned grazing is not the only form of regenerative agriculture, as we will see in subsequent chapters. What is central to gaining ecological literacy and thereby enabling landscape regeneration is that all the different forms of regenerative agriculture have four essential eco-system processes at their heart. In this book, I have added a fifth: the social or human factor. Thus, the five crucial landscape functions or processes described here are: (1) the solar-energy function (focused on maximising the capture of solar energy by fixing as many plant sugars, via photosynthesis, as possible); (2) the water cycle (focused on the maximisation of water infiltration, storage and recycling in the soil); (3) the soil-mineral cycle (focused on inculcating biologically alive and healthy soils that contain and recycle a rich lode of diverse minerals and chemicals); (4) dynamic ecosystems (focused on maximum biodiversity and health of integrated ecosystems at all levels); and (5) the human–social aspect (focused on human agency triggering landscape regeneration by working in harmony with natural systems).

The different sections of Part II are structured around these functions and how they contribute to the health and long-term productivity of our farmed landscapes. A crucial point is that all functions are dynamically intertwined and indivisibly connected via reciprocal causal links. That is, none can act alone, for all are dependent on, or linked to, the others. Nevertheless, for clarity, I will go through them one by one, in each case providing stories of some of the remarkable farmers who have thrived through working in harmony with them, before bringing them all together. In doing so, I hope to extend and promote this platform of ecological literacy.

In 1830, Scotsman John G. Robertson left his homeland for Australia with 'a light purse' – one half crown and sixpence in his pocket. He landed in Van Diemen's Land (later named Tasmania) in 1831. After working as an overseer for nine years for local farmers, he had saved £3000, and this enabled him to seek cheap land and stock across Bass Strait in the new colony of Portland in 1840. 'I had no difficulty in finding a run,' he said later, and he took up land near the present town of Casterton. He then placed 1000 ewes on the initial 11,000-acre farm he named Wando Vale. Within weeks, all the land around Robertson had been gobbled up by other squatters, each one looking for permanent water and scared to go further out 'from fear of the natives'.[2]

In a buried theme that still runs deep across the subconscious of this nation and its occupied landscapes, though Robertson professed sympathy for 'the natives', he nevertheless 'took up' the 'vacant' land they were still living on and which they had managed for millennia. This occupation was courtesy of the dirty work of others. Robertson noted:

> the first day I went over the Wando Vale Station to look at the ground I found old Maggie (that Sir Thomas Mitchell gave the tomahawk to) fishing for muscles [*sic*] with her toes, in a waterhole up to her middle, near where the Major crossed the stream ... nearly all her male relatives were killed three days before I arrived on the Wando by Whyte Brothers [because the natives had stolen some sheep] ... the Whyte Brothers' party, seven in number, surrounded and shot them all but one. Fifty-one men were killed, and the bones of the men and sheep lay mingled together bleaching in the sun at the Fighting Hills.[3]

Subsequent to their 'conquest' of this country, all of Robertson's neighbours were broken and dispersed like chaff in the strong wind of Australia's first major recession, of 1841 – except for Robertson. By 1845, he had expanded Wando Vale to nearly 12,000 acres, running 7300 sheep, and after this he kept from 8000 to 10,000 sheep 'when full stocked' on his 'pretty little station, well-watered'.[4]

At the end of his tenure, before he returned to his homeland in 1854, Robertson boasted, 'there is not a station in the Portland District better managed for its size, both as regards economy and care of man and beast on it'.[5] And yet, despite his self-proclaimed excellent management, the landscape and its functioning had dramatically collapsed around him.

We know this because in a record he wrote in 1853 for the lieutenant governor of Victoria, Charles La Trobe, Robertson laid out one of

Australia's first clear descriptions of rapid ecological collapse due to set-stocked pastoralism (the practice of grazing livestock continuously in a particular paddock) and ignorance about how landscapes functioned.

'When I arrived through the thick forest-land from Portland to the edge of the Wannon country, I cannot express the joy I felt at seeing such a splendid country before me,' he stated concerning his arrival in 1840. 'I looked amongst the 37 grasses that formed the pasture of my run. There was no silk-grass which had been destroying our V.D.L. pastures ... The sheep thrived admirably.'[6]

On such terrain, all remained well for a while. 'The few sheep at first made little impression on the face of the country for three or four years,' said Robertson, as '... all the landscape looked like a park with shade for sheep and cattle'. But then came the transformation:

> Many of our herbaceous plants began to disappear from the pasture land; the silk-grass began to show itself ... The patches have grown larger every year; herbaceous plants and grasses give way for the silk-grass and the little annuals, beneath which are annual peas, and die in our deep clay soil with a few hot days in spring, and nothing returns to supply their place until later in the winter following. The consequence is that the long deep-rooted grasses that hold our strong clay hill together have died out; the ground is now exposed to the sun, and it has cracked in all directions, and the clay hills are slipping in all directions; also the sides of precipitous creeks – long slips, taking trees and all with them. When I first came here, I knew of but two landslips, both of which I went to see; now there are hundreds found within the last three years ...
>
> One day all the creeks and little watercourses were covered with a large tussocky grass, with other grasses and plants, to the middle of every watercourse but the Glenelg and Wannon, and in many places of these rivers; now that the only soil is getting trodden hard with stock, springs of salt water are bursting out in every hollow or watercourse, and as it trickles down the watercourse in summer, the

strong tussocky grasses die before it, with all others. The clay is left perfectly bare in summer. The strong clay cracks; the winter rain washes out the clay; now mostly every little gully has a deep rut; when rain falls it runs off the hard ground, rushes down these ruts, runs into the larger creeks, and is carrying earth, trees, and all before it. Over Wannon country is now as difficult a ride as if it were fenced. Ruts, seven, eight, and ten feet deep, and as wide, are found for miles, where two years ago it was covered with tussocky grass like a land marsh. I find from the rapid strides the silk-grass has made over my run, I will not be able to keep the number of sheep the run did three years ago, and as a cattle station it will be still worse; it requires no great prophetic knowledge to see that this part of the country will not carry stock that is in at present – I mean the open downs, and every year it will get worse, as it did in V.D.L.; and after all the experiments I worked with English grasses, I have never found any of them that will replace our native sward. The day the soil is turned up, that day the pasture is gone for ever as far as I know.[7]

Here was a well-educated farmer who reckoned himself one of the best managers in the district applying long-tested British pastoral practices. Yet his entire landscape collapsed in extraordinarily rapid time under his very feet and those of the horse he rode. Clearly, even the apparently well-watered, higher-rainfall, British climate-like country that the explorer Thomas Mitchell had labelled 'Australia Felix' ('happy Australia') in 1836 behaved in a totally unanticipated manner. So, what was going on?

This new land of Australia needed to be handled very differently from the wet and green country of the Europeans who were trying to settle it. The destruction of the landscapes around Wando Vale occurred in record time: less than fourteen years, and this before the rabbit explosion that originated not far away from Robertson's farm a decade later. Strikingly, from an agricultural perspective, even the 'best' of these new Australian landscapes were incredibly fragile if managed the wrong way. Degrade the deep-rooted perennial grasses and their coevolved plant, animal, insect and microbial life and be prepared to face the consequences. For, as

Robertson's soil quality deteriorated, running water carried off the spoils, and underlying salt reared its ugly head.

Robertson had quickly seen the impact of mismanaged grass–soil–water interactions, so he decided to bail out. In 1854, just fourteen years after he developed his 'pretty little station', he bolted for home with his monetary profits courtesy of eroded natural capital.

Since first settlement, Australian farmers have repeatedly found out the hard way that grazing mismanagement can quickly trigger land degradation and desertification. This begins with gross interference with the prime healthy landscape function: that of the solar-energy cycle. Inappropriate grazing depletes a diverse grassland through first 'eating out' the most succulent, broadest-leaved, deeper-rooted perennials, native legumes, small shrubs and other key species. The less grassland there is, the less solar energy is captured, converted to carbon and deposited in the soil via photosynthesis.

Further landscape degradation then occurs via the action of another vital element: water. With no plant life to protect and penetrate the soil, it becomes more compacted, shallower and less able to retain water. Instead, rainwater runs off the land instead of being retained, meaning that precious topsoil is easily washed away. As we shall see, once the solar and water cycles are disturbed, so is the third key function, that of the soil-mineral cycle, and then also the fourth: dynamic ecosystems. And all because of the impact of the fifth landscape function: humans.

Dipping his quill in as much vinegar as ink in his poem 'Australia', A. D. Hope memorably portrayed his home nation as 'a vast parasite robber-state/Where second-hand Europeans pullulate/Timidly on the edge of alien shores'.[8]

It is hard to disagree. Over ninety per cent of Australians do crowd the littoral, living there, working there and holidaying there. Moreover, concerning this 'vast parasite' country, what shocked me as I set out on the journey of discovery and awakening I recount in this book is that those who ostensibly live close to Mother Earth – we farmers – are, with few exceptions, as illiterate as anyone else when it comes to reading our landscapes. Our Mechanical minds have made us landscape dyslexic.

Away from the coastal river valleys or the irrigated areas of the Murray–Darling Basin, and away from choice volcanic locations such as the

Liverpool Plains and Darling Downs, Australian landscapes do not match Western preconceived images of what an agricultural landscape should be. This 'should be', more often than not, involves neatly rectangular, moist, dark-soiled fields, vineyards, golden crops and verdant green irrigated rows of vegetation. For this is a far more pleasant and salving image than that of a dried dam in harsh Australian light, of dust clouds behind vast mobs of stock in shrub-dotted rangelands, or of beach-white ripened and baked native grasslands in mid-summer.

Despite some of our modernist painters such as Sidney Nolan, Russell Drysdale and Fred Williams attempting to present a different, Australian view, this 'wet country' expectation runs deep and powerful. If we then throw in the fact, as described in Chapter 1, that our ancient Gondwanan ark is both biogeochemically and functionally quite different from the young landscapes of England, Western Europe and North America, then we have a huge problem.

That is, our expectations of landscape create a disjunction between a culturally inherited ideal and Australian reality. Chapters 1 and 2 revealed that this has a long and powerful cultural tradition. Tragically, despite us immigrants having had over 200 years to apprehend this disjunction, it continues to have devastating impacts on our landscapes today. I believe this is because, rather than seeking to understand how our landscapes really function and to regenerate them, our tendency is to either do a John Robertson – leave a wasteland behind and move on to greener pastures – or else 'bodge' things up by using the short-term approach of industrial-agriculture inputs and techniques.

When I look back from my present perspective, I realise with a jolt that, over forty years ago, I viewed my landscape as if 'through a glass darkly'. To me at that time, it was still an inanimate resource to be used for grazing sheep and cattle, assisted by the occasional chemical and/or mechanical intervention with fertiliser, disc-plough and seed drill. Our land then seemed to hold no patterns nor subtleties beyond the obvious. Concepts such as ecological succession (let alone self-organising systems), positive, human-induced ecological change and the fundamental landscape functions were anathema, and I gave little thought to how animals were grazed beyond some rudimentary rotation on some of our lucerne (alfalfa) paddocks – a technique I had learnt from my father. At least half of the place was in paddocks of 150 or more acres, where animals were left at a deemed appropriate stocking rate for at least half the year, sometimes more. As a consequence of this set-stocking, thistles and the perennial exotic grass 'serrated tussock' – the so-called 'weeds' that thrived under this system – were demonised and agonised over, especially because they seemed at best never to diminish, and at worst to perpetually increase.

Our grasslands, in retrospect, had become simplified ecological systems. They definitely weren't increasing in diversity, while our monocultures of lucerne exhibited at least fifty per cent bare ground for much of the year. Because of this simplification, I wasn't aware that our original grasslands had a healthy mix of not just winter/spring-active grasses (of which we had only a few left) but also a whole raft of summer-active grasses (such as kangaroo grass – *Themeda australis*). These had once covered most of our and other regions, but because of their palatability, they had quickly been 'eaten out' under a constant set-stocking regime. As we will see later, this diversity of complementary grass types that could provide year-round green feed – the winter/spring-active grasses (what are now known as C_3 grasses) and the summer-active grasses (or C_4 grasses)[9] – are hugely relevant to this story of regenerative agriculture: whether in new holistic grazing approaches or in some remarkable regenerative cropping approaches to emerge out of Australia.

Concerning other native vegetation, I was equally blind because not only was I ignorant of the diversity we had lost but also what gum trees and shrubs remained were nearly all geriatrics. And I thought I was running a good 'show'.

Managing grazing animals in grasslands, you would think, should be pretty basic. In developed countries such as Australia and the USA (where shepherd labour is prohibitively expensive), the fence is the method of control. Once the straight lines of steel and wire are arbitrarily stamped on the landscape, managing one's animals can be done easily and without much planning, right? Just lock them in a paddock, ensure they have enough feed and water, and Bob's your uncle. And then just fiddle a bit with paddocks and moves through the year, hoping the feed won't run out on your traditional fixed number of carried stock, and just buy in the fodder to get them through if it does. That's pretty much how I operated for nearly two decades (allowing some sophistication for pregnant animals and their progeny, along with those lucerne rotations).

Today I realise I had completely overlooked the most important of all factors, the keystone of the whole operation: that our farm was a complex and dynamic series of ecological systems, and that our landscape actually functioned in specific but sensitive ways (and even, if allowed, its key landscape functions would constantly self-organise themselves towards resilient ecological health). The last things I would have considered are that, first, my decision-making on how I grazed the landscape – like John Robertson so long ago – was causing ongoing land degradation, and, second, that my grazing animals could actually be a huge ecological revitalising tool.

It is small comfort that later I realised I was not alone in my initial thinking. In fact, I shared my outlook with an overwhelming majority of landscape managers. Indeed, the presence of an alien, contrary view about managing grazing animals via a totally different approach is still not just an elephant in the room for most livestock managers, but rather it constitutes a type of invisible parallel universe. And certainly, in the first decades of my farming career, this universe was way outside my ken.

The 'five landscape functions' model is an excellent toolbox for land managers because it comprises a readily understood pathway to ecological literacy. Without such literacy, we cannot regenerate our landscapes. Now we know 'literacy' means the ability to read, and so in this context ecological literacy means the ability to read a landscape: to appraise the state of its health and how it is functioning, and thus to know how to address any issues. David Orr, professor of environmental studies at the pioneering liberal arts institution of Oberlin College in Ohio, writes insightfully and passionately about ecological literacy. He states that capable ecological literacy 'requires the more demanding capacity to observe nature with insight, a merger of landscape and mindscape'. To make his point, he quotes the extraordinary nature writer Barry Lopez, who once observed, 'The interior landscape responds to the character and subtlety of an exterior landscape; the shape of the individual mind is affected by land as it is by genes.'[10]

My view is that Lopez is correct. Therefore, this shaping of the mind can have polar-opposite impacts, depending on whether one is used to farming a heavily controlled and modified industrial landscape or a site that is re-attaining vibrant diversity under a regenerative approach.

While general literacy is driven by the search for knowledge, Orr quite rightly notes that 'ecological literacy is driven by the sense of wonder, the sheer delight in being alive in a beautiful, mysterious, bountiful world' – or what is known as 'biophilia' (the innate human affinity with living things). In short, I believe one cannot gain true ecological literacy without a great empathy with, and understanding of, nature and how it functions. Thus, one's heart also needs to be involved.

Pioneering ecological thinker Garrett Hardin once said of ecological literacy that it encompassed the ability to ask, 'What then?' But one can only ask such a question with a good knowledge of not just the fundamental basics of ecology but also the history and flow of ideas and historical development of how people and societies relate to each other and natural systems. This also means a familiarity with the processes of evolution, coevolution and self-organisation (to be addressed later in Part III); with the current predicament Earth is in (as Orr describes it, with 'the vital signs of the planet and its ecosystems'); and with the dynamics of the modern world, and thus why we as a species have become so destructive. This therefore includes questions that arise if a landscape is diagnosed as 'unhealthy'. Because 'What then?' demands that we act for renewal and restoration, for regeneration and open-ended health.[11]

In the end, David Orr encapsulates what I see as the transformative process of becoming truly ecologically literate: that an 'Earth-centered education ... is that quality of mind that seeks out connections' (not narrow specialisation); that 'the ecologically literate person has the knowledge necessary to comprehend interrelatedness, and an attitude of care or stewardship', along with 'the practical competence to act on the basis of knowledge and feeling'. Ecological literacy, says Orr, 'is to know that our health, well-being, and ultimately our survival depend on working with, not against, natural forces'. As systems-thinkers Fritjof Capra and Pier Luisi reaffirm, 'in the twenty-first century the well-being, and even survival, of humanity will depend crucially on our "ecological literacy" – our ability to understand the basic principles of ecology, or principles of sustainability, and to live accordingly'.[12]

An Indivisible, Dynamic Whole

As I have discovered in my own slow and serendipitous journey towards some degree of ecological literacy, there are clear reasons for our obliviousness to how landscapes function. The first goes back to deep cultural training (as discussed in Chapter 2). The second is that ecosystems are hard to apprehend. Not only are they complex and dynamic but they also comprise aspects that are microscopic, underground or invisible. For instance, while it is obvious that plants grow in the soil and gain nutrients there, what is not is that much of their substance is derived from either gaseous molecules in the air or else an unimaginably huge zoo of underground micro- and other organisms that can weigh ten times more than life above ground.

And so the driving systems of the landscape and its interrelated networks are invisible until evidence of their functioning or malfunctioning is physically expressed. Nothing new here, as this is an age-old conundrum – and one so clearly revealed by John Robertson. Until one's understanding and perception changes, and therefore one's mind changes, we can be blind to what is all around us. That is why understanding the five landscape functions is so important to landscape managers and, in turn, everyone on the planet.

CHAPTER 4

An Upside-Down World

Grasslands are distinctive in that they require active management.
To not act is to fail.

NICHOLAS WILLIAMS, ADRIAN MARSHALL and
JOHN MORGAN, *Land of Sweeping Plains*[1]

T he solar-energy function is a good place to start because it is the key
to all life and its systems, and to how our farmed landscapes can
work properly and regeneratively. Green plants use the abundant source of
the sun's energy to take carbon from CO_2 in the atmosphere and convert it
to carbon-based sugars. This process is called photosynthesis, and its sugars
are the fuel of everything that lives on Earth. In driving all of life's processes, the solar-energy cycle therefore also drives human economics, our
nations and civilisations.

Our management of landscapes actually determines how much of this
one-way energy flow from the sun is captured and put to good use. It is that
simple. In other words, if we can imagine plants as solar factories covered
in solar panels (chloroplasts) on broad green leaves, then clearly such green
plants are the foundation of all of life's processes on land. Therefore, to
capture more solar energy and use it more efficiently, we farmers simply
need to expand our primary base of active solar panels on green plants.

This foundation is also called the 'first trophic level', and this idea is a
fundamental ecological factor that all farmers and food consumers need to
comprehend. Briefly, a food chain in an ecosystem can be viewed as a giant

pyramid. The foundation or base trophic level is that of photosynthesising green plants – the energy fixers. Through converting the sun's rays to sugar, they produce, say, one kilogram of green grass, a hundred per cent of the energy for which has come from the sun.

The second trophic level up from plants are the primary consumers, the grass-eaters: domestic herbivores, bilbies, bandicoots, mice and so on, and even wood ducks. But this second level can be both above ground – as with the grass-eaters – or below ground. Below is that vast, seething population of microorganisms and other creatures that feed off the roots, sugars and other exudates of the plants: what are called the detrivores and decomposers (which actually feed at each trophic level). But with each trophic level after the primary energy-fixing base, there is a ninety per cent loss of energy in conversion to biomass (the rest of the energy being excreted as waste products and chemicals). So by the time a sheep finishes eating the original one kilogram of green grass and its energy, only one hundred grams of that original energy is left for whoever eats the sheep and other grass consumers.

The next trophic level up from the primary consumers are the secondary consumers: animals such as quolls, snakes, cats and foxes that prey on the mice, bilbies, and so on. Again, there is a ninety per cent loss of energy in moving through this level. Finally, at the top of the energy or eco-pyramid is the tertiary consumer trophic level: such as large carnivores, birds of prey and humans. And yet again we see a ninety per cent loss of energy in functioning at this tier.

Clearly, therefore, to get more energy into a system to drive healthy landscape function, we need maximum possible photosynthetic conversion – i.e. maximisation of the solar-energy capture. In a managed agricultural situation, this means as much year-round green leaf mass – including grass-land – as possible. This can only be achieved by three methods: (1) increased duration of plant growth; (2) increased density of plants; and (3) increased total leaf area. That is, lots of green leaves for as long as possible. Conversely, these three drivers are undermined by ploughing a paddock, overgrazing it into the ground or leaving it as bare fallow for long periods.

It was a cool, early-June winter's day some years ago. I was cruising up the New England Highway on the Northern Tablelands of New South Wales in my ute (pick-up truck). That morning, I had climbed from the head of the rich, dark-chocolate soils of the Liverpool Plains through Australia's country music capital of Tamworth, exiting through lush eucalypt forests onto the frosted, straw–khaki grasslands of the high New England Table-land. A large road sign proclaimed I was in 'Celtic Country', which I found bemusing given the Antipodean surroundings. Then another sign announced I was also in Anaiwan Country (of the Aboriginal Anaiwan people). Further along still, beside a large granite rock, was a statue of a bushranger and a proclamation that the land around there had also been bushranger country. While yet another sign proclaimed Uralla as the 'Foodie Capital of New England'. Clearly, this area had diverse landscape interpretations, if not a confused identity.

The reason for my trip was to visit a landscape manager on the western edge of the New England Tableland: Tim Wright. He was regarded by many as one of the leading exponents of the Savory system of holistic planned grazing in Australia, and I wanted to see what all the fuss was about.

Even today, most long-settled pastoral districts, especially in eastern Australia, bear the traces of a group of early settlers who, by one means or another, aggregated large acres. These 'old' families constitute the nearest thing to a British landed aristocracy in Australia – a model that some aspired to in behaviour and outward appearance, and which led to them being dubbed the 'squattocracy'. The Wrights settled early on in the New England Tableland and could be said to have been part of the New England squattocracy. However, it would be rash to prejudge the branch Tim Wright belonged to.

His grandfather Phillip (P. A.) and father, Peter, were leaders in rural education (including in natural resources and ecology) and had been instrumental in founding Australia's foremost rural-based and focused university: the University of New England in Armidale. Moreover, Tim's aunt Judith Wright was one of Australia's most famous poets, and one of the first to penetratingly write about such then unfashionable ideas as Aboriginal issues and environmental conservation.

Twenty or so kilometres west of the old township of Uralla, down a timbered road in the more atypical lighter granite country of New England, I found Tim Wright's farm entrance. Beside the cattle grid was the sign 'Lana', and beyond the grid wound an inviting road between shady eucalypts. Another sign inside the entrance proclaimed Lana as a 'Wildlife Refuge' – which Tim's father had declared back in the 1960s.

Tim Wright had pedigree both in thinking a bit differently and in the sphere of public service. I had first met him a year previously at a unique two-day gathering of a wide and eclectic group of farmers, scientists, activists and other thinkers involved in regenerative agriculture. This was held at the Australian National University's retreat of Kioloa on the New South Wales south coast. I have an abiding memory of Tim's performance at the end of that workshop when he was asked to be one of those summing up the importance of regenerative agriculture to our collective futures. He only had five minutes and a few PowerPoint slides but left a lifetime impression on me because he wept as he tried to explain the importance of what he and others were trying to do to bequeath a better world to our children and grandchildren.

I drove in a kilometre or more through a tree-lined avenue with open glades, dappled shadows and a landscape breathing 'health' to arrive at a comfortable, unpretentious homestead. Tim was in his late fifties, with a hint of greying hair, and of medium height, slim build and a handsome square face. He had the firm, callused handshake of most working farmers (in his case, he still does most of the manual labour on his farm, such as fencing and wool classing). His penetrating eyes searched me, and there was a quiet, considered confidence that came from four decades of managing varying seasons and tough droughts, but also from spectacular achievement. In addition, there was a hint of reserve towards a newcomer whose motives he was as yet uncertain of. Later I would understand this because being 'a prophet in one's home town' for a radical new ecological agriculture (and in the conservative Australian bush) is often a hard and lonely road. It is a path that leaves scars.

Tim obtained a university degree in agricultural studies at Orange Agricultural College. He took out of it the need to question the assumptions that farmers take into their day-to-day decisions. As he soon explained, a series of circumstances following the early 1980s drought led him not only to rip off his blinkers that had kept him to 'the same old sort of way of farming I inherited when I took over here' but also to avoid going down any traditional tunnels altogether.

Tim said of that drought, 'It took four years financially to get out of that one, and probably five or six years for the land, at least. After that was when I started to question everything: what I was doing here; the stuff I had learnt at college; some of the scientific stuff I was reading in journals ... I started to think, "Well, there's got to be another way. I've got to look for an alternative; otherwise, I'm not going to survive here."' The problem was, in the mid-1980s, the solution to Tim's problems hadn't yet arrived in Australia.

However, in 1989 it did arrive, when a man who would become one of the great catalysts of regenerative agriculture in Australia – Queensland agricultural scientist Terry McCosker (whose story is told in Chapter 19) – brought to Australia Allan Savory's previous business partner and educator, fellow Zimbabwean Stan Parsons. That same year, Tim attended the second ever intensive course in Australia based around the new ideas coming out of Africa under the rubric holistic planned grazing.

He travelled far north to Yeppoon in Queensland to do the course. This proved invaluable as it got him 'off his own dunghill' and into a new environment where, as he later said, 'I didn't feel threatened.' It was to change Tim Wright's life. 'I spent what I thought was a lot of money,' said Tim, 'but it's turned into a million dollars in terms of value to the land and all the rest of it. This was because light bulbs went off that first day for me. And I was open. I was looking for alternatives.'

Before going for a drive, we drank a cup of tea in Tim's spacious living room-cum-sun room. He explained the significance of the 1980s drought in triggering his change. 'In 1981, I can remember coming in and saying to Karen [his then wife], "Oh geez, I wish I had some more paddocks; I've run out of paddocks." But now I realise that wasn't the issue: it was the fact that I didn't have a plan. The secret is how you use what you've got.'

On returning from Queensland, Tim enthusiastically began applying what he had learnt at the course. But he also made plenty of mistakes, because there was as yet little collective experience about how to drive this new radical grazing and management vehicle. However, it was on reading Allan Savory's book *Holistic Resource Management* for a second time that the final pennies dropped for Tim. These involved the crucial elements of planning and setting a holistic goal; of having a structured holistic management decision-making process that one constantly reviewed and utilised; and cracking the counter-intuitive but vital component of rest and recovery of his paddocks.

By the time the next major drought came along in 1994–95, Tim had thought things through. 'In 1981, I had complained I didn't have enough paddocks,' he said. 'This time, I found them. I just fenced ahead of the stock, slowed the rotation half a day, and suddenly you've got another month's rest. Pretty simple. So I was fencing ahead of the drought, making more paddocks, creating more rest in the other paddocks and making better use of that haystack in the paddock. While our neighbours and other locals were going into town to buy hay and grain, or trucking stock out, I was passing them as I came back home with wire and poly pipe on the back of my ute.'

In short, Tim was maximising his solar-energy function through better grazing management and improved grass utilisation. His creation of more paddocks enabled him to avoid overgrazing, and this led to him maximising his photosynthesis capacity via the three methods of increasing duration of plant growth, increasing plant density and increasing total leaf area. As we shall see, this extra photosynthetic energy had profound positive effects on his other landscape functional elements, both above and below ground.

What is little appreciated, however, is that microorganisms (especially bacteria and fungi) play a crucial two-part role in making fixed solar energy available for a healthy landscape. First, soil microbiota feed off soil carbon and plant and root material derived from photosynthesis, thus releasing energy. But second, the only way cellulose (the core building material of grass and other plants) in pastures is digested is via microbes hosted in herbivore guts (including, surprisingly, in termites). Crucially, cellulose is the most common and complex sugar created by plants via photosynthesis.

The result of maximising solar-energy capture in the landscape meant that the 1994–95 drought made barely a blip on Tim Wright's radar screen. This enabled him to realise that, in all but the most prolonged cases, what the vast majority of Australian graziers call droughts are actually man-made. Tim calls them 'just dry spells'.

Today Tim has over 300 paddocks, averaging twenty acres. He started with just a few fifty-acre paddocks. He is now moving towards a twelve-acre average. Moreover, he has divided his farm into eight sub-farms (what he calls 'farmlets') for better management. 'Having stacks of paddocks is important,' he concluded, because 'the tool of stock density was really something to work with'. By this, he meant emulating the regenerative effects of vast migratory animal herds in Africa, which had been a key trigger to Allan Savory evolving his new system.

In short, Tim had immediately grasped the essence of rest and recovery for desirable plants so as to enhance his solar-energy fixing capacity. The concept 'rest' meant his plants did not run out of energy and die, and it meant the most valuable, broader-leaved, deeper-rooted, succulent perennial plants in particular were not preferentially overgrazed to the point of dying out first – which is what occurs under traditional management. This was also what happened to John Robertson's land back in the 1840s and '50s, and was why his landscape collapsed so quickly around him.

Tim now calls himself simply a holistic manager, not holistic farm manager, because, he said, 'We're managing more than a farm. We're managing ecology; we're harnessing sunlight energy through grass; we're managing soil and rehydrating the landscape; we're managing biodiversity and economics; and we're also educating other people in the world.' Tim summed this up by saying it was all about decision-making. 'The grazing part is minor in the scheme of things,' he reiterated. 'You need to plan with a clear goal.'

To Tim Wright, his own approach seems straightforward, but to ninety-nine per cent of other graziers and pastoralists, it is completely upside down and nonsensical. 'The smaller you drive your paddocks,' Tim patiently explained again, 'and the more paddocks you get with big mob density, the more stock you run and the better it works and the less fertiliser you need … and we're still leaving more feed behind in the paddocks than we probably should … We wouldn't be utilising more than half the feed … Everything is all flowing now. I have to keep splitting paddocks to keep up with the feed, then all of a sudden you've got more rest so you've suddenly got more feed, and on you go. The secret is managing both the plant and the animal.'

This sounded so back to front to me at the time that I had to pause, sipping long on my tea, to digest it all.

When Tim began really changing in the early 1990s, he had a traditional grazing operation. His father and grandfather had been pioneers of the sub clover, pasture improvement and aerially spread superphosphate revolution in New England. Tim had fully embraced this, and he annually 'supered' and ploughed up country to put in 'improved' grass and legume species. On his conversion, he largely ceased pasture improvement and began radically reducing fertiliser application. Today he hugely values native pasture species and their ecological succession. These are his most valuable pastures, he told me, and the ecological succession process, which leads to more desirable species, has accelerated under his holistic management.

As mentioned earlier, Lana is located in the western and 'lighter' (less fertile), lower rainfall part of New England. Tim calls it 'the pits of New England'. Compared with the rich basalts and higher rainfall of the eastern section, the region normally runs less than half the stock with less productivity. Today Tim has similar performance to the best areas in the east, but with extraordinary drought resilience to boot in his thirty-inch (762-millimetre) rainfall region. Prior to 1989, in an average season Lana would run around 8000 dry sheep equivalents,[2] and much less in winter. Today Lana runs an average 20,000 dry sheep equivalents, and this is increasing at around ten per cent a year. This is at the same time as fertilising and pasture 'improvement' have been all but eliminated.

In other words, with holistic planned grazing now well established, Tim told me that 'based on my grazing charts, I calculate the actual carrying and yield today has trebled since we started'.

There are also many other benefits. As nutrient concentrations became more dispersed and high densities of sheep and cattle ate both weeds (such as thistles) and their seeds, Tim was able to reduce unwanted vegetable-matter contamination in his wool. Consequently, the increase in the value of Tim's wool clip paid for his wire and poly-pipe investment. The tensile strength of the wool (a trait desired by wool processors) has vastly increased also, adding further value. Weed invasion has more generally decreased because his ground cover sits close to one hundred per cent with deep litter (thus negating the role of weeds, many of which act as constructive 'pioneer' species on bare ground). There are multiple biodiversity benefits in aspects such as parasitic wasp control of pasture grubs, and shade for livestock in hot summers due to accelerated native shrub and tree recruitment – and this in addition to the reappearance of hitherto unseen C_3 and C_4 grasses. Due to the rapid paddock rotations and healthy grass swards, animal drenching with synthetic chemicals for internal parasites has also been slashed.

Besides the obvious productivity improvement, perhaps the most startling change is the drought resilience of Lana. This is because of the way Tim's ecological management has transformed his landscape's functioning.

The day I visited Tim Wright, New England was at the end of yet another scouring national drought (what is now called the 'Millennium Drought') – a prolonged dry spell from 2003 into 2009 depending on your location in Australia. Most of Tim's neighbours had been forced to totally destock (they had sent sheep and cattle on agistment as far afield as western New South Wales), or sell most of their animals, or else, if they kept half their breeders, buy in grain and hay at great expense to keep the survivors alive. On Lana, Tim was still running record stock numbers, was not feeding and still had plenty of feed left in the paddocks. 'All we did was not increase carrying this year,' he said. But that wasn't the end of it.

All of Tim's neighbours and others nearby had dry dams – normally their main source of water because underground water in their granite country is largely unavailable. Consequently, the neighbours had taken the opportunity to clean the silt from their dam beds. By contrast, not only had Tim's dams not silted up, but at a minimum they were still over two-thirds full. Beginning with the solar-energy cycle via his grasslands, what Tim's ecological management had done was transform his soils and their absorptive capacity (that is, positively influence both his water and soil-mineral cycles). This rehydrated his landscape, which now stored huge volumes of water instead of causing it to run off. As a result, his springs were constantly recharged and dams usually full. 'Our silt scoop hasn't been used since the 1980s drought,' confirmed Tim.

My brain by now felt like scrambled eggs as it struggled to deny what I was hearing. As if not satisfied with mucking up my head, Tim decided it was time to go for a drive and finish the job on me properly.

We did a circuit of his property, and despite the 'drought' that was allegedly still ongoing, I witnessed a farm humming with life, diversity, vibrant energy, much greenness and an exudation of health. The first thing I noticed was that the entire farm was a beautifully managed landscape of linked shelter belts. This was due initially to clever selective clearing by Tim's father, and thence further judicious management of clearing and regeneration by Tim. Linking to the vegetated hills, shelter belts curved along contours and in turn linked to vegetated riparian zones (those areas adjacent to watercourses) and to open flats and their clumps of timber. As Tim said, the trees and vegetation 'are in harmony with the landscape'. One-third of Lana consisted of timber belts and forested hills. The result was magnificent biodiversity, across all organism types. Platypi lived in the creeks, and koalas, wallabies, echidnas, kangaroos, possums, marsupial gliders and a rich variety of birds and other creatures lived and moved across the rest of the farm.

'Another thing about biodiversity,' said Tim, 'is the number of insects now. There're so many insects at night, you come out on the motorbike and you've got to have bloody goggles on ... little midges, parasitic wasps. They were never there before we had all the mulch, and we're getting more winter-active native grasses also.' This observation regarding increased biodiversity has been startlingly quantified by ongoing work by ecologists from the University of New England, who, for example, have found four to five times more parasitic wasps (which are valuable because they control other pests) on Lana than on neighbouring properties. 'It's a real song out here,' Tim said proudly. 'We've got terrific mosaics and edge effects ... It's just like going into another world.'

'Too right,' I thought, 'and he even talks a different language!'

To get across to the southern portion of his farm, we drove for half a kilometre through a neighbour's place. Here, the light was harsh; there was no comfortable feeling of soft energy; the remaining grass looked grey, tired or dead; and there were large swathes of bare ground from the set-stocking. What grass there was comprised mainly annuals and weeds with very few native grasses, let alone the rich diversity I had seen on Lana.

Back in the dappled shade and comfort of Lana, Tim gave me more information as we drove. 'We get active winter growth here now,' he said. 'Our soil temperatures are at least ten degrees warmer and our plant succession we've now got means a number of winter-active perennials. And there's the flip side: in summer, our ground litter keeps the soil cooler and there's less evaporation; the whole soil is more alive, bugs function, you get germination, and so on.'

I would subsequently find with other holistic grazing managers that when they talked about new grass species emerging, and about the appearance of more desirable species in a successional event, that this was an example of nature's self-organising process. Many such new grasses had not been seen in living memory. Their seeds, having lain un-germinated at depth underground, were now being brought to the surface by a regenerated soil biology and burrowing organisms. The same was occurring on regenerating North American prairies and in Patagonia, and of course in African grasslands under such management.

I was not surprised when Tim said, 'The Indigenous people have something to tell us, how they did work with the land, even though I think they might have over-burnt.' I noticed this openness to Indigenous people and to a different world view in other leading regenerative farmers who had rejected industrial agriculture. It seems that going down this path of regenerative agriculture unlocks an entirely new way of thinking, perceiving and feeling, which is connected to a different set of values and ethics.

And I can see why, because when Tim next described how he got more with less in terms of nutrients versus fertilising, this definitely appeared back to front. That is, despite Tim radically reducing to near zero his previous expensive fertilising practices, levels of crucial elements such as phosphorus (P) and calcium (Ca) had nevertheless increased in availability. To Tim, the answer was simple: healthy ecological functioning.

'I reckon why the "P" levels are up, for example, is three reasons,' he said, ticking off his fingers one by one. 'First, it's our rest period under planned grazing. This allows lots of mulching and consequent changes in soil microbiology. Second, our animals are the fertiliser machines. They're like a tractor with a big spoon running behind. They transfer nutrients from high up in the landscape. It's part of our farm design. And third, because of the first factor, "rest", the plant roots go deeper and they're tapping into a source of nutrients that's been down there for yonks, and where soil formation is still occurring. Plus our other organisms are working – dung beetles, earthworms, et cetera – and all these factors are linked. How can you argue with that? You've got to work with the landscape.'

By now, we were nearly back at the homestead, but Tim stopped the vehicle. He wandered over to a thick stand of native pasture that was green, healthy, and diverse in colour, shape, texture and species. I enviously noted clumps of the invaluable weeping rice grass or *Microlaena* (a delicious C_4 grass). Tim dug around in the soil and exposed the thick root mass underneath. I could see dung-beetle tunnels, earthworms and other organisms.

'Notice how the country is soft, soft to walk on? And do you see any sub clover?' he asked. I nodded and then shook my head. 'No,' said Tim, 'the sub's virtually an endangered species here now. It used to dominate before I changed and stopped supering and carrying on. Our nitrogen now comes from the microbes in our healthy soils. We have unbelievably healthy pastures and I can't see any introduced legumes. Now how's that for a challenge?'

'More than enough,' I wryly concluded with what was left of any brain energy I possessed, because this was yet one more upside-down view of reality that I struggled with but that my eyes couldn't deny.

Finally we reached the homestead, and I stumbled from the vehicle and headed for the house to collect my gear. If I were a whisky drinker, I would have asked for a triple Scotch. By now, my brain wasn't just scrambled; it felt dissolved. And it was time to go.

As I drove back out to the main road, I looked anew at this beautifully balanced landscape of linked timber – to creek, to flat, to hill, to forest – at the dappled light and chiaroscuro of colour and shade, and at the gentle energising feel of the place, with its ringing cry of rosella, shrike-thrush, currawong and honeyeater.

At the front entrance, I stopped and took a photo of the 'Wildlife Refuge' sign. How true that was, I thought, and at unimaginable layers both above and below the surface. I looked at the sign on the front gate, 'Lana', and I envisaged an added shingle below: 'Welcome to Upside-Down World'.

What farmers such as Tim Wright have shown by applying holistic grazing practices is that our management of landscapes determines how much solar energy is captured and put to good or bad use.

In the case of Tim Wright's grasslands, underneath the grass trophic level was the next vital level of decomposers in the soil: all the bugs and creatures such as earthworms and nematodes (roundworms), and especially the hugely vital group of microorganisms including bacteria and fungi. So what Tim Wright was doing by leaving nearly fifty per cent of grass-vegetation matter behind and only grazing for short periods with long rest–recovery spells was maximising all three methods of increasing the density of green solar panels (leaves) – and for as long as possible.

As we shall see directly, this had positive knock-on benefits into the other four landscape functional components. Thus, due to a fully potentiated solar-energy function, more nutrients became available as soils became actively deeper and better functioning – aiding further plant growth and photosynthesis. With healthier functioning soils, more water was absorbed and made available for longer to plants – again enhancing the solar-energy cycle; and then biodiversity and ecosystem dynamics greatly escalated, as Wright indicated through things such as balanced insect predator–prey relationships.

The resulting bonus was that his grazing animals in a fully functioning landscape were healthier and more productive while playing a crucial role in enhancing landscape function (through the recycling of their excreta and also hoof action and ground disturbance). And finally there were positive-feedback loops with the human–social function in terms of the Wright family's greater farming resilience, greater income and more meaningful and satisfying life in their landscape.

John Robertson's example 160 years earlier was completely the opposite. In set-stocking his sensitive, virgin, long-coevolved grasslands, he destroyed the solar-cycling capacity of his land because the sheep quickly chewed out in particular the more palatable and deeper-rooted native perennial grasses (what is sometimes called 'eating the heart out of country'). These were replaced by shallow-rooted, ephemeral, annual exotic grasses that readily turned brown – shutting down solar-energy capture. With the solar factory wilting, the soil bugs ran out of tucker and drastically declined. This shut down the soil's mineral-cycling capacity. With no deep roots and bug-made air pockets, the water cycle collapsed, little being absorbed and most running off the increasingly bare and hard ground to cause erosion and deep gullies. With the loss of vegetation diversity, dynamic-ecosystem biodiversity in turn would have collapsed, opening the way to grass-eating grubs, pests and other corroding organisms. To cap it all, with the above functions shot to bits, active moisture transpiration and cycling in the vegetation collapsed, and so the underlying salt in Australia's ancient soil water rapidly rose to the surface, killing much of the remaining functioning vegetation. That is, the stark contrast between the two farmers came down to the fifth landscape function of the human–social: that their radical differences in landscape management entirely rested with what went on in their heads and thus how they managed.

Quite profoundly, in getting off his own dunghill back in the late 1980s, Tim Wright had wondrously changed the way his land works (its functioning) and, along with it, his family's well-being. The result was that he no longer saw a landscape through a glass darkly but had now become enabled to enhance his farming operations into a regenerative enterprise.

To the majority of traditional farmers, Wright may be operating in an upside-down world, but the difference is that he was enhancing landscape function and ecological resilience, not driving his landscape to ever lower energy levels and closer to the brink of disaster. That is, he was making his land, farm and family more resilient to droughts and flooding rains because he was fixing more solar energy; storing more in-ground water; growing more grass with deeper and year-round roots and leaves able to respond to the variable seasons; and increasing the diversity of grasses, legumes, shrubs, trees and other biodiverse ecosystem communities. This story truly is a remarkable regeneration of landscapes via a new agriculture.

Moreover, besides being instructive, Tim's story is also one of hope because the consequences of just leaving degraded ecosystems alone doesn't fix them. That is, in general if landscape functions are shot to bits then it is rare for them to self-repair, as they have gone past tipping points and into new, different states and lower stages of energy cycling and degraded landscape function. But what Tim has shown us (and as we will see in other stories in this book) is that through sensitive regenerative farming practice, caring and creative farmers can kick-start regeneration via helping to re-energise and repair different landscape functions.

CHAPTER 5

Out of Africa

Ex Africa semper aliquid novi
Always something new out of Africa

PLINY THE ELDER, in reference to Aristotle, AD 77–79

A cold, windy day in early May. There were warning signs the previous day: a broiling cloud mass creeping up over the mountains; a fish-scale sky and halo around the moon; and, late afternoon, over the mountains, a globular mother-of-pearl in the western sky. Autumn leaves are rolled, tossed and rolled again against the garden fence. I put on a beanie and lift my collar as I hop on the four-wheel motorbike, heading off to turn on a diesel pump down by the lake.

A mob of crying galahs crosses my path, torn from trees, seemingly complaining in affront, while higher up three crows arc across, buffeted by giant air-bursts. Beside me, fences are decorated in khaki and yellow filigree strands of umbrella and blow-away grass seed heads, while in the paddocks, poa tussock, *Themeda* and *Phalaris* seed heads vigorously bob and sway.

Near a bunch of blackthorns, which are crowned in a rich copper of dry seed pods that rattle in the wind, I stop the bike. For I have noticed a skylark on the barbed-wire fence. Caught by a wind gust, it has become skewered on the bony point of its wing. Marvelling at its cryptic camouflage and dumb trust, I gently release it and lay it beside a tussock – the injury slight enough to enable it to fly away when recovered.

This incident got me thinking about fences. Our natural grasslands originally knew no such impediments. Fences are human constructs, mathematically and brutally imposed – usually grid-like – upon a landscape. They interfere with how landscapes and grazing animals naturally function together as they were once coevolved to do, while also interfering with animal movement, occasionally trapping and killing both livestock and wildlife. I often discover the carcases of kangaroos that have miscalculated their leap. With their feet intersecting the top and second strands of the fence, they fall on the other side, hopelessly snared in tightly twisted wire and destined for a slow death.

But the main impact of human-built fences is that they become prisons for grazing animals under either traditional or lazy management. Such social ruminants (in our case, sheep and cattle) coevolved on grasslands and grassy woodlands to graze in dense herds for protection from predators, whose presence kept the herd ever on the move to fresh, diverse pastures. Human management – especially under a set-stocking or slow rotation regime – contravenes all this ancient ruminant coevolved behaviour. This applies first to the animals, who, in the absence of predators in their wire prison, are compelled to repeatedly graze among their own dung, ingesting intestinal parasites and re-chewing ever less palatable grassland. This leads to a dying off of the most broad-leaved, succulent and deeper-rooted species (as we saw earlier in the case of John Robertson) and ultimately to severe degradation of the solar, water, soil-mineral and dynamic-ecosystem landscape functions. But it doesn't have to be like this.

Humans couldn't have survived without grasses, as they are *the* most important family of plants for the coevolution of both mammals and hominids. Globally the area used for grazing is more than double that used for arable and permanent crops (i.e. seventy per cent of all farmed land), and as a result livestock contributes around thirty per cent of the protein in human diets. The great grasslands cover somewhere around forty per cent or more of the globe (excluding Antarctica and Greenland), thus constituting one of Earth's largest land ecosystems. They occupy the space between deserts and forests; they are central to both human evolution and ongoing survival; they are extraordinarily beautiful when in a healthy state; and yet we take them totally for granted, viewing them as some indefinitely robust and resilient resource. As a consequence – and in bush parlance – we graziers continue to 'beat the crap out of them'. Little wonder, therefore, that grassland ecosystems are some of the most threatened on Earth.[1]

Modern grasslands had more complexly evolved by twenty-five million years ago, emerging in their full diversity in the last four to five million years. In my own region of south-east Australia, grasslands have only been around since quite recently: about four million years ago. Modern Monaro grasslands would have been recognisable by 5000 or so years.[2]

The significance of grasslands is that they enabled the domestication of livestock around 10,000 years ago, when the globe entered the more climatically stable Holocene era after the prolonged ice ages of the Pleistocene. Sheep and goats, and later cattle, were the first animals domesticated (after the dog). And livestock domestication was one of the two key platforms for the foundation of agriculture (the second being the domestication of cereal plants – which in fact are annual grasses).

Following domestication, three different types of pastoral production emerged – all of which are still practised today. The first is nomadic pastoralism, where livestock are moved in a constant search for forage, and where the herders have no permanent base. The second, transhumance, involves the seasonal movement of animals and people from valley bottoms (where pastoralists have their villages) to mountain pastures in summer. Finally, there is sedentary livestock production, as practised mainly in developed countries around the world. This involves running livestock on farms and near villages year-round.[3]

The reality of all these pastoral systems is that ultimately, to one degree or another, they damage their sustaining grassland and shrubland systems. Key reasons for this are ecological illiteracy, climatic and economic pressures, and, inevitably, as part of the human condition, greed in combination with legal and political factors. As a well-known student of domesticated agriculture, Professor Carl Sauer of Berkeley, California, said back in 1956 that following the domestication of animals 'The natural land became deformed, as to biota, surface, and soil, into unstable cultural landscapes'. This, said Sauer, is primarily because the natural balance 'between plants and animals is rarely re-established under such exploitation, since man will try to save his herd rather than their range'.[4]

Historically, grazing has wrought terrible havoc. While Plato bemoaned the deforestation of Attica (famously stating in 360 BC that the hills of Attica were like the 'skeleton of a sick man, all the fat and soft earth having wasted away'), his compatriots were busily overgrazing with sheep, goats and asses. This applied to all the Mediterranean lands, and many places elsewhere in the world. And as is the case today, ignorance of landscape function was the ultimate cause (combined with greed). We should also remember that overgrazing, coupled with mismanagement of alluvial river-valley soils, has led to the collapse of virtually all the great civilisations of the past – some twenty-six at least.[5]

The worst extreme of land degradation (and primarily due to overgrazing by domestic animals) is desertification, which can occur in not just arid and semi-arid regions but even in high rainfall areas. Generally, an inappropriate response to climatic variation (such as a drought) is a key triggering factor. More frequently now, however, this is combined with political turmoil. While desertification has been going on for millennia, it massively accelerated in the nineteenth and particularly twentieth centuries. The result is that today vast areas of Earth's landscapes are affected, some to a significant degree.

At its worst extent, desertification has been starkly illustrated over recent decades when repeating droughts struck the Sahel region of North Africa: that vast transcontinental belt of nations in the semi-arid zone that lies between the Sahara in the north and equatorial forests to the south. The result has been the death of millions of people and domestic and wild animals. The harsh truth of this ongoing tragedy is that the causal factor isn't drought but humanity's response to it. It's a hard concept to accept, but droughts do not inevitably lead to desertification. As we saw in Chapter 4 with Tim Wright's story, if we can change the human–social factor, there emerges a radical, indeed counter-intuitive, approach to grazing domestic animals that can lead to regeneration rather than degradation of landscapes and societies.

An inkling of the radical potential to recalibrate the seemingly complex management factors of grazing came from an unlikely source in the late 1970s, and in the midst of the devastated African Sahel, of all places. One day, when the 1970s Sahelian drought was at its peak, the ergonomist[6] Norman H.

McLeod from the American University in Washington DC wandered into the Pentagon to resume work on examining NASA satellite photographs. At his desk, his attention was suddenly arrested by an offbeat pattern in the middle of a massively desertified section of south-western Niger – one of the worst affected areas of the 1970s. At the top of the predominantly grey-brown and white satellite photo was a series of gigantic, crescent sand dunes clearly visible from space. But at the bottom was an extraordinarily large, darker pentagon shape. Its long, straight edges revealed it was man-made. Upon investigation, McLeod discovered this was the 250,000-acre (101,000-hectare), state-owned Ekrafane ranch in Niger, which was running up to 10,000 cattle.

In 1972, at the beginning of the drought, the ranch's managers, using a barbed-wire fence, had simply divided the massive ranch into five paddocks. The key component of their grazing management was giving each paddock a long rest. In the most basic of planned grazing systems, they had grazed one paddock for a year while resting the other four. Over the next four years, they just rotated this system one paddock at a time, until over five years all had been rested for four years and grazed for one. Yet this simplest of planned grazings (along with preventing constant overgrazing by nomadic pastoralists) had yielded a startling result. Not only was the ranch not desertified, but it was preserving and even regenerating vegetation despite the drought, as was clearly visible to the Pentagon scientist. In bush lingo, this ranch in the middle of the desert stuck out 'like proverbial dog's balls' even from thousands of kilometres up in space. It was a beacon of what a mind-change could mean for vast areas of 'spaceship Earth'.

Previously, this land and its neighbouring area would have been continuously set-stocked, and as droughts worsened it too would have become desertified. But the Ekrafane ranch managers had given the solar-energy cycle a chance by instituting long rest periods to allow enhanced photosynthesis and recovery from grazing. What this quite unexpected example had confirmed was that it is not naturally occurring droughts that cause desertification but humankind's poor management.

The managers of Ekrafane were not the first to try to radically change millennia of accepted practice concerning the grazing of landscapes. For long spans of time, the nomadic herders of the Eurasian steppes, for example, had trodden lightly on their grass- and rangelands, as had skilful transhumant shepherds in parts of Europe. And in medieval Flanders,

England and other parts of Western Europe from the twelfth century, shepherds had boosted fertility for agricultural field rotations and crop production by intensely 'folding' their sheep at night (as in putting them in pens both for security and to concentrate their dung and urine).

Then, in the wake of the scientific and agricultural revolutions and European Enlightenment of the eighteenth and nineteenth centuries, the emerging caste of 'natural philosophers' or scientists likewise began to apply their minds to the key management components of animals grazing in landscapes. As early as 1760 in France, agronomic writers were describing the necessity of dividing up paddocks so a rotational grazing system could be implemented, and by 1777 the Scottish Enlightenment thinker James Anderson was outlining a management system so that sheep and cattle could 'be carried on constant rotation'.

Over the next hundred years in various European and North American botany and agronomy schools, there was progressive theorising on rotational grazing. Perhaps the most impressive of this international research (albeit in the lush, green fields of Europe) was conducted by French agronomist André Voisin, who, in a number of definitive books, brilliantly analysed the key factors in sustainable and even regenerative grazing. These hinged around rotation, length of grazing, and rest and recovery.[7]

There was, however, a fatal flaw that emerged because of the timing and location of this early thinking on rotational grazing. All of the work occurred in the reliable climates, intense agricultural areas, rich soils and humid atmospheres of the centres of northern-hemisphere agriculture. By contrast, as we have seen, most of the vast herds of grazing livestock that were busily degrading and desertifying landscapes were in arid or semi-arid zones of the world, or else in places such as temperate and subtropical Australia and parts of the Americas, where there were high evaporation rates and unreliable rainfall. In such places, the early assumptions behind rotational grazing simply didn't apply.

But then, from the 1960s, the final pieces in the puzzle of regenerative holistic grazing were put together in Africa by a number of people, including the Zimbabwean wildlife ranger and ecologist Allan Savory.

I first met Allan Savory when my wife, Fiona, and I spent three days on his 20,000-acre (8000-hectare) Zimbabwean ranch Dimbangombe in early

2014. He lives beside the Dimbangombe River in a series of simple, inter-connected, open African dwellings. On the first evening, we enjoyed a drink around a fireplace and grate designed to discourage mosquitoes, chatting on an open stoep (verandah) out the front of the huts overlooking the river. The sounds of an African night drifted in, as Allan, not feeling well, quietly listened to our conversation with his wife, Jody. Next day, we drove around the ranch with Allan, walking to selected points of interest.

Allan generally goes barefoot: the better to feel the softness of the earth, its temperature, degree of ground cover and energy. That way, Allan told me, 'it's a huge advantage; you're reading the ground every damn step you take, and in long grass when you can't see, you can still detect any bare ground or litter'. It was not a totally relaxed drive, however, as Allan doesn't suffer fools. Between stories about points of interest, he targeted rapid questions at me, testing both my observations and my understanding of the principles of holistic management.

In the ensuing days, I spent my time walking extensively with a Mata-bele hunting guide called Elijah. Not long previously, I had seen the now famous TED talk where Allan Savory explains how his methods can arrest and turn around desertification. I made sure I went to all the photo-points shown in his presentation, and was agreeably surprised: the landscape's transformation was even better than in the video.

The reason is that instead of unplanned, haphazard grazing – with its resultant overgrazing and especially over-resting – Dimbangombe's 500-plus cows are corralled each night with a portable canvas cloth for protection from lions in an area of an acre or so. The cloth, nearly two metres high, is unrolled and strung around trees and stakes in a long, rect-angular shape, with one guarded entrance. Then, during the day, with up to four stockmen and small dogs to ward off lions, the cattle are grazed intensively according to a plan. This ensures no land is grazed for more than three days and that all areas have up to two or more months of recov-ery. After a week or so, the corral is moved to a new site and more pasture, thus enabling grazing without fences. The result in this classic tropical monsoon savannah landscape, which receives large amounts of rainfall in short periods (in their case, 24 inches/600 millimetres in the four months from November to February), is remarkable recovery in landscape func-tion via increased ground cover, species succession, more carbon via mulch into the soil, and so on. Again, via managing the way the vegetation is grazed, this functional regeneration begins with the solar-energy cycle.

On our drive, Allan pointed out the Dimbangombe stream, which now ran year-round. It had been dry for decades when he first bought the land. He indicated the flood level on the thickly and richly revegetated banks – some two feet high – and where the excess water had lain down the copious vegetation beside the stream without any erosion damage.

We then went to a point where the Dimbangombe River joined another stream: the Tsitsingombe. The catchment of the latter is similar to that of the Dimbangombe, but it is managed by the National Parks Service, which doesn't allow any cattle-grazing impact. The flood level there was over fourteen feet high, with eroded banks. It was a dramatic lesson in the difference between healthy landscape functions working together to absorb more moisture and the alternative dysfunctionality.

In short time, Allan Savory's new grazing approach had turned the soils of much of the catchment of the Dimbangombe River into a giant sponge – one capable of holding over a thousand times more water in the landscape. Pointing out the difference between the two rivers, Allan concluded, 'If you watch the off-flow of water at the lowest point of your land, it tells you most about your land and its environment, because if you get your water cycle right, then everything else comes right.'

Born in 1935, Allan grew up in what was then the self-governing British colony of Southern Rhodesia (now Zimbabwe). He was clearly impacted by the physical beauty of Zimbabwe's rich land, flora and fauna, accompanying his civil engineer father into the wilderness to inspect projects and spending time on an uncle's ranch riding horses and learning to hunt.

On leaving school, Allan went to Natal University, where he majored in botany and zoology while studying ecology. However, he found the reductionist separation between disciplines highly frustrating. Inspired by a godfather who was a game ranger, at the age of twenty Allan left to become a research biologist and game warden in Northern Rhodesia (now Zambia).

Out in the wild, Allan found the game parks and grassland savannahs rapidly deteriorating because of the proliferation of man-made desertification. He blamed both excess cattle and also elephant numbers. This led to one of his searing early learning experiences that would prove cathartic. In charge of a game park by now, Allan concluded that the deterioration of the park was due to excess elephants, so he set about instituting a massive culling program. Some 40,000 were killed, which Allan would subsequently regard as one of the greatest regrets of his life. Later he came to realise his mistake and became determined to dedicate his life to finding solutions.

Allan quit his job as a ranger and began servicing farming clients (for, by the 1960s, he had changed his views that all farmers and their cattle should be banished). Flying over parts of Africa in his own plane was also instructive as he witnessed the long-term impacts of overgrazing and injudicious burning. Along the way, he developed a benchmarking tool called the 'brittleness scale'[8] to aid understanding and management of different types of country in different climates.

A further crucial factor in his thinking was when, in the 1960s, Allan encountered government officials and scientists from Zimbabwe, Botswana and South Africa arguing that the cause of desertification was an increasingly unreliable rainfall, or that 'the rainfall had changed'. At that time, Zimbabwe was in drought, but downstream on the Limpopo River Mozambique was having floods. Allan recalled, 'Clearly to me as a government research officer, Zimbabwe's drought was Mozambique's flood because water that should have been soaking into our soil was flooding Mozambique.'[9]

These floods culminated in a seminal cathartic moment in 1960, which confirmed the change in Allan's life's direction. At the time, he had moved to the south of the future Zimbabwe as a research officer in the game department. Two things then came together. Allan had grown deeply concerned about the process of human-caused desertification. He realised that this in turn, through increasing the degree of bare and less-absorbent ground, was leading to excess flooding. His observations and conclusions were a major step in evolving his holistic management process because, in his own words, he realised that 'desertification begins when the available rainfall becomes less effective'.

This led to an epiphany one day in 1960 on the banks of the Umzingwane River: a tributary of the lower Limpopo, where it runs from Zimbabwe into Mozambique. In front of Allan this day, a raging flood was in progress: 'A lot of soil, dead goats, vegetable material, trees and so on were sweeping past. And suddenly I began to cry.' Evidence of his nation being destroyed was passing before his eyes. 'The sudden emotion,' said Allan, 'caused me to ask, "What am I going to do about it?" Then I reasoned that if I was in the army, I would be prepared to give my life for my country. But what I was witnessing was a threatening event to my life, land and nation. So I asked myself, "Am I prepared to give my life, devote my life to my land and nation?" The answer was "yes", and that cathartic moment explains my dogged persistence in the fifty-five years since in trying to solve this problem.'

However, no one wanted to listen to Allan's argument that desertification was occurring 'because of the massive bare ground and non-effectiveness of their rainfall'. Unfortunately, Allan had made himself the enemy of a powerful clique of rangeland academics – who have continued in violent opposition to him ever since.

In the meantime, another formative event helped shape Allan's thinking. On the outbreak of the Rhodesian Bush War (or Second Chimurenga War of Liberation) – a subset of the Rhodesian civil war from July 1964–80 – in response to communist terrorist insurgency, the Rhodesian Army established a crack counter-insurgency tracker combat unit under Allan's command (utilising his bush knowledge). This later became the famous Selous Scouts.

His time training and commanding army trackers had an unexpected impact on his thinking. 'We could not light fires at night, and after tracking all day I had much time to ponder on what I saw and to think,' Allan told me. 'I noticed how tracking was either easy or difficult over country under different management. Where livestock were grazing on commercial ranches was very different from where they grazed on communal lands, and the tracking was again different in national parks and hunting areas where only big game was on the land. The management made an enormous difference to the amount of bare soil on which tracking was easiest. So over long nights I began to work out why there were such differences. All of this helped greatly to unravel the mystery of desertification that has been misunderstood for centuries.'[10] By this time also, Allan had begun going barefoot, and this further helped in his intimate reading of the land.

As his understanding and ideas developed, a number of other key thinkers helped to inform Allan's ideas. One important influence came at a time when Allan had begun formulating his new management approach but realised that key pieces were still missing. He therefore paid a visit to an extraordinary South African botanist called John Acocks, author of the 1953 book *Veld Types of South Africa*.

After studying botany at Cape Town University, the twenty-five-year-old Acocks had started work as a pasture ecologist in 1936, conducting botanical surveys of research stations. His major work began when in 1948 he was permanently posted to the Grootfontein College of Agriculture in Middelburg – smack bang in the middle of South Africa's key semi-arid grazing country of the Little Karoo and Greater Karoo. People who knew Acocks said the botanist was a shy man and something of a loner, quiet and unassuming. In many ways, this suited him to his self-driven task. Acocks undertook decades of walking hundreds of miles across all terrains and in different environments to methodically survey the grass and shrublands of South Africa. In the process, he put together the story of how they had deteriorated and what they had originally been like – a story penned in his definitive book. It was through this effort that Acocks realised how poor veld management had hugely degraded diverse grasslands that were long coevolved to be grazed by large herds of migrating animals. This eventually led to him promulgating a solution to the ongoing degradation.

In the introduction to his landmark work, Acocks makes a statement that sums up the entire story of grassland degradation worldwide, including in South Africa's sister Gondwanan nation of Australia and in the Americas:

> The near-desert and semi-desert conditions that have developed in some parts arise out of the fact that the grasses best fitted to cover and hold the soil, for the building-up of which they are also largely responsible, are at the same time the most palatable; so that under a form of grazing management which recognised no need for knowledge or understanding of vegetation and gave no thought to it, it is precisely that doubly valuable species that have tended to be selectively grazed to extinction, and it is only because less palatable species have taken their place to some extent that desert has been prevented from developing.[11]

In his beautiful analysis of the grassland ecology of southern Africa over thirty years, and after identifying over 700 grass species, Acocks was able to paint a picture of what the grasslands had been like. This provided a blue-print for what visionary farmers could work towards restoring. The 'all-important grasses' Acocks called 'master grasses', and the less import-ant ones (though still performing a useful function) he called 'caretaker grasses'. In short, Acocks was describing a vision of a healthy mix of the most palatable winter active (C_3) grasses and summer active (C_4) grasses – which had quickly been lost.

'[I]f soil erosion is to be halted and the full carrying capacity of the veld realised,' said Acocks, the master grasses 'must be restored to their old posi-tion of dominance'. These master grasses, he added, were what once 'covered the whole country with a sufficient mat of foliage to protect the soil'.

One of Acocks' key insights was to point out that 'overgrazing' and 'overstocking' were not synonymous terms. As he explained, 'the overall stocking rate on paper may seem very low, but the actual local stocking rate, as regards the few species or the small patches of veldt which are actu-ally being grazed may be excessively high.' That is, he pointed out, 'The system responsible for the deterioration of the veldt is the continuous grazing system, the traditional system which causes the vegetation to be overgrazed to eventual extinction even while it is understocked.'

Part of Acocks' legacy was to encourage farmers to visualise what the original ecosystem had once been, thus revealing a pathway to how, with good, imaginative management, they might restore the condition of their earlier grasslands. So Acocks had come to understand both the causes of South African range- and grassland degradation and the solutions. But it wasn't until 1966 that he publicly presented in a paper his pithy diagnosis, which would come, via Allan Savory's work, to reverberate worldwide. He simply said, 'South Africa is overgrazed and under-stocked.'

Deeply concerned at what he had discovered, Acocks developed what he called 'Non-Selective Grazing' or 'NSG'. This comprised the approach of short periods of heavy grazing of all plants so that none were favoured by not being grazed, followed by long rests. By 1966, he had begun work-ing in particular with Len and Denise Howell of Hillside Farm, Springfontein, in the semi-arid southern Free State.

He then extended this work, primarily with sheep farmers and then later with cattle operators. In varied landscapes, he was able to increase stock numbers and create more paddocks, thereby increasing stock density.

He gave the grazed paddocks longer rest periods so the plants could recover. Albeit a far cry from tens of millions of wildebeast, springbok and other animals in vast migratory herds, it was nevertheless a first step to mimicking the natural grazing patterns of the predator-driven African animals in the wild. Unsurprisingly, the results were startling. The veld began regenerating, ground cover improved, 'master' and 'caretaker' grasses began to appear, other key landscape functional elements such as the water and soil-mineral cycles and biodiversity improved, and before long pastoralists were lifting stocking rates.

Acocks' insights threatened rangeland academics and South African Department of Agriculture personnel. Predictably, he was quickly silenced. And so the baton was passed to Allan Savory.[12] It was at this early phase of NSG that Allan travelled to South Africa in 1966 to talk to Acocks and the Howells, among others.

While on the Howells' ranch one day, Allan got out of the vehicle to open a gate. At that moment, he observed a small part of the land in a nearby fenced corner showing obvious improvement. He rushed over there, to be joined by the Howells, who asked him why he was so excited. Allan pointed out that where the sheep had bunched as a herd (followed by sufficient recovery time for the pastures), he was seeing the same changes he had observed where large buffalo herds were still intact and were being pushed to be constantly on the move by accompanying pack-hunting predators (such as the big cats). In both cases, there was more vegetative soil cover, more seedlings, greener colour in the grass, and so on. Accidentally, those sheep, which had crowded in adverse weather, led to Allan instantly realising that livestock could be used to mimic wild herds if only he could work out how to do so consistently.

Thus, Allan added one more vital component to his evolving approach: what he called the 'herd effect'. Holistic planned grazing should incorporate practices that caused livestock to bunch together as a herd, so their concentrated dung, urine and hooves could help to break up hard soil surfaces while laying soil-covering plant material or litter. This was especially important in the more seasonal rainfall environment where desertification mostly occurs.

With this next element apprehended, Allan returned to Rhodesia to further refine his coalescing thoughts and to continue conducting experiments with ranchers there. Crucial in the progression of his thinking was another key piece of the puzzle. This came from the famous French agronomist André Voisin's classic book *Grass Productivity*. There, Voisin had outlined the supreme importance of building flexible rest and recovery periods into a simple planning process he called Rational Grazing. Voisin thus gave Allan two vital concepts. First, as Allan told me recently, 'Voisin clearly explained why rotational grazing over centuries in Europe had led to degraded pastures and thus how this could be overcome by replacing rotational grazing systems with a planning process. And second, Voisin showed that overgrazing of plants was a function of time of exposure and re-exposure to animals and not due to animal numbers.'[13]

Armed with these insights, Allan first tried Voisin's rational grazing planning but found it did not cater for the greater complexity of African grasslands with all the wildlife, nor the integration with crops in a far harsher environment.

Allan then realised that Voisin wasn't wrong but that a more sophisticated way of planning grazing was needed. Again, serendipity stepped in, resulting in two major consequences. The first was when he became part of a famous seven-year international trial in Rhodesia in the early 1970s (called the Charter Estates Trial). Though Allan only applied a simplified prototype system (what he then called 'Short Duration Grazing'), he was able to sustainably carry twice the number of cows over those seven years compared with traditional grazing approaches.[14]

What Allan proved was that inappropriate management, rather than too many animals, was the cause of land degradation. But in the process Allan had begun to prove an already defensive group of scientists and academics wrong. Thus, it was after this trial that he was blacklisted by African and then American rangeland academics, not to mention the Rhodesian and South African governments.[15]

Because the Charter Estates Trial had been set up to resolve what had become a divisive impasse in academic and government circles, the Rhodesian government appointed a trained scientist to do the economic evaluation of the trial. His name was Dr Stan Parsons, and thus began a thirteen-year partnership with Allan Savory that further helped the final development of today's HM approach and teaching – and indeed helped spread the principles of holistic grazing management worldwide.

Like Allan, Stan Parsons was raised in Rhodesia and educated in South Africa. Stan's father, a far-sighted man, prophetically told Stan that one day all he may take from that country was an education. Stan took his father's advice, gaining first class honours in his agricultural science degree at the University of Natal and thence completing a PhD in reproductive physiology. He then went to Purdue University in the USA to study economics for two years, where he gained a Masters degree.[16]

Parsons' economics degree allowed him to become a lecturer at the university in Rhodesia. It was these skills that led to Parsons being chosen to analyse the Charter Estates Trial results: results clearly in favour of the Allan Savory approach, but much to the chagrin of rangeland academics who vehemently disagreed with Allan's assertion that it was domestic stock and their management that were destroying the landscape.

Parsons was a prodigious reader and he introduced Allan to Jan Smuts' seminal work on holism. This in turn profoundly influenced the refinement of the HM system. In his 1926 book *Holism and Evolution*, Jan Smuts – the South African general and statesman – outlined his extraordinarily prescient vision on holism, which would influence ecological and philosophical thinking across the globe post the 1960s. For Allan, of relevance were ideas about the way nature functioned in wholes and patterns, and also about managing complex situations while using all available science. This approach was clearly opposed to the Western mechanistic view that complex systems consisted of linear, machine-like entities (even if somewhat complicated).

Parsons and Allan, working together and through trial and error, developed much of the original thinking around the interaction between grazing and ecology. Allan was the ecologist while Parsons had the production and economics background. Parsons did not just bring a sophisticated financial planning approach and financial skills to the table along with his PhD in animal nutrition. He also had exceptional teaching skills from which Allan learnt considerably. They worked as a team from 1970–83 when the partnership broke up. They worked through Zimbabwe initially, then Namibia, South Africa and the USA from the late 1970s on. According to Terry McCosker (the Australian grassland and rangeland grazing educator who worked with Parsons), 'The world will be a better place because of what these combined intellects achieved.'[17]

The final piece in the puzzle of the Savory HM system was how to crack a more sophisticated system of planned grazing. Because of his work with Parsons, and after seeking possible answers in business planning schools and other professions, Allan told me that he 'found the answer in centuries of experience in military planning in Europe'. Accordingly, he said, 'I simply adapted the British military academy of Sandhurst's process of planning in ever-changing immediate battlefield situations.' That is, realising ranchers in Africa faced even greater complexity and needed to plan for far longer periods of time, Allan adapted the military planning process to develop a grazing planning chart on which several dimensions could be expressed over many months or even years.

By now, Allan was advising many ranchers across four countries. However, while he found their management results were successful, these results became increasingly erratic. Clearly something was still missing. Upon further analysis and discussions with Parsons, Allan found this was due to factors he had failed to account for – social and economic ones. It was at this point that the final piece of the puzzle was put in place, for Allan had realised that he needed to include Smuts' theory of holism and thus bring all social, cultural and economic factors into the management process. And thus was born the holistic framework in the early 1980s and the grazing planning that became the holistic planned grazing of today.

In its essence, Allan had apprehended that grazing should occur only in short periods, combined with long rest periods (but also that too much rest was also deleterious, as it degraded the solar-energy cycle – when grasslands get rank, old dead leaves shade out the solar-energy capture and tie up nutrients in dead tissues). Like Acocks, he realised that sheep and cattle could replicate the effects of the large, migratory herds of buffalo, wildebeest, springbok and such, while lions and cheetahs could be replaced by humans utilising planned management. Thus, with refinement, holistic grazing management was born.

It was after his 1983 split with Parsons that Allan Savory finalised the modern form of his Holistic Planned Grazing and Holistic Management organisation. To this platform, he then added a further element based on his army experience: a specific process that became the famous holistic management step sequence of decision-making. Meanwhile, Stan Parsons would play a crucial role in bringing the new thinking to Australia from

1989. His particular strength was a focus on greater profitability from animal performance, which, as a longstanding client of Allan Savory's once told me, was 'Allan's blind spot'.

Today, practitioners of Savory's holistic management approach to grazing and business operations can be found managing an estimated seventy-four million acres (thirty million hectares) of land in every grazable location from Tasmania to Patagonia, the Monaro to the short- and tall-grass prairies of the USA, New Zealand to southern Africa, and even the uplands of Sweden and the hills of Turkey. Tim Wright is but one excellent example of a well-applied and individually adapted application of Savory's holistic management in practice.

In the next chapter, I will cover some more examples, embracing South Africa, Australia and Patagonia. These comprise one of the most longstanding current practitioners and three quite remarkable new ones.

But Allan Savory's radical and groundbreaking regenerative grazing approach has also helped spawn three quite extraordinary and equally radical Australian regenerative innovations in broad-acre cropping: those of pasture cropping and no-kill cropping (see Chapter 10), and of natural intelligence agriculture (see Chapter 17). I could fill up this book with wonderful case studies derived from Allan Savory's and Stan Parsons's work and that of their offshoot extension bodies. In these, through holistic management practised with varying degrees of skill and experience, whole landscapes, farming families and their businesses have been resurrected and set on a pathway of growth, regeneration and well-being – and this multiplied in every grazing continent.

Moreover, like so many biological and other elements in this story (including the story of humanity), this radically different approach to grazing, and thus cropping and landscape management, has come 'out of Africa'.

CHAPTER 6

Make Mistakes
but Don't Do Nothing

An agriculture cannot survive long at the expense of the natural systems that support it and that provide it with models ... We can build one system only within another. We can have agriculture only within nature, and culture only within agriculture. At certain critical points these systems have to conform with one another or destroy one another.

Under the discipline of unity, knowledge and morality come together.

WENDELL BERRY, *The Unsettling of America*[1]

In its fundamental aspects, enhancing the solar-energy landscape function means growing more green leaves for longer periods on a wide variety of plants. Yet, as we have seen, until the late twentieth century, various grazing regimes around the world simplified and diminished this process. The result was widespread land degradation. But since the 1970s, the new holistic grazing regimes that came out of Africa via the work of Allan Savory, Stan Parsons and others have been increasingly found to reverse the damage.

After visiting Allan and his wife, Jody, at their Dimbangombe ranch in Zimbabwe in 2014, my wife, Fiona, and I travelled 2000 kilometres south to the famous Great Karoo and Stormberg regions of South Africa. Our purpose was to visit Norman Kroon and his Australian-born wife, Jennie, who live in a comfortable but unpretentious house fifteen minutes out of Graaff Reinet in the Kamdeboo region of the Great Karoo: challenging, semi-arid, twelve-inch (300-millimetre) rainfall country. Norman was then in his seventies. He had the meat-axe, scarred hands of a life spent outdoors, and the open, kindly, weather-beaten face of a resourceful individual whose regenerative work allows him to look forward to each day.

Norman did not come from a traditional farming background (probably one reason why he is such an open thinker), and after his father – a financier who also owned a butcher's shop – took up country in the drier Karoo, Norman decided to take it on.

Norman's brother Malcolm was Allan Savory's first Namibian grazing client in the late 1960s. A near-neighbour to Norman was a farmer called Smith Edwards, who worked with John Acocks and was the first in the area to practise Acocks' NSG approach. So, with these influences, and with his father's encouragement to 'go and make mistakes, but don't do nothing' ringing in his ears, Norman began a holistic grazing approach in 1972.

For Norman, the appeal of the new approach was that 'it adopts a holistic approach to farming by considering the plant, the animal, the management and economics, and the results are judged in all four facets jointly'. All grazing systems will fail, he said, 'if they cannot satisfy these factors.'

Norman first took this new approach on his 30,000-acre (12,000-hectare) farm Kariegasfontein in the seven-and-a-half-inch (190-millimetre) rainfall country of the Great Karoo, where he had found a legacy of shocking, livestock-induced degradation. There were thousands of acres of hard-pans: areas of exposed subsoil of high-clay content that is dense, compact and slowly permeable, and which has been exposed by total wind-blown soil loss due to overgrazing in droughts. There were also gullies up to fifteen metres deep, and capped soil (i.e. with a hard crust on top) – so that after two inches (fifty millimetres) of rain there was 'tremendous run-off'.

Early Dutch and English records reveal that centuries ago this Karoo country had once been a healthy, hydrated landscape, where rivers hosted mobs of hippopotami and whose dense, diverse, healthy grasslands were traversed by vast hordes of migrating wildlife – constantly on the move and driven by groups of large predators. But in the now post-European 'managed' landscapes of a settler economy, Norman first began by expensive strip-ploughing of the hard-pans, six feet apart to retard the run-off. This had limited results, he said, 'as it only treated the symptom.'

Consequently, Norman then implemented holistic grazing management with large flocks of sheep, short grazings and long rests – and the place began to transform. The tough hooves of stock ('animal impact') chipped the hard-pan ground, creating seed-germinating spaces, and all paddocks began to grow more feed, with plant recruitment occurring in bare ground. In time, new, more desirable grass species began appearing – some of Acocks' 'master' and 'key' grasses. 'The condition of the existing palatable plants changed from short overgrazed sticks to healthy unhedged plants with long shoots,' said Norman. A key area of 1230 acres (500 hectares) he called Rivercamp had been unusable as it was a dust bowl that, after rain, grew so many weeds he couldn't stock the land with sheep as the wool became contaminated with dust and vegetable material. Today, production on this block has trebled and with little wool contamination.

There was a rapid rise in effective rainfall due to these ecological changes and increased litter on the ground, which impeded run-off and enhanced water infiltration into the soil. The entire water cycle on the farm then began to change, with increased plant cover (enhanced solar cycle and soil-mineral cycle). After rain, sand-rivers now flowed for nearly four days instead of twelve hours, and the water table lifted between twenty and fifty feet (six to fifteen metres). And finally, as Norman's grassland ecology and variety burgeoned, his animal performance also dramatically lifted as their nutrition improved – and due to what Norman calls 'the miracle of the move'.

I have a photo of the boundary fence today between Norman's Kariegasfontein and the farm of a traditionally grazed neighbour, which looks exactly like Norman's did forty years ago. As Norman recalled, when he first started, you had to walk miles to find a perennial grass plant and a shrub that wasn't a depleted stick. On the neighbour's farm, there is ninety-five per cent bare ground. On Norman's, immediately over the boundary fence, there is an entirely different landscape. Ground cover has increased enormously with no visible bare ground, and there is a riot of colour and diversity, with plants at all different heights and compositions, and in different stages of flowering, seeding and mulching. From vehicle tyre tracks, you can see that his land is soft and absorbent. One picture truly does tell a thousand stories, but, more importantly, also speaks to forty million years of landscape coevolution.

On the 7500-acre (3000-hectare) home farm of Watervlei, the same story has been repeated. At the time of our visit, Norman was running two 'flerds' (mobs of sheep with cattle), rotating through sixty-four paddocks from two to four times a year. Owing to short grazings, these animals had reverted to their ancient behaviour when under the pressure of predators – where they would have kept bunching and moving off their own dung and urine. With no plants overgrazed now, Norman was experiencing the return of former grasses and increased biodiversity. In such broad-acre brittle environments of South Africa (and similarly in much of Australia and North and South America), under traditional management such lands continue desertifying. This was what John Acocks meant by the South African veld being 'overgrazed and under-stocked': that desertification is due to set-stocking and thus the most attractive, nourishing and ecologically valuable plants are being constantly overgrazed when on 'the range' (or 'the veld'). The second factor is that, due to low stock densities in this broad-acre country, the landscape was being over-rested by having livestock too scattered and thus of too low a density for extended periods (i.e. with no 'herd effect' to quickly but only partially eat down all grass species without selectively damaging the best, while impacting through dunging, urination and soil disturbance via hooves – as occurs in the big, wild migratory herds). But that wasn't the case with Norman.

Moreover, on all his farms, he had eschewed fire and focused on grazing management. Once asked by a neighbour, 'Why don't you burn?' Norman had simply replied, 'I prefer to see my product go off as wool and meat and not smoke.'

After forty years of holistic grazing management on Norman's farms, the ecological and financial results were hugely impressive. On the dry Kariegasfontein farm, between 1973 and 2010, long-term carrying capacity went from forty acres (sixteen hectares) per livestock unit to twenty-two acres (nine hectares). On his farm Vrijnsfontein (which he bought in 1984), carrying had tripled. This allowed Norman to pay off this most recently purchased farm in only three years. On top of this, drought resilience had hugely escalated, while he maintained healthy lambings and calvings irrespective of the seasons.

The morning before Fiona and I departed to continue our magical journey among lateral-thinking pastoralists in this toughest of all grazing countries, I hopped on the back of a ute with a friendly African kid who was keen to come and do the gates. In the front were Norman and his son-in-law, catching up on farm-management changes and the season while I took a good look about.

Norman headed to a paddock of healthy, head-high shrubs and variedly coloured grasses in between. Two days before, there had been half an inch (13 millimetres) of rain, and the ground and air smelt fresh. Despite the large flerd, the vegetation was so thick that the fat, shiny animals were initially hard to spot. We frequently stopped to examine new shrub and grass recruitment in hoof-print indentations in the soil. These were the result of Norman running a big flerd immediately after rain in the previous year so as to break up residual hard-pans and create a healthy seedling environment. We stopped to examine a remaining patch of bare ground, perhaps lounge room-sized. It was covered in sheep dung. 'They like to camp on bare ground here, and this will start successional change,' Norman observed. 'Our biggest change,' he concluded, 'is that our effective rainfall has increased enormously.' By this, he meant that his maximisation of solar-energy capture via regenerative holistic grazing had made his soils far more water absorbent (instead of water repellent, as usually occurs in poorly managed landscapes).

In short, it was as if he had moved into a wetter climatic zone – and all because of this radically different approach to grazing.

While holistic management and related grazing regimes may have begun in Africa – and on what are perceived as either tough and brittle, semi-arid rangelands or else equally brittle savannah grass- and woodlands – this new regenerative grazing approach has also caught on internationally (as we saw in Chapter 4 with Tim Wright in New South Wales). Moreover, it is constantly having its boundaries, practices and thinking expanded: an evolution spurred by encounters with questioning minds embedded in different ecosystems.

The next story comes from the broad horizons, big skies and often merciless sun-drenched landscapes of the true outback Australia – a story that is mind-boggling in daring and scope. It concerns the Dunnicliff and Armstrong families' conglomeration on the Barkly Tableland, between Alice Springs and Katherine in the Northern Territory. And if anyone embodies the precept of Norman Kroon's father – of 'having a go' – these people certainly do.

I got to know John Dunnicliff relatively early in his career, after he began farming in his own right in northern New South Wales, aged just sixteen in 1956. From there, he traded his way up to ever larger holdings. Accompanied and supported by his resilient wife, Trish, he then went on a perambulating journey through many of Australia's great landscapes over many years. This saw management of the famous Isis Downs near Longreach in the mid-1980s; of a blue-ribbon property near Holbrook, New South Wales; of an Angus cattle and dairying farm on King Island; and then of a cattle station on the big pastoral holding of Cherrabun in the Kimberley. And it was this combined experience that enabled John to realise the huge potential of the Barkly Tableland.

In the meantime, he had encountered Terry McCosker and the concept of a holistic grazing management regime, which he first began implementing on his Holbrook property. Later, the Kimberley country taught him how to develop big pastoral stations and deliver a low cost of production, and so by the turn of the millennium he was well prepared for the big move.

In 2002, the Dunnicliff family bought the perpetual pastoral leases of Beetaloo station, OT Downs and Mungabroom: over 2.6 million acres (1,054,700 hectares) of beautiful downs country, with nice lancewood (an acacia) and other timber and shrubs on sandy and red country, and with coolabah on the wetter country (the latter a eucalypt of the Australian inland, usually found in or near watercourses). When Fiona and I stayed with the Dunnicliffs in May 2015, the property was also being run with daughter Jane and her husband Scott Armstrong.

The combined place is 180 kilometres across, and this is country, John told me, 'where our trees and shrubs play a big role: shade, fodder, biodiversity – it all fits'. With lakes, billabongs and swamps in the wet, the place is a haven for water and other birds: plains bustards, magpie geese, brolgas and more, totalling over 200 species.

Crucially, and due to the prohibitive expense of tapping and reticulating available water, John had immediately seen that less than ten per cent of the land was effectively watered, and so only around fifteen per cent of country had been utilised (OT Downs, for example, virtually comprised one paddock of half a million acres/202,000 hectares). That land was seriously degraded, and the un-grazed land meant a high incidence of damaging, late, dry-season fires.

John's experience told him that for effective grazing and best animal performance, a cow shouldn't walk more than two to two-and-a-half kilometres a day to water. So he did the maths. To maximise both the ecology and production he would need to divide the country into a grid pattern and have 600 water points (bores, infrastructure, et cetera) plus fencing and poly pipe. It added up to $40 million. The maths also told him that, with an effective though flexible holistic grazing regime, he could take stocking from a start of 20,000 head to near 100,000. And he knew he could get cost of beef production down to 30 cents a kilo or below.

To make it work, John convinced Brett Blundy to tip in the money: on a convertible share basis if the plan were successful (Blundy being the Singapore-based ex-Australian farm boy who became one of Australia's leading retailers). Along with other cattle interests, Blundy has also invested in the adjoining 783,000-acre (317,000-hectare) property Amungee Mungee, owned by John and Trish's other daughter, Emma, and their son-in-law Adrian Brown.

By May 2015, John's extraordinary vision was well on track to realisation in what was the largest operation of its particular type anywhere in the world. With over ninety per cent of bores and troughs installed (seventy per cent of these only four kilometres apart), stock numbers had grown to 80,000 head in an operation redolent of constant, flexible innovation. John then was also well on the way to splitting some of their rectangular paddocks of 4000 acres (1600 hectares) into triangles so as to enhance ecological and animal impact. But with 5000 flexible holistically grazed livestock units (LSUs, with a cow rated at 1.5) on average to the 3950-acre (1600-hectare) paddocks, the land was already responding. 'It's open-ended,' John told me. 'We don't yet know where the limit is.'

In keeping with the unique land and climate, Beetaloo was flexibly run, with two main types of holistic grazing regimes, which were still being refined, and others in experimentation. For it was the land and climate that dictated practical, sensible and profitable ecological management. The combined landholding is comprised, first, of red country, which is higher, and second of low-lying black soil country. In the wet season, much of the latter is either under water or else isn't suitable for optimum intensive rotational management or stock performance, as the cattle pug the soil, ruining soil structure and ecology. Consequently, the bulk of the herd (the breeders) are shifted out onto the red country in October–November, just before the wet. Here, in mobs of 7000 to 8000, they are managed in a loose rotational system: what Jane Armstrong described to me as a 'seasonal rough rotation, where "rest" is crucial – what is known as a seasonal resting program.'[2]

The cows with calves are then brought back down onto the black country (where yards and infrastructure are located) at the start of the dry season in March–April, and the red country is then locked up for extended rest. It is on their return that weaning and other management operations occur, and when the first drafts of animals for live export are taken off.

Conversely, the second cattle component of their operation is up to 12,000 bulls – the basis of their livestock export approach to Indonesia. After the best draft are sent to market, the remaining bull weaners are put into large bull mobs, where older lead bulls quieten them down and lead them around: management of testosterone-driven males that can be achieved with Brahmans but not with a British breed like an Angus. These bull mobs are judiciously run in suitable black country all year (including on the flood plain in the dry) in mobs of 3000 or so, and in a modified holistic grazing management system. Here they use rotational blocks of forty paddocks of 990 acres (400 hectares) each, moved every three days or so, with every water point within two kilometres of the animals. For ecological enhancement, in a four-year cycle a different block is rested every year while the other four are utilised.

Typical of the Dunnicliff family's 'can-do' approach, they can now develop a cell of one hundred paddocks for $5 million, running 3200 bulls, moved every three days. At current calving rates, they can pay this off in six years – and one person can run three such units.

In 2015 John Dunnicliff told a reporter: 'I think long term, if you want to utilise the land, you have to take the water to the cattle, you can't take the cattle to the water. It doesn't work.' Moreover, he explained, 'Cattle do two things, they make energy or they make product. You can't sell energy, you only sell the product. If you don't take the water to the cattle, you'll burn up too much production, and that is not economic.'[3]

However as a postscript to this and as other station owners begin looking to increase watering points to enhance production, daughter Jane

Armstrong (who with husband, Scott, was John's right-hand person in his absences) added a cautionary ecological message in a recent conversation with me. 'Having been through the development of this new watering system,' she said, 'if anyone asked me, I would tell them that, when they put up their first tank, they should already have in mind how they will manage their country – because most people don't have that in mind. You see,' she concluded, 'much of the Territory is still relatively pristine, and it's urgent we learn to "see" and manage this country.'

While in 2011 the Dunnicliffs flew Allan Savory out from Zimbabwe for advice, they use Terry McCosker on a consultancy basis, and ensure all the family and staff do one of Terry's courses plus the associated low-stress stock-handling course run by Jim Lindsay. Cattle are trained under this regime so as to quietly respond to both chopper and motorbike.

In every respect, the Dunnicliffs have upended the old conservative thinking that still pervades this big cattle country. Bulls run with mobs the whole year, and Beetaloo only produces bull calves for specialist Indonesian markets that John has cultivated – again delivering huge cost savings in mustering and management. This bull market matches Indonesian culture, the beef ending up in traditional bakso balls.

Following difficulty in obtaining reliable quality poly pipe, John's son-in-law Adrian Brown established his own poly-pipe factory in Katherine, which now churns out 1300 kilometres of pipe a year – not only for Beetaloo and Amungee Mungee but also for a growing clientele as holistic grazing slowly catches on.

Moreover, the cattle are a base of hardy, climate-adapted and tick-resistant but previously lower fertility Brahmans that are now being crossed with a mix of Senepols (a short-haired Senegalese breed with tick resistance), plus other African strains, along with a judicious mix of better muscled *Bos taurus* breeds. By running bulls all the year and culling dry heifers, the Dunnicliffs place pressure on environmental adaptation and fertility. Aiding this is the fact that the long rest periods for pasture under a holistic grazing regime has broken the tick's life cycle.

And finally, as with all good land and human management, the Dunnicliff example of ongoing cycles of innovation presents a huge and positive social revolution for the large areas of Australia's remote and sparsely settled northern cattle country. The Beetaloo operation is now on course to reach 100,000 head. This five-fold increase means that six or more family or business units (plus all the associated labour and services) can now fit where there was once one. The approach applies to both non-Indigenous and Indigenous enterprises alike, and this without calculating the knock-on effects in the regional community due to these extra families through job creation and demand for infrastructure and other investments. Such impacts are already manifest due to the Beetaloo operation's growth in the Northern Territory. The double benefit of this extraordinary social revolution is startling ecological regeneration and biodiversity, which in turn opens up additional enterprise options (such as ecotourism). As far as the so-called triple bottom line (social, environmental and financial) is concerned, this approach seems a no-brainer.

Thus the Dunnicliff operation is a startling testimony to what happens when landscape managers work with, and not against, natural ecosystem function. In this case, the solar-energy cycle was regenerated through effective holistic grazing of the magnificent grasslands of the Barkly Tableland. From this, in indivisible and synergistic fashion, regeneration of the other four landscape functions flowed.

Sadly, however, with his vision, research projects and plan nearing completion, John prematurely died at the age of seventy-six in March 2016. His OAM nomination (he was actually presented with the medal by his long-time accountant and friend Mick Boyce only four days before he died) states that 'his contribution to the grazing industry – especially beef cattle in the north of Australia in the Kimberley and Northern Territory – is one of the most significant of the modern era'.[4]

By the time of his death, in the words of Mick Boyce, John had instituted 'the most extensive economically and environmentally sustainable development program ever seen in the northern Australian beef industry ... a proven and economically viable and simple to operate method of cattle production that is recognised by his peers, land conservationists and governments as a future model for the beef industry in northern Australia'. Moreover, said the nomination (and hinting at the accompanying implicit social revolution), John's new approach 'is a major breakthrough for an industry that has been struggling to find a way to utilise the vast areas of land available in northern Australia to support sustainable agricultural pursuits'.[5] As much of this land is also under Aboriginal ownership or direction, this new approach to pastoralism could enable a different reality for contemporary Indigenous livelihoods on pastoral country.

The old saying that 'the only thing that has ever changed the world is an individual' rings supremely true here. John Dunnicliff's brother-in-law, Professor Sir Gustav Nossal, is one of Australia's leading scientists, with considerable knowledge of the agricultural industries. He recently stated that the social, economic and environmental revolution John Dunnicliff instigated could only have been undertaken 'by a person of staunch character, wide experience and high motivation' requiring 'a quite unusual mix of courage, enterprise, vision and plain hard work'.[6]

So, despite John's premature death, through his family, ideas and example, his work of bravely pioneering a radical approach to regenerating land and society in Australia's broad savannah and other lands lives on and continues to grow.

However, it seems now that his legacy will almost certainly be taken on by others outside his family, for in August 2016 Beetaloo and Mungabroom stations (trading as The Barkly Pastoral Co.) were placed on the market – conservatively estimated to be worth $200 million. This move always seemed inevitable, given the convertible share arrangement negotiated with Brett Blundy, and as John's family and Blundy judged that the business had moved from the development phase to an operational phase, comprising 'one of the most improved large-scale cattle production businesses in Australia.'[7]

Aside from sound ecological literacy, though financial capital is the limiting factor in the rapid spread of similar visionary approaches to that of the Dunnicliff–Blundy enterprises, the looming shortage in world protein plus a desire for more ecologically produced food and ethical investments is nevertheless leading to similar developments elsewhere in Australia and further afield. Some 1600 kilometres to the south-east of the Dunnicliff family operations is an equally ambitious scheme, also based on ecological-agriculture principles and a holistic grazing regime, but this time jointly funded through foreign corporate equity. Again the key point of difference to traditional agriculture is the capturing of the potential of regenerative agriculture to create a much bigger landscape painting, and also with major social implications.

This example is led by a company called Sustainable Livestock Management Australia. In 2007, the leading and popular holistic management educator Bruce Ward (who prematurely died in 2012) partnered with accountant and businessman Tony Lovell to dream big and form the investment vehicle known by the acronym of SLM, which was eventually launched in 2012. Based on the Savory system of a holistic grazing regime, and along with other partners in the company, the business has now raised over $70 million and purchased fifteen properties totalling 1.2 million acres (480,000 hectares) in the semi-arid grazing lands of western Queensland and northern New South Wales (grouped in geographical hubs around Cunnamulla, Quilpie and Blackall). This area is typical of much of the Australian outback: landscape health, function and thus productivity have been constantly declining, and therefore so has employment and in turn rural and social infrastructure and the human population.

To run this aggregation (grouped into seven management units) which comprises breeding and finishing country with rainfall between 14 and 18 inches (350 and 475 millimetres), CEO Tony Lovell from SLM engaged the experienced western New South Wales sheep and cattleman Graham Finlayson as livestock manager.

Back in 2010, I had visited Graham and his strong partner and wife, Cathy, on their 22,000-acre (9000-hectare) property Bokhara Plains in the Western Division of New South Wales beyond Brewarrina. By that time, after being influenced by both Allan Savory and Terry McCosker, and following a life-changing Nuffield scholarship trip that took them overseas to visit many leading holistic grazers and practitioners of low-stress stock-handling, they had been among the first in Australia to publicly demonstrate the extraordinary potential of a holistic grazing regime in semi-arid, brittle grassland and rangeland country. Indeed, what Graham and Cathy had seen in similar country overseas had revealed the extraordinary potential of this approach to regenerate vast swathes of degraded land across the world – transforming the grazing base by first regenerating the solar-energy cycle. Consequently, they had more than doubled their stocking rate on Bokhara Plains while at the same time healing clay-pans and regenerating landscape function and biodiversity. The move by SLM was only an extension of this work by the Finlaysons, coupled with learning from the Dunnicliff operation north-west of them across the Northern Territory border.

Today the SLM business model is based on a few key points of difference compared to traditional operations, which is why they have attracted the big Danish pension fund Pensionskassernes Administration (PKA) as a lead investor in the $75 million deal. SLM's platform rests on: (1) a sustainable, indeed regenerative, low-input management regime adapted to Australia's variable climate; (2) a low-cost beef production system, environmentally green and delivering healthy food; and (3) a grazing regime that in time lifts production and builds resilience into ecosystems and businesses. This leads to greater investment in infrastructure and increased employment – with subsequent positive social impacts.

The large investment is based on the same principles pioneered by the Dunnicliff–Blundy investment model: a massive infrastructure investment in five key sets of cattle yards, 400 kilometres of water pipes linking 150 new water points, along with 3000 kilometres of single-wire electric fencing – the enclosures follow a grid-like pattern. The result is over 1200 paddocks for holistically grazing cattle.

The basis of this investment is the never-ending, interrelated regeneration of the five key landscape functions. The aim is to increase ground cover and healthy soil so that the solar-energy cycle is maximised. This in turn means still healthier soils and improved soil-mineral and water cycles, and in time, dynamic-ecosystem function.

In short, and as with the Dunnicliff–Blundy operation on the Barkly Tableland, this is a new development in big-business, broad-acre agriculture, as it is based on regenerating ecological health. Thus the main distinction between companies such as SLM and conventional agriculture is that the former are eliminating (or at least part-controlling) four of the largest risks in such operations: rising input costs, volatile weather, environmental degradation and global warming (which they combat by storing greater soil carbon).[8] Added to this is an eye on increasing demand for healthy, nutritious, ethically produced food as we continue to creep into the Anthropocene era.

In a further move in 2015, SLM partners began seeking $100 million for a holistically grazed sheep-meat and wool operation on over 247,000 acres (100,000 hectares) of Chile's Patagonia region. This, says financial journalist Chuck Stanley, 'is part of SLM's commitment to the World Resources Institute's *Initiative 20x20*, which aims to restore twenty million hectares [fifty million acres] of degraded land in Latin America and the Caribbean by 2020.'[9]

Across the Andes in Argentina, Allan Savory's own organisation – The Savory Institute, also in Patagonia – parallels this involvement. In 2014, the Savory group linked with an ecologically based sheep management organisation, Ovis XXI, which is located in the Chubut Province. Scientist and agronomist Pablo Borrelli and farmer and large estancia (landholding) owner Ricardo Fenton started Ovis XXI in 2003 with the aim of regenerating thirty-seven million acres (fifteen million hectares) of Argentina's fifty-odd million acres (twenty-odd million hectares) of degraded grasslands. Holistic grazing management was begun in 2008. So far, in collaboration with 160 farmers, they have regenerated over 3.2 million acres (1.3 million hectares) while more than doubling wool and meat production. A further key partner in this arrangement is the innovative organisation The Nature Conservancy, through collaboration from 2013 onwards in the project – called GRASS (Grassland Regeneration and Sustainability Standard).[10]

As Graham Finlayson concluded in his *Nuffield Report*, he has 'the underlying belief that all of agriculture has much more to gain by focussing on *management intensive principles and values that are flexible and resilient*'. He adds that, above all, 'Individual management and skill have much more influence on our results than all external forces, and focussing on what we can change is much more empowering and positive for the psyche of the Australian farmer in all regions, all environments and through all situations.'[11]

While most regenerative agriculture involves small to moderate sized landholdings or shared systems, the large operations described in this chapter excitingly point to how broad-scale land degradation across tens of millions of acres can be addressed. However, such operations tangibly capture the great tension in large-acre contemporary regenerative agriculture: that of using an ecological approach within the parameters of a modern global financial and economic system – and a system devised and endlessly refined by modern purveyors of the Mechanical mind.

The irony here, therefore, is that such regenerative-ecological farming systems and practices are not only in high tension and even contestation with the dominating and extractive approach of the Mechanical mind and its predication on a neo-liberal, unsustainable 'endless growth' philosophy, but that they are ultimately subversive of this very system. But – as we will see throughout this book – this juxtaposition is one of the great conundrums and yet avenues of hope for us all: that as a broken system and its components (whether in agriculture, economics, human health, food and its distribution system, or Earth systems) races towards an inevitably painful denouement, a counter-reaction is increasingly being promulgated.

In this 'subversion and undutiful behaviour to destabilize and deconstruct the authority of the inevitable' lies the only hope for Mother Earth and its multitudinous and varied inhabitants and life forms.[12] For as Wendell Berry says at the start of this chapter, 'An agriculture cannot survive long at the expense of the natural systems that support it and that provide it with models … At certain critical points these systems have to conform with one another or destroy one another.'

There is, however, only one truly long-lasting, healthy, regenerative and self-renewing, self-organising pathway and this is the one that works with and not against nature and its systems. And so, while the Dunnicliffs, Finlaysons, SLM, Ovis XXI and others described in this and the previous two chapters operate in sometimes harsh, sunburnt and windswept country, they have also learnt to work with these natural functions – and particularly with the key solar-energy cycle. In this, their recognition of the prime importance of the sun for the functions of life is not unlike the inhabitants of most previous human civilisations. Virtually all such civilisations, in one form or another, sought to worship, appease or encourage the sun through various religions and belief systems that enshrined sun gods and goddesses. It could even be said that the purveyors of modern economic rationalism in a way worship the fossil fuel-embedded solar energy that drives our unsustainable growth economies and has thus tipped us into the Anthropocene.

But the regenerative farmers I have just discussed revere the sun for other reasons. In recognising the role and power of the sun as the prime energy source, they have sought to learn how to use its blessings to positive effect. This smarter way of farming, via storing more long-lived carbon in the soil and rejuvenating soil health, is regenerating Earth and potentially its human communities.

As we will see in the remaining chapters of Part II, the other four landscape functions can likewise be positively regenerated in synergistic fashion.

CHAPTER 7

Water, Water Everywhere

Examine each question in terms of what is ethically and aestheti-
cally right, as well as what is economically expedient. A thing is
right when it tends to preserve the integrity, stability, and beauty of
the biotic community. It is wrong when it tends otherwise.

ALDO LEOPOLD, 'The Land Ethic',
A Sand Country Almanac[1]

Water seems so abundant but also so integral to life that we take it for granted. In the form of steady, soaking rain, it can be soul-replenishing to a drought-stricken farmer; heart-stoppingly beautiful on an early morning as mist rises off a river dappled by damsel-flies or concealing a platypus that bobs and ripples the secret shadow-surface; incredibly sensuous as one slides into a hot bath or lets the frigid cascade of a waterfall pound and prickle one's skin on a blazing-hot day; hellishly destructive when raging in torrents or coursing across exposed soil; and chameleon-like as it mysteriously shifts between liquid, steam and ice. To boot, when clean it is utterly transparent, tasteless and odourless.

We know that much of our universe's water is a by-product of star formation and has been around since not long after the 'Big Bang'. And we know that life on Earth evolved from primitive microorganisms that developed in water. As groundbreaking writers on the microcosmic world Lynn Margulis and Dorion Sagan put it, our cells live in a medium of water and salts that mirrors the composition of the early seas, so much so that our bodies, 'like those of all life, preserve the environment of an earlier Earth'.[2]

The reason water is vital to us and all life resides in its physical and chemical nature. At the macro level, because water exists in all three states of gas, liquid and solid, it plays a major role in climate control. It lessens climate extremes because it absorbs and releases energy during its transformation between its different phases. Water therefore enables the transfer of energy around the globe via ocean currents and vast storm systems. Equally importantly, water vapour, along with carbon dioxide, is a major component of Earth's temperature buffer in the atmosphere, thus crucial to the maintenance of Earth's steady-state temperature, which is so vital to today's life on the planet.[3]

Water's chemical nature and the strong attractive forces between molecules gives it its high surface tension. Just watch a water boatman on a pond's or river's surface: she skates around like a coxless four in thin treacle. It is surface tension that controls drop formation in rain clouds, without which we wouldn't have agriculture and human society. But the same molecular chemistry in water also gives rise to capillary forces: the very factor that allows water to move up a narrow tube in the face of gravity, and thus the crucial property enabling all vascular plants such as trees, shrubs and grasses to function and sustain agriculture and life. In fact, without water, both respiration and photosynthesis could not occur in life's organisms.

We know water covers nearly seventy-one per cent of the Earth's surface, though only a tiny 2.5 per cent of our planet's water is fresh. Around seventy per cent of all fresh water is used in agriculture, yet many of us farmers squander its abundance and yearn for it in alleged drought times. As John Robertson so painfully found after 1840, water is also a great energy carrier, and this, combined with its solvent capacity, can do huge damage to mismanaged landscapes, in the form of either erosion or dissolved salt.

As I was learning from Tim Wright and others, if one manages the solar-energy cycle and regenerates soil to health, huge ecological and production benefits accrue. Lose soil moisture and you risk losing diverse soil- and other life forms, and thus extra and longer active plant growth. This echoes the catch-cry of regenerative agriculturalists, 'Bare soil is dead soil', especially in a harsh Australian environment.

Over the decades since Europeans came to Australia, there have been individuals ahead of their time who attempted to throw off their previous cultural training concerning water and soil. One such person was Percival Alfred Yeomans, who came to have a worldwide influence on the use of water in the landscape.

Yeomans' very surname is apt, given that the old English word means a 'self-sufficient family farmer who lives close to the land'. He was born in the New South Wales wheat–sheep town of Harden in 1905. His father was a train driver and close friend of fellow train driver and later prime minister Ben Chifley. Like many in the depressed days of the late 1920s and '30s, Yeomans struggled to find work and tried many jobs to sustain himself and his young family. Eventually he gravitated to the mining industry, where he built a reputation as a reliable and trustworthy assayer.[4]

Not only did Yeomans drag his wife, Rita, and young family widely through remote areas of eastern Australia and New Guinea, but in the process, as land-use historians Martin Mulligan and Stuart Hill recount, 'he developed a sensitive eye for landscapes'. This meant he learnt to detect patterns in the landscape, such as being able to associate plant communities with particular mineral-carrying country, and 'was able to observe a diversity of landscapes and appreciate the importance of access to water, so essential for mining as well as agriculture'.[5]

The extraordinary architect-cum-artist-cum-philosopher Christopher Alexander was able to recognise that nature, like the built environment, has a 'deep geometric reality of order', an order he called 'pattern language'. It is clear that Yeomans could read this language. That is, his sensitivity to the broader landscape allowed him to be able to grasp and perceive its distinct patterns and their meanings.

As with many innovators, Yeomans' personal development was spurred on by a series of sudden mental and physical shocks, which changed his perception. The first occurred in 1931 when the twenty-six-year-old Yeomans was on a family camping trip in the western Queensland desert. One day, his eldest son, three-year-old Neville, wandered off and became lost. With the boy close to death, it was an Aboriginal man who tracked and found Neville, and then, using his desert knowledge, he dug for water and gently nursed him to health (later aided by Aboriginal women). Such an experience could not but have a profound impact on a parent. Researcher into the Yeomans family Les Spencer states that the life work of both the son, Neville, and his father, P. A., 'was informed and guided by a ... familiarity with Australian Aboriginal and Torres Strait Islander wisdom about the social and natural life-worlds', including their 'loving and affectionate relating to Earth as their mother who nurtures them'. This was a vastly different relationship to non-Aboriginal people, who saw 'Australia as a harsh and hostile place to be conquered and tamed'.[6]

Les Spencer further related that P. A. Yeomans 'was deeply impressed by the Aboriginal tracker's profound knowledge of the minutiae of his local land' and of 'minute traces left as evidence of the movements of a little boy that would not be made by other creatures or natural phenomena'. That is, the Aboriginal man was reading a pattern language, and he did the same in being able to find and then dig for life-saving water.

Stuart Hill, after interviewing Neville Yeomans, stated, 'According to Neville, it was probably this incident that gave his father his enduring interest in the movement of water through Australian landscapes. He could see that an understanding of this would be a huge advantage for people living in the driest continent on Earth.'[7]

The second influential traumatic incident in P. A. Yeomans' development occurred in 1944, by which time his career had moved into farming and earthworks. By the late 1930s, Yeomans had become a successful earth-moving contractor. This experience added to his knowledge of soil, geology and water, and of how to engineer landscapes. In 1943, Yeomans bought 1000 acres (400 hectares) of run-down, 'light' country at North Richmond, just out of Sydney. This comprised two farms, which he called Yobarnie and Nevallan.

Yeomans' brother-in-law, Jim Barnes, became farm manager, and he, with Yeomans' guidance, began implementing the popular soil conservation practices promulgated by the New South Wales Soil Conservation Service. This approach, said Yeomans' second son, Allan, had 'unfortunately originated with the agriculturally illogical practices "invented" by the United States Corps of Engineers, guided and advised by US Army construction officers' in the wake of the 1930s American dust bowl. In effect, it was a badly thought through, land-dominating, engineering-based thinking that took no account of landscape function, seeking as it did to limit damage ('conservation'), and not regenerate soil and landscapes. But before the folly of this approach could be fully exposed on Yobarnie, tragedy struck in December 1944.[8]

Barnes, along with Neville (then aged sixteen), were out on the farm when a fast-moving grassfire swept in. Neville survived by squeezing into a hollow tree, but his uncle was killed. This second trauma compelled Yeomans to fire-proof the farm through effective water use. Already dissatisfied with traditional soil conservation approaches, he now began experimenting in earnest. His knowledge and experience told him that holding water in the landscape, not hurrying it off, was key to successful farming in Australia.

Yeomans developed his regenerative agriculture practices through the late 1940s and into the '50s, bringing his unique attributes and experience together in a process of trial and error conducted on his own farm. This also involved extensive reading of leading ecological agriculturalists combined with international travel, plus consorting with a rare few open-minded academics.

By 1954, he was able to prove that the accepted belief that it took 1000 years to build an inch of topsoil was false. He showed that you could turn nutrient-poor, light-coloured Australian soil into rich, dark, healthy and biologically alive, self-sustaining topsoil to the depth of many inches in only three years. He saw that water was the key to this and that, if done properly, farmers could set themselves on a path of ongoing improvement and regeneration that would lead to dispensing with industrial artificial fertilisers. In short, Yeomans' approach was to regenerate the solar, water and soil-mineral landscape functions first, and then later the dynamic-ecosystem function. He also believed he could fire-proof and drought-proof a farm with the astute capture and use of water – water that, in the driest inhabited continent on Earth, often fell quickly and hard and raced off a landscape, taking with it impoverished, degraded and non-absorbent soils.

Yeomans was so fascinated with running water that whenever there was a big rain he would go out to see what water was doing on his land and how his new engineering constructions were faring at any time, day or night. Stuart Hill, who knew Yeomans and studied his work, relates the crucial moment on one of Yeomans' water-studying excursions in heavy rain: '[Yeomans] had noticed a reflective band running across some of the hillsides where they change from being convex above to concave below.' This was Yeomans' 'aha!' moment. Said Hill, 'He later called this line the "key-line", and the points where it crosses the drainage lines within primary valleys the "key points".'[9]

From this flash of insight, based on pattern perception, Yeomans cracked the secret to his Keyline system of regenerative agriculture, as he set out in his 1954 book *The Keyline Plan*. This 'whole of landscape' approach started with the key point, where one could build the first dam, or from where you began cultivating and designing. Yeomans' son Allan, who continues his father's work, said that starting from the keyline 'and cultivating parallel to it, both above the line, and below the line, produced off contour furrows, which selectively drifted water out of the erosion vulnerable valley' and towards the ridges.[10]

Every component of Yeomans' plan fitted together. He introduced tined cultivation implements into Australia – later developing designs of his own. These multi, chisel-like implements didn't invert the topsoil but, on progressive workings over three years, went deeper and deeper, allowing water and plant-root penetration and the creation of necessary air pockets to develop healthy soil, while at the same time drifting water out of the gullies towards the ridges. This, combined with the introduction of rotational grazing or else mulch-mowing, allowed litter and dead root material to feed microbial life in the soil. And it was this combination of practices that allowed Yeomans to show that farmers can build healthy soil in only three years. Furthermore, based on his deep reading, Yeomans also saw that a healthy soil and landscape translated into healthier animals and people.[11]

Yeomans built a series of contour channels across his landscape, beginning along the keyline, which intercepted water and channelled it into storage and flood irrigation dams: all designed to check erosion and allow maximum soil penetration of water or storage for later. What he was doing was enhancing the water cycle by slowing movement of water over his land. His contour channels were sited at intervals down a slope with tree-breaks below them so that a completed farm could look like a forested landscape, where the tops of trees were level with the bottom of trees in the break above.

And all of this fitted into what he called 'The Keyline Scale of Permanence'. In some respects, this conceptual extension of the original Keyline Plan anticipated by decades Allan Savory's approach to a regenerative decision-making approach for farmers. Yeomans developed a list of 'permanent' factors that needed to be considered in a priority sequence in the planning, development and management process if a successful Keyline Plan was to be implemented. In his own words, these key factors provided 'a yard stick or guide to every type and kind of decision that has to be made' in a Keyline Plan's execution.[12]

In order of priority, these permanent factors included: climate, landscape, water supply, farm roads, trees, permanent buildings, subdivision fences and soil. While he rated soil fertility of utmost importance, Yeomans placed it last of the eight factors 'because the fertility of the soil can be lost in less time than a line of fence posts will rot', while 'a poor soil may be converted into rich fertile soils in a tenth of this time'.[13]

In his last book, *The City Forest* (1971), Yeomans explained how his plan also applied to sustainable urban planning: enhancing aesthetics and energy use while recycling water, sewage and waste, and growing food and timber crops. He was way ahead of his time, and in effect his strategy involved triggering all the landscape functions: himself epitomising the best of the fifth function, the human–social.

One of Yeomans' strongest supporters was nutritionist Professor Sir Cedric Stanton Hicks. Hicks stated in 1955 that Yeomans' landscape design 'appeals to me as the basis for renaissance in Australian land use'. However, like most major innovations in agriculture and elsewhere that require a paradigm shift to be fully embraced, only a few early adopters persisted with his 'whole of landscape' transformation. There are a number of reasons for this, the first being that he really was ahead of his time in respect of holistic and systems-thinking.[14]

Stuart Hill aptly summarised the issue. Opposition to Yeomans 'trying to change farming practices', he said, 'relates to the fact that his ideas posed a direct threat to the powerful pharmaceutical and chemical companies that were increasingly becoming the major players in world agriculture … A demonstration of increased production without the need for purchased inputs [such as chemicals, fertilisers and implements] would certainly have been viewed as bad for agribusiness. As such, it would have been subjected to all the usual "damage control" strategies by the threatened industries, government departments and "colonised" academics, i.e., ignore as long as possible, then ridicule it, describe it inaccurately, conduct fake experiments, make false comparisons and then dismiss it.'[15]

Nevertheless, Yeomans' ideas not only lived on but they had a profound impact worldwide and, in an unexpected way, played a major role in influencing one of the greatest Australian contributions to a sustainable farming movement: that of permaculture (as we shall see in Chapter 16).

As I have indicated already, when it came to me overthrowing traditional and deeply inculcated beliefs and practices concerning farming in a total landscape, I was a slow and hesitant adopter.

I had the chance in the late 1970s to adopt Keyline landscape design and management while Yeomans was still alive. Not long out of university and still enthusiastic as to how one could rapidly change the world in environmental matters, I was a founding member of a small, eclectic group that comprised concerned farmers, foresters and agronomists on the Monaro. We gave ourselves the sexy name 'Monaro Tree Decline Committee'. This had come about because of the shocking and sudden mass die-off of eucalypts on the Northern Tablelands of New South Wales about this time (see Chapter 13 for this story). We had recognised that the Monaro potentially had similar issues.

As a result of this, I became friendly with the other farmer on the committee, a sensitive, deep-thinking man called Rix Wright. Rix's father was 'tree and nature sympathetic', and Rix had grown up as an only child on an isolated farm in the midst of thick and vibrant bush. So, being of creative spirit, he was incredibly sensitive to nature. The paddocks around their house are dominated by beautiful candlebark gums and as a child and later, Rix had a name for each one.[16]

He was also a talented artist – his mother, Hilda Rix Nicholas, was a famous Australian painter of the early twentieth century – who sculpted all his life while also dabbling in play writing. I was able to visit Rix on his farm in the southern Monaro near the lower Snowy River north of Delegate. He had lovingly designed and sculptured his property according to Keyline design principles after reading Yeomans' books and also briefly working with Yeomans in 1957. On an inspection one day, I discovered a series of vegetation belts and channels across his landscape that followed contours that fed into dams. I mistakenly thought these dams were set too high in the landscape.

Rix was disgusted at the Department of Soil Conservation's 'dismal attempts' to control erosion, and at their myopic philosophy. He said they saw 'run-off water … as a problem to be got rid of "safely" down grassed waterways to eventually leave the farm unused'. So Rix, in designing his own Keyline system, said he wanted to 'harness and not waste' all water. Keyline, he said, allowed him to 'sustain and develop land to produce the maximum capacity while nurturing, rather than stressing it in the process'. To the artistic Rix, as with many regenerative farmers, Keyline regeneration represented much more. For him, healthy, moving water was beautiful, and conserving water on the farm was creative and constructive. He said that Keyline Planning was 'aesthetically beautiful, being in harmony with nature and the landscape. Tree belts wind their way round hills and gullies in contours acting as wind-breaks, firebreaks, erosion control and wildlife corridors. The erosion is halted and gullies healed.'[17]

Regrettably, at this time of befriending Rix, I didn't fully grasp the deeper landscape function issues behind his approach. Nevertheless, in 1983, after the drought broke, I set about Keyline-tilling a 200-acre (eighty-hectare) paddock according to Yeomans' principles. I used a small dobby-level tacked onto the front of our old 1950s Fordson tractor with a Graham Hoeme chisel plough and small seed-box behind, and tilled along the contours of the hills.

But due to my lack of sufficient ecological literacy and thus an inability to see the wider picture, this effort was wasted. As I got busy with the intellectual challenge of stud merino breeding, and seeing no immediate beneficial results of this early Keyline tillage attempt (benefits almost certainly nullified by my set-stocking regime), I became a lapsed adopter.

It wasn't until fifteen or more years later that I returned to Yeomans' principles and approach. Friends in Western Australia, Mervyn and Carol Hardie of the renowned Wallinar merino stud, asked me over to talk at a field day on their farm in 2001. They lived in Western Australia's leading agricultural belt, the Great Southern, near the town of Broomehill. It was country that had been over-cleared in the pioneering phase and, like much of Western Australia, had a landscape underlaid with ancient salt. A rising water table and salinity began to appear in the 1960s in the Great Southern. By 1986 signs of soil salinity in the bottom country of Wallinar were still appearing despite the planting of over 100,000 trees and shrubs. So in 1989 Mervyn and Carol (each concerned about aspects of environmental health) had engaged a Western Australian 'whole of landscape' designer to lay down a restorative, integrated farm plan for them. This person's name was Ron Watkins.[18]

Having driven south from Perth through denuded, over-cleared and drab-looking landscapes, my wife, Fiona, and I were profoundly impacted by the healthy and vibrant nature of that part of Wallinar where the Hardies had implemented their landscape design plan on a recently purchased 600-acre (243-hectare) block that had been totally cleared and was prone to high wind erosion. I immediately recognised that I was looking at a partly matured Keyline Plan. Over a cup of tea, the Hardies, who are professional managers and business people, explained they had revegetated sixteen per cent of their land in this particular planned block and yet had increased production by over forty per cent; their average stocking rate having gone from 900 to 1700 sheep.

The Hardies said that, prior to the revegetation, their landscape was silent. When I went for a walk at dawn, with a gentle mist rising off the hydrated soil, the valleys and hills were a cacophony of birdsong. And to cap the experience, the landscape was restfully beautiful, with its secret, sheltered nooks and crannies and harmonious design redolent with varied shapes, vegetation, colour and shadow. I remember Mervyn pulling up in his ute at the head of a gully. It was a windy day, but we were cocooned by a crescent of layered vegetation and active bird energy.

Weeks of cogitation and research followed before I picked up the phone to the Hardies' intriguing landscape designer in Western Australia, Ron Watkins. Four months later, I found myself at Canberra airport, trying to imagine what this bloke who had executed that beautiful Keyline-type design looked like.

In fact, I nearly missed him, but then I noticed the determined walk and the alert, observant mien that subsequently would become familiar. After a few predictably awkward moments, I further assessed Ron as we waited for his luggage. He was of medium height, slightly round faced and neatly moustached. What struck me was his energy, combined with a cheeky grin and penetrating gaze that contained both wariness and warmth.

I subsequently realised Ron was also closely assessing me. The only reason he had hopped on the plane was that I'd sent him maps showing how we controlled the water on our farm. (We were able to do this because we owned the high country and weren't dependent on neighbours above us.) But having come this far, he still needed to know we were not dabblers but people committed to a challenging journey.

In no time, I found myself carrying an assorted and awkward bundle of surveying equipment, laser instruments and other paraphernalia. Over the next two years, this airport scene would be repeated as his plan was implemented on our high country, and as I lined up graders and bulldozers to execute his modified Yeomans system. Within ten years, the contour lines became Yeomans channels, with thick tree-breaks below, which then merged into a twelve-acre (five-hectare) central gully of vegetation that had by then filled with parrots, thrushes and a variety of insectivorous and other birds, not to mention encamped wallabies, kangaroos, echidnas, snakes, wombats and other creatures.

A number of years passed before I returned alone to Western Australia in 2010, this time on a research trip. I got to the farm of Ron Watkins and his wife, Sue, after ten days and 3500 kilometres of travelling through that state's agricultural lands. The journey left me with two powerful impressions. The first was how rapidly those crucial water transpiration pumps and soil conditioners – the vast swathes of vegetation (trees, shrubs, grasses) – had been cleared to make way for wheat and sheep. The second emphasised the consequence of the first: I was shocked at the impact of the white cancer of salinisation across huge areas. Abundant were stark, grey and white dead trees, broad areas of bare soil, white, salt-encrusted pans and dead, eroded creeks with rust-coloured water. I now understood why one of Australia's leading intellectual 'pastoral' poets, John Kinsella, wrote with such perception about the degradation of his native landscape. Clearly, those salted landscapes looked like war zones, and, as my daughter Alison commented, 'it was because they were indeed war zones'.

It was reading Kinsella's poem 'Why They Stripped the Last Trees from the Banks of the Creek' that put in my mind those famous lines from Samuel Taylor Coleridge's *The Rime of the Ancient Mariner*: 'Water, water, every where/Nor any drop to drink.' For here was the ultimate irony: in attempting to make a living in this driest of all lands, the actions of farmers (encouraged by successive state and federal governments) had indeed brought water to the land (rising upwards once the vegetation pumps were gone), but this salty water was killing the land in the process. I would soon realise how influential this issue of salt was in shaping the work of Ron Watkins.

Ron was one of six children growing up on a small family farm in one of the wetter, 'safer' agricultural areas in Western Australia, near Frankland in the south-west and 320 kilometres south-east of Perth. He was the third generation to farm their 1364-acre (552-hectare) property of Payneham Vale, his grandfather having 'taken up' the place in 1908 when it was all 'native bush'. Ron's parents took over the farm in 1947 and began seriously clearing vegetation. It was a time (the late 1940s and '50s) when, with government assistance and the widespread arrival of bulldozers, most vegetation clearing was done in that and many other parts of Western Australia.

On his marriage in 1973, when he in turn took over the farm, Ron carried out the last of the clearing – some 300 acres (120 hectares). This was because Ron was brought up in the heroic post–Second World War period when agricultural research had inculcated the view that you couldn't survive on 'native country'. The problem was that almost no one had foreseen the consequences of over-clearing deep-rooted vegetation in the ancient, often sandy and salt-laden landscapes of Western Australia.

Soon after taking over the farm's management, in May 1973 Ron was alerted that all was not well with his farm environment. It was the house dam that rang the alarm bell. 'Mum noticed that her prized Washington navel orange trees began to die, and that the vegie garden was less productive,' he recalled. 'On investigation, we discovered the creek and the dam had begun to go saline ... So salinity was the catalyst for me to ask myself, "What am I going to do about it?"' This would prove a life-changing moment.

However, Ron said he quickly moved away from just focusing on salinity. This was because he realised the 'white cancer' problem was caused by water. So he thought, 'How can I stop its destructive effects and turn it into a useful resource?' That's where P. A. Yeomans came in. 'At ag. college we were given one lecture on this Keyline Plan stuff, and despite a bad lecturer the message stuck,' he recalled.

The upshot was that, in October 1980, Ron and a colleague funded P. A. Yeomans to visit for a week. 'He was willing to let me pick his mind,' said Ron. 'We would stay up until midnight. I just quizzed him and quizzed him. "Water was the key," Yeomans said, "and energy from gravity, and the shape of the land." He taught me how to identify and recognise features in the landscape.'

What Yeomans didn't need to teach Ron was something that can't be taught, but that each of them possessed in some measure: a capacity for independent and original thinking. While this trait challenged the status quo, it sought to connect information in different, more holistic ways that often resulted in breakthrough solutions to problems.

Ron realised that he had to adapt the Keyline system to Western Australia's different soils and its saline landscapes. But he didn't yet know how. One excruciating day was to prove the catalyst.

At the time, a local farmer with an engineering bent claimed to have a solution to the salt problem by building water-intervention banks across the landscape with the idea of holding back rainwater and hopefully letting it infiltrate into the soil instead of running down into lower soakage areas that became saline. This farmer had formed a consultancy group, and Ron joined up. However, he quickly realised the interceptor banks weren't working in some areas, as farmers were still getting saline soaks below the drains. This alerted Ron to the fact that there must have been another source of water entering the landscape.

He soon apprehended that the Western Australian soils were part of the problem, as many of them were 'duplex soils'. Soils of this make-up sit on a hard, concrete-like rock base but are intersected by an often moist, plasticine-like clay layer and then a quartz layer at various depths. After Ron organised a backhoe to dig on his farm, he discovered that the extra water (beyond that from the clay layer) was travelling through the quartz layer.

Matters came to a head on a three-day field course with the consultancy group, when Ron began challenging some of the leader's methodology and logic from the first day. What happened the next day is seared into his mind. When they first drove out to do the paddock inspection, the farmers attending the course were equally distributed on the backs of three utes (one of them Ron's). But after Ron persisted in expressing his concerns, he returned with an empty ute, bar one polite friend in the front, while the other two utes were crowded to the gunwales. Ron immediately withdrew from the organisation, and he and Sue have never forgotten that feeling of peer rejection.

So Ron returned to his farm, the place where he did his thinking and testing, determined to explore his hunch. 'I need to test it here, on my country, country I know, that I've walked, because all my life I've lived this thing … It's part of me, and I'll pick up the nuances, the missing causes,' he said.

Ron hired an excavator to dig holes twenty-three feet (seven metres) deep across his farm. This detailed work, in addition to verifying a rising water table, also confirmed his theory that water did travel in the quartz layer as well as in the clay. Ron discovered he had eight different underground streams or 'preferred pathways of water' running through the quartz layer. 'These streams could be coming from anywhere from a level contour up slope, by gravity, but could all be being forced into this one area,' said Ron. He realised that other people's poor management was being off-loaded onto his farm, and his own management onto others, and so on down the catchment.

Suddenly it hit Ron. If he could refine a Yeomans-type landscape design system to address this complex issue, then it could have enormous regenerative impact on the Western Australian landscape and the livelihoods of farmers, their communities and landscapes further afield. What he evolved was a system, translatable across Australia, that refined what Yeomans was aiming to do but could also address and regenerate problem soils like those in the West.

The next breakthrough came from the ancient practice of geomancy: of being able to divine subtle-energy[19] flows in the landscape. Ron had learnt to accurately water-divine from an old uncle. What he found was that the eight underground streams wandered all over his farming landscape in a kind of mosaic. He knew he had to get his drains into the clay and its water source; and he knew he had to start high in the landscape, intercept rainfall and store it in dams, while slowing the excess and directing it out onto ridges and retarding it in gullies.

Ron reasoned that tiered contour drains would cross all the underground streams and intersect the clay layer also. If he planted wide, deep-rooted vegetation belts below the contour banks, they could cross the mosaic of water flowing across his farm. If he also planted vegetation in a wide corridor down the main gully, further water interception would occur. In addition, the native tree belts, via natural transpiration, acted as nature's pumps – reducing excess underground water recharge. Trees in turn would encourage biodiversity and create sheltered microclimates for livestock, while preventing desiccation of pastures and cropland. Moreover, tree belts following contours like Yeomans had originally designed (and as opposed to standard straight-line tree-breaks on fence lines) catered for varied wind direction and were more aesthetically pleasing and, in beauty and form, fitted the landscape – as Fiona and I had experienced on the Hardies' farm.

The result was a series of corridors linking to fifty-acre (twenty-hectare) native bush areas in the central gully and main drainage creek. This interconnected system addressed the salt and wind issues while protecting the creek by naturally pumping out and transpiring its water. What Ron was doing, in addition to addressing elements of the solar, soil-mineral and water landscape functions, was encouraging a healthier dynamic-ecosystems function. Ron's plan was thus a 'whole of landscape' design, shaped by the Western Australian environment – a design that used nature's principles in working with the landscape. Reflecting on his efforts, Ron said, 'I'd like to think that old P. A. Yeomans would be proud of what we've done here.'

Ron began working with clients from 1984, the Hardies being among the earliest. In 1988, he travelled overseas on a Churchill Fellowship to New Zealand, the USA, the UK, the Netherlands and Israel, examining agroforestry and water-use management. By 1995, Ron Watkins' work was being recognised internationally when he won the United Nations Environment Programme's prestigious Global 500 Award (known as 'the environmental Oscars'), which was presented to him in South Africa by Nelson Mandela. He also received in the same year the first Saving the Drylands Certificate at an international conference on desertification, in Kazakhstan. However, and typical of most agricultural innovators, Ron's work is much less recognised in his home country.

On my visit in 2010, Ron and I went for a drive around his farm, landscaped by him since the early 1980s. I found a maturing environmental haven. We stopped to examine native bush full of multiple wild orchid varieties. The paddocks, secure and protected from any wind, were sausaged around the slope between mature tree-breaks. So neat was the planning to align the tops of the trees with the bottom of the tree-break above that, from some angles, the whole landscape looked like a forest. In between, Ron showed me spots where salty patches had disappeared. We passed contented livestock lazily chewing their cud, and a mob of hundreds of chooks in a moveable pen, with a self-sufficient solar system for the electrified protective fence, and a roof over their roosts that caught water for the birds: part of his organic chook and egg set-up.

I found out that, soon after refining his home system, Ron was irrigating crops from his dams. He now successfully grows canola crops without chemicals (crops that, in a traditional industrial-agriculture system, are very prone to insect predation and often need repeated insecticide sprays) and has moved his farm to full organic status. He delivers eggs and vegetable produce weekly to a farmers' market in Albany to the south.

Reflecting on our discussions over a cup of tea as we gazed out over his tiered and timbered landscape, Ron attempted to sum up his philosophy. 'The umpire is what this environment says,' he told me. 'My philosophy is that God put all of this together, and there is just so much diversity and interaction that we as humans would never be able to come up with the design ... the system works because of all the principles that are instituted in it ... We need to look at the whole, but few people can do this.'

Translated, Ron was saying that farmers needed to be thinking ecologically, and the key was working to enhance all the landscape functions – functions that are indivisibly connected. Ron had realised that, in his environment, the landscape function in the direst need of restoration was the water cycle. So that's where he had started.

But he was also saying something even more profound while alluding to an issue that other regenerative agricultural innovators I had interviewed frequently touched on. This can be encapsulated as, 'It's best if we farmers get out of the way and let nature get on with it. She knows best.' I came to realise that, by this, they meant the propensity of natural systems to self-organise (a major concept to be discussed later in Chapter 18, and one crucial to this whole story).

So, what are the key features of a healthy water function in landscapes? For starters, Ron Watkins and others such as Yeomans have demonstrated (as with the solar-energy function and more green leaves) that the water function is key to all life and its systems. It follows a cyclical pattern in the landscape: falling as rain to ideally soak into the ground, where it is held, or to run off into rivers and creeks and away to the ocean, swamps or evaporation pans in salt lakes, or to be evaporated off bare or unprotected ground in poorly functioning landscapes or transpired by plants in a more healthily functioning landscape.

An ineffective water cycle means increased run-off and erosion, and less water penetration of the soil. When the soil consequently loses the protection of deep-rooted vegetation and ground cover, the result is greater water evaporation at high temperatures and less plant growth and production. An ineffective landscape water function then clearly means less stable rivers and creeks after floods.

Conversely, an effective water cycle – through good landscape management – means trapping, holding and recycling the maximum possible amount of rain falling on a landscape. This delivers the maximum amount of water to plants, to help drive the solar, soil-mineral and dynamic-ecosystem functions. It means less evaporation and less run-off. It allows less organic matter to be carried off by escaping water, which in turn maintains healthy, porous soils and so a good air-to-water balance in soils, which, to complete the cycle, allows greater water absorption by plants. It also means in some regions (as we will see in the next chapter) a crucial freshwater layer or lens that sits atop natural underground saline water in Australia's ancient soils.

Other notable features of a healthy water cycle are: increased amounts of water going to the atmosphere via plant transpiration (which aids climate control); greater recharge in underground supplies; less soil-capping and greater vegetative ground cover; and increased growth of deep-rooted plants (whose root cavities – biopores – become conduits for greater water penetration and absorption). Overall, this results in an enhanced solar-energy cycle, through keeping plants in a longer growing season.

All these features of a healthy water cycle have one thing in common: they are dependent on active and skilful human management. And if farmers get this part right, their management pays dividends. Research by soil scientist Glenn Morris now shows that, in the top twelve inches (thirty centimetres) of a healthy soil, by adding just one extra per cent of soil organic carbon through a healthy solar-energy cycle and thus nurturing spongy, healthy soils, farmers can store an extra 144,000 litres of water per 2.5 acres (one hectare) in their ground.[20] This is prodigious and has enormous synergistic impact on all the other landscape functions.

———

Ron Watkins' last visit to our farm in 2003 coincided with a bulldozer and grader implementing the contour-channel design Ron had laid out. The two contractors were suspicious of this seemingly crazy idea, cheekily answering to inquisitive mates over their two-way radios that they were building another Snowy Mountains Scheme. Nevertheless they worked with Ron, who, confident and authoritative because of his engineering experience, revealed he understood the nuances of land-shaping with earth-moving machinery. That didn't prevent their scepticism increasing throughout the day, because some of Ron's laser-pegged drainage lines, to the naked eye, looked like they were running uphill into the main gully.

Then, at the end of the day when the work was nearly complete, a small thunderstorm broke over the top of the main hill above the earthworks, delivering three-quarters of an inch (twenty millimetres) of rain. We all waited on the result: Ron, me and two disbelieving earth-moving contractors. Then, after half an hour, water began trickling around the contour from two kilometres away and into the main gully. The contractors were visibly impressed. I turned to see Ron's expression and noticed tears running down his cheeks, which he surreptitiously brushed away.

CHAPTER 8

Call of the Reed Warbler

The land is a living, breathing entity. If you love the land, the earth, it'll love you back. It's just the way it's always been … there's no big secret to it. If you want a relationship with the earth, you just have to love it … We're not exclusive from nature. We're part of it! … part of everything around us.

ARCHIE ROACH, 'Archie Roach sings from the soul'[1]

It is winter again and I am out early in the frost to catch a mob of sheep coming off where they camp. I let the sheepdogs out and wait for them to 'empty' by cocking their legs on favourite marking points. They spray seemingly limitless volumes of urine while also investigating the night's smells (perhaps a wandering fox checking out the chook pen or the fresh diggings of a wombat). I patiently sit on my motorbike, rugged up in my thick gloves, beanie and woollen 'Bluey' coat, while taking in the vibrantly clear morning.

In the west, a deep rose lights up the snow on the main range, running in sparkling glints across the royal blue foothills, whose tiered ranks and folds, deep shadows, humps and ridges seem like the living, breathing flanks of ancient megafauna monsters. Above these are purple-pink snow clouds that already are creeping over to smother the high range in a billowing blanket.

From the hill above our house, flocks of currawongs have begun descending in their peculiar looping-glide flight – a parabola down, wing-beat up, then a loop down again. They will spend the day on *Cotoneaster* berries in tree-breaks near the garden. Our crows have long been awake, nesting pairs beginning to caw before dawn. Soon, foraging squadrons of them will move out into the paddocks – what I call our unpaid pest controllers.

For us, the real winter starts after the shortest day of the year: 21 June. My father's litany has great truth: as the days get longer, the cold gets stronger. The six weeks after the shortest day to early August are the tough times on the Monaro. They often entail clear skies and big frosts, once reaching as low as minus sixteen degrees Celsius in the low parts of our farm, when the dogs' water became pure ice six inches (150 millimetres) deep, and pipes burst in ceiling and paddock alike.

Eventually the dogs are ready and looking expectantly at me. We set out on the four-wheel motorbike, the dogs jostling for position on the back as my exposed cheeks start to burn. I skim through showers of ice crystals spraying up from the long grass that glistens in the early light. Later I know we will be bathed in lukewarm sunshine after the frost has gone and I have had more time to marvel at mid-winter life going about its business.

This winter time is the local Aboriginal Ngarigo season for ice, not water. Yet ice is water and water is essential to life. Without it, life could not exist, and, as the dogs at my back and a myriad of spiderwebs sparkling in the early sun remind me, even at this coldest and driest of times life exists due to ubiquitous, precious water.

In this vast, usually parched continent, finding water has been a national obsession of white settlers, who dreamt of inland seas and giant rivers of abundance: what Michael Cathcart called 'water-dreamers'.[2] But in the end, dreams don't make water. The basics surrounding the water-cycle landscape function are consistent worldwide. Mess with these – as Australian and North American land managers and a myriad of other pastoralists have done – and pay the consequences.

While there are many pathways to regenerating a landscape's water cycle, all – as we saw in Chapter 7 – require change and regeneration in the other landscape functional elements (not least the human–social). Bluntly, it comes down to one's view of how the world and its natural systems function, which in turn determines one's management approach. Unfortunately, many farmers are simply not aware of the concept of landscape functions, as they have little ecological literacy.

The way we pasture our animals is the greatest determinant of a healthy water cycle in Australia. In Chapter 4, Tim Wright's replete, spring-fed dams (by comparison with his neighbours' dry dams) were a stark testimony to how holistic grazing enhances a water cycle even in drought. Norman Kroon's rising water table and rehydrated Karoo landscape described in Chapter 6 was further testimony. Short, intense, high-density grazing and nutrient spreading and disturbance, in combination with adequate rest for the most palatable plants, completely regenerated the landscapes of both these farmers – built around an enhanced solar function putting more carbon in the ground. With increased desirable perennial plants, roots went deeper; a healthy, vibrant soil meant more burrowing organisms and more soil air pockets (or biopores); and this meant vastly more water stored in the soil. All landscape functions had worked in synergy.

This was dramatically brought home to me one day in my own district. I was in town when a big rain was setting in. On my drive home, I passed two farms on identical red basalt and granite country. The first was set-stocked, and while from the road it looked to have a good cover of tall but low-grade native *Stipa*, or corkscrew grass, I knew from when I had walked through it once that about fifty per cent of the soil was bare. The second farm was holistically grazed and had near one hundred per cent ground cover of a healthy mix of perennials, annuals and forbs (broad-leaved, herbaceous plants other than grasses).

On getting home in the still-pouring rain, I tipped out the rain gauge, as farmers are wont to do: it showed two inches (fifty millimetres). I thought back to my drive, and that first, set-stocked farm I passed had already been running vast sheets of red water. The dead soil (due to over-grazing and resultant shallow roots) simply could not hold the water. Consequently, the water had begun running off, taking with it precious topsoil. Needless to say, there was no water running off the paddocks on the second farm because it had all been readily absorbed by a healthy, functioning soil with deeper and more varied plant roots.

That rain continued for another twelve hours, tripling the first measure to six inches (150 millimetres). Going for a drive, I discovered that only at this end point did the second farm begin trickling clear water into a few gullies. The lesson was stark: in the driest continent on Earth, with often irregular rainfall that can come in big downpours, the second farmer had effectively tripled his rainfall capture compared with his neighbour in that twenty-four-hour period. I have subsequently heard similar stories repeated all over Australia.

The thirst for abundant water in this nation's harsh climate regularly spawns diverse, sometimes eccentric, water-dreamers.

Many of these dreamers witness drought-breaking rains or violent thunderstorms after a dry, and then watch the valuable water pour off their land and disappear downstream, downhill, and more often down man-made gullies. Hence many water-dreamers have devised methods to capture as much water as they can with each scarce downfall. Widespread and various approaches to water-ponding are one such example in different regions across Australia (led by pioneers such as Bob Purvis in the Northern Territory and a group of pastoralists around Cobar in New South Wales). And a sophisticated version is P. A. Yeomans' approach to the same challenge (discussed in Chapter 7).

However, two of the most recent examples of such 'hold and capture' approaches have come out of the school of farming hard knocks. Both work on relatively similar principles. The first has gained wide notoriety, while the second is virtually unknown. Yet the second approach, in my opinion, has far wider application to Australian and even wider landscapes and is supported by extensive scientific and engineering knowledge. The originator of the latter innovative approach – loosely described as 'Successional Land Repair' – is Peter Marshall.

Originator of the first approach – called Natural Sequence Farming – is Peter Andrews: who can sometimes appear as prickly and inconsistent, yet at times insightful and clever water maverick-cum-visionary.

While I have met Peter Andrews on a couple of occasions, it was an unexpected encounter with an early adopter of his that revealed to me he was on to something. I had been asked to address a Landcare[3] group at Sutton, near Canberra, and I arrived mid-afternoon at the farm of my hosts, David and Jane Vincent. David then took me for a drive to view some of his creek reclamation work. He runs a mob of 150 or so Angus cows, holistically grazed, and on the drive out I saw dense tree-breaks of tree lucerne, neatly trimmed back to the fence and browse-line by the last mob of cattle in the paddock.

After a drive up and down steep hills, we came to a valley, passing on the way the upstream paddock of a neighbour that drained into the valley's creek. It was a dry spring, and much of the hill country on David's block had browned off. The set-stocked neighbour's place was in far worse shape. There was little ground cover, and that which existed was a yellowy-brown. The neighbour's creek was eroded, with large areas of bare ground. David informed me that a large mob of merino wethers (castrated male sheep) were set-stocked there all the year and that there were outbreaks of dryland salinity also.

Then we crested a foreground ridge and looked down on the same creek now running through David's farm. I was met with a stunning sight. Not only were there no erosion scars or banks visible, but to a width of fully 100–200 metres either side of the creek there was a bright-green swathe of grass, including large tussocks and *Phalaris* up to half a metre high. On closer inspection, there was also a variety of broad-leaved native grasses and forbs, plus sub and white clover. The contrast to the neighbour's land only half a kilometre upstream was dramatic.

As we parked and walked down to the creek, David told me he had heard Peter Andrews talk once, and then, with a small group, he had driven up to the Hunter River and Andrews' home farm, Tarwyn Park, to see for himself. While there, he and his group also visited the farm of the famous retailer Gerry Harvey. Gerry had employed Andrews to apply his pioneering water works. David said the visit was mind-changing because, in a drought, Andrews had converted the entire valley to a lush green compared with neighbours' overgrazed brown. There were even green patches high up steep hillsides. I was now experiencing a similar scenario on David's place – even though he had been applying this work for only nine years.

The first thing that struck me was that, unlike on his neighbour's place above, here the creek was readily running again (the landscape was releasing stored water). Moreover, there were sizeable, weed-cloaked ponds and bunches of the large perennial grass *Phragmites australis* (or common reed), along with tall cumbungi reeds (*Typha*) and other succulent water plants. Furthermore, the creek banks had largely healed (aided earlier by cattle hoof impact before David fenced off the creek), and there was no sign of salt scalds. I then stood and looked back, outwards from the creek, and could follow how the water that was now clearly being held in the creek vicinity had begun to rehydrate the lateral landscape.

David showed me his creek crossing points (narrow trackways for vehicles and livestock), where he had built up slightly elevated banks with tyres and rocks, which impeded and backed up the water while allowing it still to gently flow through. This was a famous Andrews design called a 'leaky weir'.

Then a sound caught my attention: the unmistakable high-pitched song of a reed warbler, high up and swaying sideways in a rocking reed bunch. I had seen many such birds in bigger reed clumps on various expeditions along river courses in the past. I was surprised, therefore, that the bird could be found in this small founding patch of reeds that David had helped to introduce. Clearly it knew there would be a much larger reed bed very soon.

Later, over a cup of tea, as David and Jane checked out the bird and its call on an impressive app that Jane had on her iPhone, I pondered the likelihood that these locally rare birds probably hadn't been seen in this creek for more than 130 years. The next morning, as I left Canberra heading home, with filaments of dawn mist rising off dams and cloaking irrigated flats of lucerne (alfalfa) beside the Murrumbidgee River, I concluded that the lovely reed warbler could only be a talisman of a watercourse and landscape function on the path to healthy regeneration.

Not unlike a number of others in this story (including P. A. Yeomans) who were able to see Australia and how its landscapes functioned through fresh eyes, Peter Andrews claims to have been strongly influenced by Aboriginal people. Over the years, Andrews engaged with such people to learn about their perception of landscapes and to study their artistic, almost abstract aerial views of landscape.

Andrews grew up in hot, water-scarce sheep country in the New South Wales Western Division, eighty kilometres from Broken Hill. It was that pastoral country of New South Wales and South Australia that had been overgrazed and denuded for decades, and which consequently was repeatedly swept by dust-storms: country that in the 1930s the CSIRO scientist Francis Ratcliffe had called 'The Kingdom of the Dust'. Andrews' father, Archibald, ran three family properties out in this 'Desert Kingdom' while his brothers were fighting in the Second World War.[4]

One day, when aged three, Andrews and his brother and sister were packed into a cool underground room to sit out another dust-storm. After three hours, the children emerged to witness a previously grass-covered landscape now stripped bare by the storm's corrosive particles. 'This made a deep impression on me,' said Andrews. 'I am sure it was the start of what proved to be an intense lifelong interest in how the landscape functioned.'[5]

Andrews grew up with stock: first sheep, then cattle and always horses. When his father, on experiencing a heart attack, moved to Gawler in South Australia (forty-four kilometres north of Adelaide), Andrews left boarding school in Adelaide and went back to the 59,280-acre (24,000-hectare) farm at Broken Hill to help his brother. What makes a natural stockman is acute observational skills, and Andrews, at the age of just sixteen, proceeded to fine-tune his. He became a top horseman and judge of a horse, and in time an outstanding breeder and manager.

In time, so as to further his horse-breeding ambitions on better country, Andrews moved to the upper Hunter Valley in New South Wales, in the Bylong Valley near Rylstone. The 865-acre (350-hectare) farm, called Tarwyn Park, became his home: the farm already having a reputation for breeding fine racehorses. Here Andrews began to develop his water capturing techniques as he also continued to travel widely in the Australian landscape, observing patterns, questioning. Along the way, he came to understand some of the basics of landscape function, which, when integrated, resulted in him evolving original theories about Australian landscapes and water function (which we will shortly see). Furthermore, he designed a series of interventionist steps that would allow some components of degraded landscapes to begin returning to healthier function.

Early in his first book, *Back from the Brink: How Australia's Landscape Can Be Saved*, Andrews laid down the key thrust of his life's work. The degraded landscapes he was faced with were a result of European settlement and practices and the introduction of rabbits in particular. Therefore, said Andrews, 'since the Australian landscape functioned perfectly well on its own for millions of years, we ought to be able to solve the landscape's current problems by somehow reinstating whatever it was that enabled the landscape to function so efficiently then.'[6]

But first he had to put some of the puzzle pieces together, and Tarwyn Park was his proving ground. His main focus became water in Australia's landscape – moreover, underground water, not the more obvious above-ground water. 'In-ground water is what kept, and keeps, the Australian landscape alive,' said Andrews. 'It's the ultimate key to our landscape's survival.'[7]

At Tarwyn Park, Andrews found a badly managed farm that had lost much landscape function. Salinity was rampant, the creek was eroded, and the pastures were unhealthy and covered in tall weeds. Using the health of his horses to monitor landscape health, and by mechanically slashing with a tractor and other non-chemical interventions (partly because he had no money), Andrews began to restore his property. 'A farmer gets three things for nothing,' he explained, 'air, sunlight and water. Everything else farmers pay for. You would think, then, that the logical thing for farmers to do would be to concentrate on making the most of the things they get for nothing.'[8]

After over a decade of experimenting, combined with wide reading, talking to scientists and travelling through the Australian landscape, Andrews developed the belief that the solution to restoring the landscape functions was twofold: 'revegetate the land and reinstate the hydrology system that once kept salinity suppressed'. Andrews realised you needed to start with the solar function: deep-rooted, photosynthesising plants and their solar panels. He came to recognise that soil organic carbon represented soil fertility, but that this in turn came from plants. 'The green surface area of a landscape,' he said, 'represents the productivity of that landscape. This basic fact is the key to sustainable farming.' His mantra was 'Plants run the landscape'.

So, seeing deep-rooted weeds as allies was crucial. This is because their nature is to heal by bringing up nutrients, breaking open the soil, preventing erosion, having deep tap-roots that allow water to penetrate as they decay, and so on. To this end, the sometimes cantankerous Andrews made a point of challenging a still deeply held belief in Australian agriculture: that tap-rooted weeds are bad. Andrews wasn't the first regenerative farming innovator to challenge this view. But it was total heresy to a whole cohort of industrial farmers and attendant education and research institutions, plus chemical suppliers dependent on weeds being seen as a dangerous landscape element.

Andrews also part perceived the importance of dynamic ecosystems: what he summarised as 'biodiversity'. He realised the crucial significance of having not just diversity in one's pastures, but also 'trees, scrub and weeds' so as 'to produce the extra fertility that pastures require'. While not the first to understand the importance of these elements to healthy landscape function, and that birds, insects, animals and others recycled nutrients in a landscape, part of his farm planning was to have tree-breaks ('hedgerows', he called them) so as to 'link the various components of the landscape that ensure biodiversity'. Significantly, in regard to trees and shrubs, Andrews didn't care whether this biodiversity was native or exotic, and he soon became a controversial champion of willow trees.

As discussed in Chapter 7, Christopher Alexander revealed that nature works in, and is made up of, patterns. This was Andrews' breakthrough also. 'We can see what's happening in the landscape,' he said, 'and [it] falls into certain patterns.' One of these was that most of Australia's inland rivers (unlike the traditional view of rivers in the geologically younger and wetter areas of Europe and parts of the USA, for example) formed 'chains of ponds' in between and around wetlands. However, Andrews claimed that ninety per cent of these had disappeared since white settlement. The result of this naturally occurring pattern in our dry landscape was that much of the water stayed 'in-ground'. Andrews drove 150,000 kilometres around Australia verifying his thinking. 'The answers were written in the landscape,' he said. '... Once you were able to recognise them, you would see them everywhere.' He added that scientists miss these, and he provocatively concluded, 'Their mistake is not looking at the whole.'[9]

Confirming Andrews' observation are diaries of early white explorers such as Thomas Mitchell and John Oxley, in which the term 'chain of ponds' frequently appears. Others, such as Charles Sturt and Ludwig Leichhardt, gave excellent descriptions of chains of ponds and vast wetlands and reed beds. Of great relevance is that various Aboriginal artists (such as Tim Leura Tjapaltjarri), with their unique aerial view, beautifully capture the series of steps in valleys that appear as chains of ponds: as seen in Tim Leura Tjapaltjarri's *Water Dreaming* and similar paintings. There is even a town in South Australia called Chain of Ponds.

Andrews found that many of Australia's ancient flood plains comprised huge wetlands, perhaps fifty kilometres across, located every 200 kilometres or so down the bigger water flow valleys (I hesitate to call them rivers). The chain-of-ponds pattern was a brilliantly evolved, self-organising system suited to Australia's unique physical and climatic environment. This was an environment where landscapes experienced long dry periods and then sudden floods. In such a situation, it is an advantage (for landscape health) that as much as possible of this sudden influx of water is captured. The wetlands acted as plugs to the whole system, thereby preventing the valuable water draining out to what later developed (after land degradation) into gullied drainpipes. This series of steps I had first clearly perceived that day of my visit to David Vincent, when the reed warbler announced the coming of a future wetland plugging up with *Phragmites* and cumbungi reeds.[10]

Moreover, the chain-of-pond streams, like much of Australia's landscape activity, appeared counter-intuitive: the wetlands and bulk of the streams were higher than the lateral valley country, which in floods then received the over-wash and its fertility. This also meant that, outside flood time, the landscape's in-ground water was constantly on the move in a vast arterial system comprised of sandy trails.

Unsurprisingly, the vegetation – such as giant reed beds – played a key role in both the maintenance and creation of the chains-of-ponds set-ups. Reed-plugged marshes took the energy out of the system; they trapped mulch and prevented erosion, and in a flood event the reeds blocked the flow of water and forced it out onto the higher ground – where the sandy recharge areas were also located.

Having insightfully analysed this ingenious Australian-adapted system for keeping water in the landscape, Andrews then set about developing a series of techniques to regenerate and restore function to eroded creek and river valleys: the results of which – marked by a broad green band – I witnessed that day with David Vincent.

First, Andrews realised he couldn't change things overnight and so had to start slowly. This meant slowing the speed of the water flowing in his creek at Tarwyn Park so he could hold more of it in the landscape. Then he worked on trying to recreate, however simply, some form of chain of ponds that would rehydrate his landscape.

Various impediments to flow were used, such as rocks and logs. Water was diverted either into more separated flows or onto trees or rocks to dissipate energy; or else a stream was diverted and via various means directed back so at the head-on meeting of the two or more parts, further energy was dissipated. This led him to using fast-growing willows as a major impediment tool – and much of this was aimed at the worst erosion effect in a creek, which is back-cutting of head walls (that part of an eroding creek at its highest level where water actively drops over an edge and becomes the site of high-energy erosion – forever working its way further upstream). Once he blocked this and put in impediments, the flowing water lost energy, and so silt began to settle and the creek level lifted.

All this fiddling led to the development of his famous 'leaky weirs'. Here, and sometimes used as a creek crossing point but usually at the bottom of a pool, one placed a slightly raised structure made of rocks, logs or any solid impediment material. But water was still allowed to trickle through, so that too much erosive energy was not built up, as with an impervious dam. As energy was lost and water blocked, reeds, other weedy-type vegetation, rushes, willows, flood debris and so on could take hold, and the retained water could begin seeping laterally into the landscape. Over time, as the natural functions took over, the creek of its own accord began to move towards a modified chain-of-ponds set-up. In no time, Andrews had part-reinstated ponded wetlands, as I saw David Vincent do in less than nine years.

Having proven he could mend the eroded main water channel by turning it into a chain of ponds, Andrews next concentrated on getting as much of that water in-ground as he could. He thus looked at a modified contour plan, which he saw as a simulation of ancient landscape systems. He combined this with fertility pits, filled with organic material, to which the channels were linked so the fertility could also be moved into the landscape. Later, he implemented a system of double contour banks higher in the landscape. In total, Andrews' whole-of-landscape approach is now known as Natural Sequence Farming.

Peter Andrews is vehement that there are fundamental differences between Natural Sequence Farming and the Yeomans' Keyline Plan. He contends that his approach to managing water in the landscape, 'far from resembling the Keyline approach, is in several important respects the very opposite'. This is because Andrews believes, first, that water can't be stored efficiently in the Australian landscape (by which he means in dam storages), and, second, that water can't be flushed across the land 'without running down the fertility of the soil'.[11]

I think there are more similarities than differences between the two major innovative approaches. Keyline, for example, is a far more complex and holistic approach than just what Andrews calls 'the installation of dams and the flushing of water across the land'. Like Natural Sequence Farming, it is based on restoring landscape function in key areas, including regenerating grassland and forest, soil health, biodiversity and so on. Some of Andrews' ideas on engineering intervention of contours high in a landscape have much in common with those of Yeomans. On our home farm, via Ron Watkins (as seen in Chapter 7), we partially implemented a modified landscape design based on Yeomans' principles. It quickly morphed into an ecological approach that blended the best of both Keyline and Natural Sequence Farming.

However, I have major concerns with some of the upbeat promotion of the broad benefits of Natural Sequence Farming. I believe that Andrews' breakthrough vision of restoring an original chain-of-ponds hydrology simply doesn't apply to those vast areas of the Australian landscape that have no running creeks or watercourses, dry or wet, or which have steeper catchments where run-off water needs to be slowed, diverted and captured so as to rehydrate valley systems. The fact is that the basic approach of well-managed practices like holistic grazing and integrated agroforestry are proven to, firstly, more efficiently restore (over large areas) healthy grasslands, rangelands and woodlands; then, second, to allow regeneration in the water cycle and other functions. This is more effective in rapidly restoring broad-scale landscape function than any other known approach. In their focus on reinstating hydrology, Andrews and key supporters (a number at the highest levels) seem to thus have missed the efficacy and wider potential of other regenerative landscape management techniques. For example, regarding deeply eroded country, I believe the work of Peter Marshall has greater application – as we shall see directly.

Nevertheless, it is hard to disagree with Andrews' criticism of over-reliance on farm dams, particularly in regions with accessible underground water. Far better to aim for a true rehydration of in-ground water, as the landscape has coevolved to operate with this. As Andrews concludes, in this hot, dry climate and land, 'the only efficient method of storing water … is the method that the natural landscape employed'.

There is yet another contentious aspect to Andrews' work: dealing with Australia's salinity. On taking over a degraded Tarwyn Park, Andrews found he had also inherited a salinity problem. As he restored landscape function, his salt problem disappeared. But, in typical fashion, he confounded scientists and others by arguing that the cause of salinity in his and many other landscapes wasn't a simple matter of tree-clearing leading to the rising of a salty water table. Instead, he argued dryland salinity was a more complex problem, to which Natural Sequence Farming provided a solution.

Crucially, Andrews revealed (now part-supported by some soil scientists and hydrologists) that if the landscape is functioning healthily, then a layer or lens of fresh water sits atop the saline water. The secret to restoring land and turning around salination is restoring that freshwater lens.

To Andrews, the solution is a simple two-step process. 'Salinity has nothing to do with water tables, but it has everything to do with plants – or rather the absence of them,' he says. 'Revegetate the land and reinstate the hydrology system.' This revegetation should encompass not just trees but plants of all kinds, and especially a healthy grassland replete with so-called weeds. We saw in Chapter 7 that a key part of Ron Watkins' solution to the dire salting problem in Western Australia was returning landscape function especially via plants. Andrews arrived at the same answer.[12]

In such discussions it needs to be borne in mind that much recharge of aquifers and underground water in Australia occurs on the once-vegetated ridges, and that revegetating these with deep-rooted trees and shrubs is a major solution. As we will see in Chapter 20, another leading regenerator of landscapes, John Ive of Talaheni, Yass in New South Wales, has demonstrated the huge benefits of revegetating ridgelines.

Evidence from Peter Andrews' clients' farms shows that, concerning salinity, he appears to be right. The self-organising capacity of the landscape (via its chains of ponds) has been shown to regulate excess salt build-up through a reed die-off and saltwater release cycle, followed by a reed regrowth phase.

Over time, the sometimes-difficult maverick of Peter Andrews has gained powerful support in high places, including with government funds. However, compared to the next water innovator, Peter Marshall, who uses similar though more sophisticated and scientifically researched principles, Andrews' system appears to have limited application. First, it really only applies where there is running water (a minute proportion of the Australian landscape); and second, it doesn't work if the surrounding soils are so compacted that they cannot transmit water laterally and away from water courses. Most clay-containing soils compress and bake under inappropriate hoofed grazing and lose their permeability. Rain runs in sheets off such ceramic soils, taking the fertility with it. Underground water cannot move through the tight profile, either laterally or vertically.

Practices like Yeomans' ripping, holistic grazing management or integrated agroforestry can restore soil pore spaces. The result of NSF, if soil porosity is not restored, is a thin ribbon of green in a brown, ill-functioning broader landscape.

Nevertheless, through his perambulating, eclectic information gathering, acute observations, and peripatetic mind excursions, brainstorming and teaching, Peter Andrews has received support from a number of respected and courageous leading scientists, including Dr John Williams (ex-chief of CSIRO's Division of Land and Water), David Mitchell of Charles Sturt University (a wetland scientist), and Polish limnologist[13] Wilhelm Ripl of the Berlin Technical University. These people appear to have helped refine Andrews' thinking.

As John Williams pointed out, it was Jared Diamond in his book *Collapse* who showed that 'societies that can listen to the small voices and take that and change accordingly are the societies that survive change and can accommodate environmental consequences. Those societies that can't hear the small voices and smother them are the ones that have a problem.[14]

It is not an exaggeration to say that if we keep on current track, Western and global society (like all civilisations before us that have trashed their soils, water, environment and landscape functions) is destined for an almighty collapse. So now, of all moments, is when we desperately need those 'small voices' John Williams referred to – of the Yeomanses, Watkinses, Andrewses, Marshalls, Savorys and similar brave pioneers to be heard by a wider society. Only then will we be able to hear the songs of a multitude of reed warblers echoing throughout the valleys of this vast nation. This next story is a dramatic illustration of this.

Hitherto, Peter Andrews is a small voice that has been loudly heard. Yet a parallel system evolving at the same time as Natural Sequence Farming has gone virtually unrecognised. To my mind, however, this approach offers great potential in restoring broader hydrology and other landscape functions to not just Australian but wider degraded landscapes globally.

It was after the first Australian edition of this book was published that I encountered these other visionary water innovators: Peter and Kate Marshall on their farm near Braidwood on the Southern Tablelands of New South Wales. In the midst of thousands of hectares of degraded farms and catchments, using broad experience in various disciplines and a combination of long-rotation and self-organizing agroforestry aligned with an original water-capture and diversion system, the Marshalls have entirely rehydrated and revegetated their land to abundance. Today, they now have permanent water, vast reed beds and singing reed-warblers, plus a mixture of productive oak and other exotic and native forests, and an award winning truffle farm.

I first met Peter Marshall in the early colonial town of Braidwood, over 120 kilometres east of Canberra. A highly reflective, original and independent thinker, his conversation is brisk and exhibits a broad and eclectic knowledge across an impressive breadth of scientific and other disciplines. Whilst he has another business, farming forestry is his love and passion, and the 630-acre farm he runs with his wife Kate is twenty minutes southeast of Braidwood. Tucked nicely on the western side of the hills of Monga National Park, the farm (receiving between 600 and 800 millimetres of rain) is on a watershed between the Shoalhaven and Deua rivers, resting atop a landscape that was once part of an ancient river system. It is a farm like nothing else I have encountered: a hint of unique difference being given on first arrival by welcoming truffle dogs, seed collecting and growing apparatus, a profusion of native and exotic trees, and a large biochar retort near a shed.

Like many key innovators, early life history powerfully shaped Marshall's life direction. He told me there were two main influences. His forbears were gold miners at Hill End, New South Wales: active in both alluvial and hard rock mining. Says Marshall: 'They destroyed huge areas of Australia's countryside, and destroyed people's health'. Marshall recounts the excoriating experience of when, as a boy, his job on Sundays was to sit with dying miners in the Bathurst Base Hospital as they suffered from silicosis: their lungs choked with mining silicates. 'These men were my heroes for their work ethic and skills with tools,' said Marshall. 'But they were "dusted" – for their lungs had turned to cement'.[15]

Marshall's father, Alex, didn't want to run the family mines, as he had seen too much human suffering. He took a horticulture apprenticeship instead. This led to Marshall's second life-shaping experience, for Alex contracted with the large British chemical company Burroughs Wellcome to trial and grow *Digitalis*, the source of the heart drug digoxin. In those days 'good farmers' were encouraged to use the 'modern' agricultural chemicals coming on the market. Under pressure to increase drug yield, Alex tried them all. Marshall calls it 'the Green Revolution in compressed time: a test case for new herbicides, pesticides and fungicides, with my father and our family as the guinea pigs and victims.' In the process Marshall witnessed widespread environmental destruction on the farm, and later a savage human cost.

The chemical company eventually synthesised the drug in their laboratories, and the farm closed down. The Marshall family tried organic vegetable growing but before there were any markets, and went broke again. Worse, in time Marshall's father and family members died of a range of chemically induced diseases. The net impact on Marshall was an aversion to agrochemicals; a desire to combine forestry and farming; and a strong drive to evolve biological or mechanical solutions to farming's environmental challenges.

So, with early experience helping plant trees on 'wrecked mine-sites'; of working with visiting scientists on his father's farm; of a mentor who taught him geology; and with an uncle who had been a sapper in the Second World War and had 'returned to the mines, and could move water around mountains', Peter Marshall gained 'a reasonable forestry education' via a science degree. In 1990 Peter and Kate bought their first farm near Braidwood – despite enormous opposition from some locals. Kate, while a city girl, had loved visiting her uncle's sheep farm, and suspended a promising career as a textile artist to help Peter with what he calls 'my obsession with fixing land'. Peter Marshall says Kate has now become 'a top practical ecologist with an eye for reading landscapes'.[16]

The farm they bought was a run-down dairy farm: 'The only things holding it together', said Marshall, 'were blackberries and Scotch broom (and thank God for them!)'. It was wasteland after decades of mismanagement. The soils were compacted, and erosion gullies were six metres deep. 'Soil compaction is the greatest unnoticed disaster in the Australian landscape. We had hundreds of exposed clays, B and C soil horizons, that were baked as hard as tiled floors', said Marshall. Down in the low lands gold dredging had ripped out and diverted streams, and billabongs were filled with sediment from the erosion upstream.

But the local climate was good. With experience in mining and farming, and an engineering and geological orientation plus a reading habit, Marshall felt that 'there had been water here once, an old river system, and our task was to try and put it back, discover where the billabongs were, and bring them back to life'.

Marshall hated the land degradation that had resulted in deep gullied erosion and particularly compacted soil (making the latter what he called 'water repellent'), so his first steps were two-fold – first steps were to slow the water, allow it to soak in, and thus to reengage the water cycle. First came deep ripping with a Yeomans plough so as to crack the ceramic surface and invite rain and air to revitalise the soil. Then came re-stepping the landform and restoring the old, self-levelling reed beds long eaten or mined out. Deeply incised gullies were dealt with by 'biogeotextile wrapped fascines'. Sediment-filled ancient billabongs were identified from aerial photos and brought back to open water by helpers and 'land artists' Andy 'Kiwi' Whaiapu and Jason Griggs, ace excavator operators with an eye for levels.

The property had only come on the market because declared noxious weeds had defeated the previous owners. The Marshalls, using biological and mechanical means, then set about eliminating and controlling the Scotch broom and blackberries. The Scotch broom required lateral thinking, as decades of herbicide spraying (the preferred method of local councillors) had just stimulated germination and killed soil biota. Goats were drafted to eating the flowers and prevent seed-set. Deep-ripping buried the seed bank where ants could store seeds, and thus carbon, deep. The blackberries would take a long time, as they were valued by the Marshalls as the only organism which held soil together in many areas (a reason early explorer-botanist Baron Ferdinand von Mueller had spread their seed over a century ago). Again, goats did the initial work, and now the Marshalls were cutting and piling dead canes so as to shield establishment of native trees and shrubs, which then shade-out the blackberries.

Through trial and error, plus eclectic reading, research and interaction with water, engineering, geological, fungal, forestry and other experts across the globe, the Marshalls in time evolved a unique farm. In a hitherto dry catchment, this farm now has permanently running water and comprises an extraordinarily evolved form of water capture, repair and reticulation. Partnering this rehydrated landscape with forestry and the truffle enterprise has created a financially and ecologically resilient farm.

But it is the water capture and water cycle regeneration that stands unique. While it shares some basic principles with Peter Andrews' NSF (though Marshall's system evolved in isolation), the execution and management style is more sophisticated; is easier to understand and more cost-effective on the farm scale; and is more grounded in modern, multi-disciplinary science and technical aspects. I believe the Marshalls' approach offers wider scope in regenerating landscape functions – and indeed has global applicability.

When the Marshalls began with their degraded, soil-compacted farm and deep gullies, they had no permanently running water to deal with. Their original problem was the 4,000 acres of mismanaged, set-stocked (and thus poor ground-covered), compacted soil catchment above them: off which, in any serious rain event, millions of gallons of water and silt poured into and down through their land. Marshall told me, 'We originally had floods running 18 inches deep over the farm, but none of the water soaked-in. Two days later the system was dry again. And that's typical of most degraded catchments, as they can't absorb water, be it rain, fog, dew or snow'. Initially, low strainers of rocks and logs were placed in the base of gullies to impede flow and dissipate energy. *Phragmites* reed was grown as 'tiles' and laid over these structures to bind them strongly. Ancient technology was vital, like the old Roman military technique of fascines: large, tight bundles of parallel sticks or poles placed parallel to the current so water could still trickle and pass through, leaving sediment trapped in biofilms. Live revetments of bamboo are used – and indeed if you go for a walk with Peter Marshall then one has to expect a gentle trickle of such technical terms across broad disciplines. 'We don't call them "gullies" any more, we call them "creeks"', says Marshall. 'They have come back to life.'

Today on the Marshall farm, all one can see is a green and forested landscape of lush, slashed grass, diverse oak, poplar, eucalypt, wattle and other tree species; of multiple shallow running creeks densely covered in native shrubs and trees, while the entire watercourse system of multiple channels comprises a green and hydrated valley. The main creek now is a series of stepped, earthen arrangements that impede and hold the water in moderately sized ponds, often with islands in the middle for waterfowl, water dragons and other fauna now protected from foxes and other predators.

Particularly noticeable are immense *Phragmites* reed beds throughout the ponded system, often with rafts of water lilies. The water is clear, healthy and functional, and there is a lot of bird and other activity. The ground over the fifty or more acres of this system is moist and spongy, and dotted by more recent plantings of mainly oaks. Spiralling out of the meandering, ponded system (which itself is designed in zig-zag fashion) are cleverly arranged side ponds. Marshall designed the lateral pondings so that rushing flood water was deflected, using its own kinetic energy, out into ponds where islands further dissipate the water's energy. Some of the ponds, in playful design, have long, rectangular islands which Marshall calls his 'Chinampas': named after the floating garden food beds of ancient Aztec Indians on Mexican freshwater lakes.

Early life experience in mined rivers, close observation and an earlier phase as a rafting and kayaking guide helped shape Marshall's design approach. 'Kayaking and rafting teaches you how water moves', Marshall told me recently. 'The flow of water and objects in it is the same at different scales. A twelve-foot river raft and a floating wombat scat behave the same. Boats need to avoid "strainers" and "stoppers"; you can trap sediment and debris by building such things'. He concluded: 'It is the water that gives you the lead every time, and all we're doing is using the dynamics of fluid movement, air and water'.[17]

The upshot of all this design and excavator work, says Marshall, even in a big flood today, is millions of gallons of water pouring onto their land, loaded with a huge tonnage of silt. The planned, stepped, ponded, meandering, energy-dissipating system then traps the valuable topsoil, and by midway through their one-kilometre system the inflow water is already running clear. What were once six-metre deep gullies are now shallow creeks with clear water, shaded and dense vegetation laterally out to many metres, and diverse volunteer ferns, *Lomandra*, tea tree and blackwood – all surrounded by a diversity of bird, insect, frog and other life.

The measure of the system's functional rehydration of their valley is clear. In a big rain the compacted catchments uphill deliver thousands of tonnes of turbid water. It leaves the Marshall's farm crystal clear. During the non-flood period (much of the year), their inflow creek is dry, but at the bottom of their system the creek permanently runs with a gentle clear flow. That is, their part of the valley is functioning again as it once used to before European settlement – the valley's hydrological system retaining and spreading out violent downpours, rehydrating the entire system. In the process they have stimulated a mosaic of environmental niches, and are delivering water twelve months of the year.

The crucial difference between the Marshalls' approach and some of the NSF work, therefore, is that not only do the Marshalls not need permanently running water to regenerate a catchment, but they specifically also work on regenerating the lateral landscape to become water absorbent. This occurs via strategic ripping with a Yeomans plough; with careful manipulation of clay, soil and creek bed structures; and through farm forestry and judicious slashing. The contrast to some NSF work is that the Marshalls have created an entire rehydrated land system versus a narrow ribbon of green surrounded by compacted, brown dry land: a pattern sometimes seen in some well-known NSF approaches. David and Jane Vincent's rehydrated lateral land cited earlier in this chapter is evidence of how one must work not just on regenerating a creek system but also the lateral landscape – which the Vincents achieved with holistically grazed cattle.

Key to this broad rehydration, therefore, is enablement and regeneration of the riverine system's 'hyporheic' zone. This is the biome where the stream's water body interfaces with the side and bed of the stream, and where there occurs the mixing of shallow ground water and surface water. The interactions of these two different types of water is crucial to a stream's physical and ecological health: that is, for water quality and riparian health and habitat.[18] The Marshalls therefore pay great attention to disturbing the impervious clay layers of eroded banks, and to enabling riparian connection which fosters a healthy hyporheic zone – a zone that can extend many metres laterally under the ground away from the creek or river. If need be, if the lateral soil is extremely hard and degraded, using his mining knowledge, Marshall is not averse to drilling a connected series of holes, placing explosives in them, and firing a timed sequence. Says Marshall: 'Blasts all go off at once, shockwaves bounce between charges and fracturing breaks the blockages', (including compacted soil and sealed, incised banks), 'thus rebuilding a hyporheic zone' and developing connectivity and absorption in what he calls 'capillary tubes'. Therefore, he concludes, 'You now have lateral movement from aquifer and then capillary tubes lifting water up to the field'.[19]

Midway through his evolving landscape regeneration and rehydration work, Peter Marshall stumbled on two key features that he realised tied his entire approach together. Each idea came out of North America, and particularly its Southwest which he visited, beginning with that great water engineer, the beaver. These animals are known as a keystone species (analogous to a keystone in an arch). A keystone species has a disproportionately large (if not determining) effect on its environment and thus other organisms within that system – and a large effect relative to its abundance. That is, such species play key roles in evolving and/or then maintaining the structure of an ecological community, thus affecting and determining many other organisms and their types and numbers in their ecosystem and/or other ecosystems. Crucially, without keystone species, such ecosystems either would be different or wouldn't exist. Keystone species, which can be plant or animal, are often dominant predators – like a wolf, jaguar or eagle – controlling prey populations, or else ecosystem engineers that create, maintain or change an environment (such as Pacific otters, corals or elephants). Crucially, they impact how an ecosystem functions.

We now know how beavers play a keystone role, being co-evolved to impact the hydro-morphology of riparian systems and to re-engineer entire riparian systems with their dams and 'lodges'. That is, they shape their and other organisms' environment. They re-create a rehydrated riparian habitat that supports other organisms and sub-ecosystems (i.e. restoring wetland diversity). Their ecosystem engineering thus leads to flood mitigation, the recharging of aquifers and removal of pollutants from surface and ground water, drought protection, decreased erosion and repair of incised and damaged stream channels and watersheds, maintenance of stream flow, and so on.[20]

Marshall planted *Phragmites* reeds on his farm in the beginning, and then noticed that waterfowl did the propagating job for him (as had occurred on David Vincent's farm). He still shifts rhizomes to new sites, nowadays in one tonne lots by excavator. He came to recognise this reed as a keystone species and ecosystem engineer. It was this species that was determining the nature of his landscape. *Phragmites australis*, in different variants, is found around the world and thrives in the right conditions in Australia.[21] As we observed another big patch of reeds covering nearly half an acre, Marshall told me: 'They're not "leaky weirs". Weirs are a negative word to a canoeist. They are not dams, as we access the water from underground to start, not from overland flow. We call them "beaver berms" because beavers self-level and regulate the porosity of the berm very carefully, trapping sediment effectively, dissipating energy, absorbing shock and preventing overtopping. Our *Phragmites* does the same thing, it throws out its rhizomes sideways, forming a composite material with our biogeotextile. When covered by sediment it grows upwards and puts out another horizontal sheet of rhizomes.'

He concluded with the observation: 'And so we have *Phragmites* berms, sort of hydraulic pillows. These large masses, these pillows, they just lift up and absorb the shock of the hydrostatic head of a big flood. That's how it worked for thousands of years – but cows and horses craved the nutrients in the reeds and walked in, and their hooves punched holes in those pillows and destroyed them.'

The second crucial 'aha' moment that reinforced Marshall's thinking was when he was in the American Southwest on a business trip and saw how the Native American Hopi and Zuni peoples valued and used a simi-lar, naturally evolved water system to what he had created – though a system with some key differences. These systems are called *Ciénegas*. Effec-tively, they are aerated water meadows. Most are largely destroyed now following the introduction by Europeans of grazing animals, which degraded their long coevolved functions.

Ciénegas were usually associated with seeps or springs, and have evolved over long time periods into alkaline, freshwater, spongy wet meadows with shallow gradient, and permanently saturated soils in dry landscapes like those of New Mexico and Colorado. They have strong integration with their hyporheic zones. Occupying the entire width of valley bottoms, water was able to slowly migrate through these features via large-scale mats of thick, sponge-like wetland sod. Such wetlands played a crucial biodiversity role in dry areas, providing rich habitat for birds, animals, plants and other organisms – many of which are migratory or threatened species.[22]

These systems, for Marshall, reinforced how, given the chance, nature will evolve similar, though variant, systems to hold, control and reticulate scarce water in dry continents, and to act as flood mitigators in absorbing hydrostatic flood shocks and laterally spreading flood pulses while protect-ing soft surface sediments and trapping sediment loads.

So after nearly twenty years of learning to understand the nature of their ancient catchment and river system; of constant trial-and-error and refinement; of ongoing excavator development, corrections and experimentation, the Marshalls not only have a rehydrated farm that is far more resilient and productive, but have delivered a template for landscape regeneration of degraded catchments that can apply to not just Australia but many parts of the world.

To walk or drive on demarked tracks with Peter Marshall through their green, ponded, tree-, shrub- and *Phragmites*-lush landscape is to be regaled with a lifetime's research in multiple disciplines but also relevant technical talk. One can visit a pond that contains gorgeous pink water lilies whose ancestors were 'borrowed' by a First World War Australian 'digger' from Monet's garden at Giverny. Walking, Peter will carry a penetrometer to measure soil compaction (their levels have dropped from 1000 p.s.i. initially to around 200), or else a bucket of grass seed to be spread on areas bared by recent backhoe excavations; and sometimes is a little hand-augur to plant punnets of oaks they have raised in old toilet paper rolls. He showed us a huge pond, ringed by *Phragmites* reeds, which he calls 'Swan Lake'. So abundant is the water in his landscape now that he was able to build this large pond of sufficient length for swans to do their ponderous, awkward, water-running take-off flight. Within months of the pond filling, swans had occupied the site. He calls it 'opening a window to the ground water'.

It was while my wife Fiona, daughter Tanya and I admired this wetland that was alive with waterhen, reeds and throngs of birds in nearby vegetation, that I pointed out to Peter the penetrating song of a nearby reed-warbler. In fact it was one of dozens I had heard that day scattered across the farm landscape. 'Oh, is that what they are called?' said Peter. 'They don't seem to mind it here'.

CHAPTER 9

From Stardust to Stardust

All come from dust, and to dust all return.

ECCLESIASTES 3:20

The soil is the great connector of lives, the source and destination of all. It is the healer and restorer and resurrector, by which disease passes into health, age into youth, death into life. Without proper care for it we can have no life.

WENDELL BERRY, *The Unsettling of America*[1]

In the 1938 classic tome *Soils and Men*, head USDA soil scientist Charles Kellogg stated, 'Essentially, all life depends upon the soil ... There can be no life without soil and no soil without life; they have evolved together.' On the cusp of the massive unrolling of industrial agriculture, this was a brave statement, but Kellogg was absolutely spot on. The prescience of his words has been increasingly confirmed and increasingly ignored since.[2]

As we saw in Chapter 2, when hatching the rise of industrial agriculture Justus von Liebig and fellow soil scientists kidnapped the concept of a living soil. They turned it into a toolkit: one that could be ploughed and knocked around like a Meccano set, or else simply treated as a chemical set by adding industrial ingredients such as nitrogen, phosphorus and potassium – and later herbicides, pesticides, wetting agents and the like. That is, this ancient substrate had become a plaything of the Mechanical mind.

I studied traditional soil science at two universities, some thirty-seven years apart. However, nothing had substantially changed the second time, and while both experiences were informative, I found them boring, focused as they were on soil chemistry and physics. Such knowledge is of course valuable for farmers. It partly explains what soil is made of and some aspects about how it functions and why things can go wrong if badly managed. Yet in both courses the most vital ingredient of all was missing: the diversity and dynamism of a healthy soil and its complex, living biota (a biota largely hidden from view and mainly microscopic). Without this second knowledge component, we cannot grasp the implications for not just preventing land degradation but also regenerating soils, landscapes, life and animal and human health.

A hint of the invisible, living soil world and its little understood connections to plants and the biologically driven processes that make essential nutrients available for animals and humans alike, was given to us by water-dreamer Peter Andrews. Early in his career Andrews was a serious racehorse-breeder, and during this time he received a major lesson in the field of environment and nutrition.

This came when Andrews visited England's leading stud of Beech House, Newmarket, owned by Lord Derby. It was one of the most successful studs in the world in terms of winners and genetics. The stud manager told Andrews he didn't believe in superphosphate and only used some fish-meal on his pastures every three years or so. Moreover, he greatly valued pasture diversity, including a variety of so-called 'weeds', and a general rule of thumb was that good horse paddocks needed at least eighty different plant species. In talking to other leading English horse people, Andrews also discovered 'there was a traditional belief... that once a field was ploughed and reseeded the pasture wouldn't be any good for young horses for five years and was unlikely to produce a Group One winner for ten years'. Andrews never forgot this wonderful lesson of vibrant, green diversity encouraging healthy landscape functions.[3]

What also makes a living soil hard to study (let alone comprehend) is that it comprises a constantly adapting, self-organising system. This, to a Mechanically minded scientist, is like trying to pin down a will-o'-the-wisp – so these scientists stick with what can be comprehended and taught. In doing so, the true essence of a living soil disappears from view and mind.

This dichotomy goes to the heart of the agricultural world. On the one side is a dominating industrial agriculture based on the Mechanistic mind. It is supported by all the embedded paradigms of the modern industrial world and its driving power base of government bodies and transnational corporations. On the other, the regenerative agriculture movement is largely eschewing industrial inputs and mechanical practices, and is thus issuing a serious world-view challenge to the dominant sector.

This is not to decry the basics of soil chemistry and physics. But we also need to know about Albrecht von Thaer's humus (that healthy mix of vibrant microbial life and dark, healthy soil organic matter) as a key ingredient. There are a number of reasons why humus is vital (instead of just endlessly adding replacement chemicals via fertilisers, as with industrial agriculture). The first is that the chemical chains of good humus have a large surface area. These carry electrical charges that attract and hold mineral particles. Second, as Jeff Lowenfels and Wayne Lewis describe in their excellent book for organic gardeners *Teaming with Microbes*, the molecular structure of the long carbon chains of humus 'resembles a sponge – lots of nooks and crannies that serve as veritable condominiums for soil microbes'.[4]

But a third crucial reason relates to the chemical composition of humus. That is, healthy humus is composed of 'humates': various forms of carbon-rich, humified organic matter and humic substances that are highly resistant to further biodegradation. Such humates are the real secret and most productive input in regenerative agriculture, as they contain humic and fulvic acids. While humates in general lead to heightened nutrient absorption and enhanced plant-root growth (besides having a multiplicity of other functions that enhance healthy soil structure and function), humic and fulvic acids are so valuable because they powerfully promote healthy fungi and make available crucial elements such as phosphorus. Furthermore, while they are also a product of soil microbial metabolism, they are powerful organic electrolytes: meaning they are soluble in water and can promote electrochemical balance in plant and animal cells. These abilities also mean they are very active in dissolving minerals, metals and other nutrients (along with vitamins, coenzymes, hormones, natural antibodies and such), thereby making them more available and more readily absorbable to plants – and through plants to animals and humans. In humans, for example, these acids further enhance health by helping to catalyse enzyme reactions, increasing assimilation of key nutrients, stimulating metabolism, detoxifying pollutants and so on.[5]

So, instead of 'either/or', a healthy agriculture requires from its practitioners both the chemical–physical knowledge *and* the organic. However, there seems a vast gulf between the two. Some regenerative farmers call the industrial fertilising and chemical approach to agriculture (and the soil tests that accompany them) the 'more-on' approach. This is because they believe the application of synthetic chemical fertilisers tends to be based on narrow soil tests (tests, they say, that have been developed by industrial chemists in laboratories to serve both industrial agriculture and the transnational companies that foster high fertiliser and chemical inputs).

However, in soils such as those of Australia, where microbial and plant life have adapted and coevolved to rapidly recycle and make use of microunits of scarce elements like phosphorus, the addition of huge dollops of P, S (sulphur), N and K leads to massive overdoses of these elements and them thus being rendered inert or left in solution to be rapidly leached out. Even worse, the huge concentration of chemicals and the acidic nature of other elements such as heavy metals in modern fertilisers kill off the soil biology, also rendering soils inert and making them more dependent on further fertiliser applications, thus exacerbating the 'more-on' approach. All this is to the healthy profits of the transnational behemoths and middle men but to the detriment of farmers and a healthy, living soil. Such a relationship is little different from cynical drug-pushers and their dependent addicts.

The key difference between industrial and regenerative agriculture is that in the former humans generally believe they can control the chemical inputs while ignoring the biology. In the latter, the biology comes first and is encouraged to self-organize itself. Regenerative farmers, and organic farmers and gardeners of all descriptions, believe humans should largely get out of the way and let nature do what she does best, no matter how unbelievably complex that may be. The assumption of the latter approach is that, by allowing self-organising processes in nature to work, the chemical and physical side will look after itself (in most cases with little help). The second key difference is that in the former approach (and by most measures of a healthy soil) industrial agricultural soils are degrading (sometimes slowly, sometimes quickly), while in true regenerative agriculture they are improving – and often rapidly, through an accelerated process of soil formation.

To begin to wrap my mind around the origin and importance of a living soil, I met microbiologist Walter Jehne on a hot January day. We chatted under the shade of cottonwood trees outside University House at the Australian National University (ANU) in Canberra.

A man in his mid-sixties who was meant to be semi-retired, Walter now devotes himself full-time in acting as a catalyst to change societal, farming and political thinking about the role and function of soil. He is perpetually optimistic and enthusiastic. His red face, blue eyes, ginger hair and whiskers, and constantly cheeky grin somehow reminded me of the avuncular Fozzie Bear in *The Muppet Show*. But this appearance hid an acutely sharp mind and prodigious memory. When I said to him that he was a walking encyclopaedia on the subject of soil, human health and global climate issues, he refuted it swiftly with a quip: 'Not at all. Encyclopaedias are redundant on the shelf from the day of printing.'

Walter Jehne's whole life had prepared him for his ongoing role as a national and international healthy soil advocate. This also includes a role as one of the scientific advisors to former governor general Michael Jeffery, who heads up the nine-year-old non-governmental organisation Soils for Life. In 2013, Jeffery was appointed the prime minister's advisor on soils, and Soils for Life was both a lobby and extension organisation, gaining access to the highest levels of government while at the same time publishing case studies on regenerative agriculture. Walter played a key role in the founding phase of Soils for Life.

Having earlier escaped across the border from Jena in East Germany in 1956, Walter and his parents emigrated to Australia. But Walter's formative experience was in being lucky to grow up in the bush in the Blue Mountains west of Sydney. As an eight-year-old experiencing the 1957 bushfires, he recalled that the dynamics of regeneration in the Australian bush nevertheless 'just got me'. This, combined with being a Boy Scout and plenty of bush tucker foraging, set him on a path towards forestry and a love of the natural world.

Two-year stints at both Sydney University and the ANU Forestry School had him hooked. Eventually he gravitated to the CSIRO, to work on forest diseases and protection. It was while researching alpine ash (*Eucalyptus regnans*) dieback and fungal pathogens in Tasmania that he began to understand just how fundamental microbiology was to all life. 'What I came to realise,' said Walter, 'is that life in the soil is the dominant driver of macro ecosystems.' Ironically, it was through looking at the microscopic world that Jehne began to think macro-holistically.

This was aided by him moving north in 1976 with CSIRO to Brisbane and then to the Queensland tropics. In the process, he joined the world-class CSIRO team on soil formation, which looked at a wide variety of soil environments and the role of microbial and mycorrhizal (root fungus) symbionts. Walter found this crucial to his deeper understanding of the real drivers of healthy soils. To him, the mycorrhizal fungi were the engine of the whole system.

Following a varied career (including in the Federal Department of Agriculture and Forestry in Canberra and working for Federal Agriculture Minister John Kerin), Walter returned to soils and microbiology in 2004 – disillusioned with the lack of strategic direction in government. However, this serendipitous journey had prepared him for his current role as a change agent attempting to turn around Australian farming practices to deliver solutions to such issues as land degradation, climate change and helping people be healthier via healthier food.

More recently, I asked Walter to lecture students at a university course I was teaching in sustainable rural systems at ANU. For fifty minutes, without notes, but while drawing diagrams on a whiteboard, he kept the students mesmerised with his unrolling story of soil formation and its significance to agriculture. He began with the fact that planet Earth, all life and modern soils are derived from stardust and an exploding supernova some 13.7 to 13.8 billion years ago (the 'Big Bang'). The microbiologist in him also pointed out that a huge evolutionary step then occurred some time before 460 million years ago – via microorganisms – when fungi moved out of the ocean onto the mineral detritus on land. Here they were able to dissolve into digestible solution untapped nutrients from nearby land-based rocks and move these nutrients back to their original host colony. That is, they had begun the first step in soil creation.

These fungi then formed symbiotic (mutually interdependent) associations with blue-green algae: the algae providing energy; the fungi the nutrients. Lichens had thus evolved, and they now began breaking down rock to form early soil substrate. The second ingredient they released was the residual organic detritus from their cell walls, and now only recently known as 'glomalin': the initial source and still a major component of most soil organic matter and an amazing soil-builder.[6] So began the first living soil and thus the capacity to support further microbial and, in time, plant life. This event soon became an ever-increasing, positive-feedback process as life's evolution now began to move into overdrive.[7]

Bringing the students from a half-billion years ago right up to the present day, Walter's message boiled down to the fact that at the core of all ongoing, coevolving ecosystems are the microbial and fungal worlds. It is these that continue to deliver in extraordinary ways the complex carbon chains of soil organic matter (that is, solar energy stored in a carbon matrix), and these chains are what drive healthy soils, most life, and agriculture.[8]

Yet it is this very process, as we shall see, that industrial agriculture seeks to destroy by replacing it with human-made chemicals and destructive mechanical intervention.

Walter Jehne makes no bones about his belief that the secret to regenerative agriculture and to restoring the world's living skin – its functioning soils – is to recreate the four healthy steps of soil formation. The first step is the photosynthetic fixation of carbon dioxide from the air: the task of green plants capturing energy from the sun. This – as we saw earlier – produces sugar and biomass to provide the substrate and the building blocks for most terrestrial life. Then comes the second part: the oxidation of the photosynthetically produced biomass to rerelease and dissipate the solar energy stored in the biomass. But this step is a double-edged sword and needs to be limited by good soil and landscape management.

'That oxidation is just crazy,' concluded Walter, 'given that the only basis through which we influence the geo-biosphere is in what happens to each molecule of carbon fixed by photosynthesis. This is because we need to remember that nature conserved at least sixty per cent of photosynthates as "C" via managing its soil microbial ecology. We can and must do this as well.'[9]

The third key element is the creation of biomass by microbes of stable soil organic matter so as to store that fixed solar energy for up to millions of years. This requires creating suitable microbial ecologies and conditions in both soils and the guts of our grazing ruminants. As we will see in later chapters, this is a crucial element in the area of healthy food, as it relates to our own internal and external microbial world and our coevolved sensory mechanisms of taste, sight and smell. Finally, the fourth key element in soil formation is the generation of a series of positive-feedback processes that can significantly aid the level of photosynthetic fixation of carbon dioxide in step one. These processes involve improved water and nutrient retention and thus improved production of food supplies for humans, plants and animals from the soil organic matter laid down in step three above.[10]

Put simply, regenerative agriculture is about maximising steps one, three and four, and limiting the possible overproduction of processes in step two. Poor land-use practices that lead to land degradation – and this includes most of industrial agriculture – continue to ignore steps one, three and four while exacerbating step two. That is, the chemical and mechanical treatments of the soil in industrial agriculture turn the oxidation in step two into the dominant process. Shaking his head in disbelief for the students' benefit, Walter said, 'We currently drive one hundred and twenty per cent plus of photosynthesis – fixed carbon to be burnt!' This leaves soils in the dead mineral state that reigned on the planet before active soil creation began around 420 million years ago, which in turn accelerates land degradation.

Walter Jehne then succinctly summed up this entire story by explaining the significance of regenerative agriculture to the future of the planet. 'Through regenerating suitable soil microbial ecologies that aid this natural sequence of soil creation processes,' he told my students, 'innovative farmers throughout Australia have demonstrated that they can readily and rapidly improve soil carbon levels, the structure, hydrology and nutrition of soils and thereby the productivity and resilience of their plant biosystems, animal production and soil health.'[11] 'That is,' added Walter, 'if we manage our soil microbial ecology, then this investment in "C" drives the very powerful series of positive-feedback processes and force multipliers that exponentially drive soil fertility improvement and get more carbon fixation by plants.'

To this I would add (as discussed later in Chapters 20 and 21) that regenerating soils also ameliorates climate change (via less carbon released, more carbon fixed, and via more water retained in the landscape) and enhances human health. 'In other words,' as Walter told me recently, 'the balance of oxidation and carbon fixation is entirely in our own hands via farming. So too, therefore, is our future.'[12]

The idea of a vitally dynamic living soil was late in developing because of slow developments in microscopy, which delayed our better understanding of the microbiological world. Some forty-two centuries ago, the Chinese made a schematic soil map of their country for taxation and administrative purposes. As F. H. King in his famous 1911 book *Farmers of Forty Centuries* showed, the Chinese had been awake to the value of humus and soil organic matter for over 4000 years even without being able to use microscopy.

Intuitively Rudolf Steiner, in establishing biodynamics in 1924 (as we will see in Chapter 11), clearly understood that the power and energy of his special 'preparations' came from an abundant microbial life. 'Healthy soil' pioneers such as Sir Albert Howard and Lady Eve Balfour equally understood this. After the Second World War, in the USA, Jerome Rodale and successors began to highlight the significance of soil biology. Then, from the 1960s, we got a fuller articulation of soil biology – led, for example, by Nobel Prize–winning scientist Selman Waksman. In the meantime, industrial agriculture, the Mechanical mind, and university and agricultural college teaching of a reductionist, physics- and chemistry-based soil science appeared to have won the day. However, like earthworms and microbes under the soil, a group of insurgents had been slowly at work.

Besides the pioneers of an early 'organic' tradition and an emerging regenerative agriculture, by the early twentieth century a number of scientists (in areas such as microbiology, soil ecology and energy flow) had begun unravelling the vital physical, chemical and biological functions of healthy soil and soil organic matter. What became clear was that soil organic matter was now seen as a key component of ecosystems: notwithstanding it was a hugely complex natural material.

In the footsteps of thinkers and communicators such as the brilliant soil scientist Dr William Albrecht (who helped to introduce the idea of trophic levels, food chains and food webs), others now stepped forward to begin spruiking a different gospel to that of von Liebig. As we shall see, some innovators emerged out of Australia too. But before them, one of the most influential pioneers from the late twentieth century to play a key role in Australia was an American soil scientist, Dr Elaine Ingham.

Ingham gained her PhD in 1981 at Colorado State University, examining soil microbiology. After varied postdoc work, in 1986 she moved to

Oregon State University, where, by 1991, she had opened a soil-testing facility specialising in examining soil microbial issues. By this time, Ingham had realised that farmers and others utilising landscapes and gardens needed informing about healthy soil biology in the face of the onslaught of soil-harming industrial agriculture and its inputs and practices. So she began popularising her work, and in 1995 set up the private commercial organisation Soil Food Web Inc. In addition to research, this organisation began to focus on disseminating soil-health knowledge and developing courses in the concept of the soil food web.

While specifically focusing on soils, this concept came out of holistic systems-thinking and ecology, which revealed how entire ecosystems are linked in complex ways in terms of energy exchange, nutrient and mineral exchange, and interactions between all organisms in the system, across all trophic levels. But for Ingham, the answer to a healthy ecosystem was clear: the major engine driver (after the solar and other basic cycles) was the microbial and other biological worlds in the soil. Her teaching on this subject led to leading innovator Terry McCosker inviting her to Australia – where she came to strongly influence many Australian regenerative farmers.

Authors Jeff Lowenfels and Wayne Lewis, who have popularised Ingham's work, summarise its key message as follows: 'Since some organisms eat more than one food chain or are eaten by more than one type of predator, the chains are linked into webs – soil food webs.'[13]

A healthy soil food web contains a gobsmacking plethora of life. Oft quoted are the statistics of what is found in a teaspoon of soil, such as many billions of bacteria in some 25,000 different species. But in any small volume of healthy soil can also be found perhaps up to 10,000 fungal species and many tens of kilometres of their invisible fungal hyphae (their long, thin feeding tubes), thousands of protozoa and dozens of nematodes, not to mention tens of other, larger creatures (the ecosystem engineers) such as earthworms, centipedes, springtails, ants, slugs, ladybird beetle larvae, and bigger, burrowing and excavating creatures such as lizards, snakes, mice, prairie dogs, bandicoots, bilbies, badgers and even wombats. All of these are active in recycling and mixing soil, creating structure and organic waste material, and so on.

We now know that in a healthy soil around eighty-five per cent of all plant nutrients need first to be cycled through microorganisms. However, for this crucial microbial component – and therefore others down the food chain – what has happened with industrial agriculture is that modern chemical fertilisers, pesticides, insecticides and fungicides, along with mechanical intervention (tillage) and heavy traffic, have wiped out or severely modified this living foundation of a healthy soil. This occurs via direct toxic poisoning, by warding off beneficial organisms, or by changing their environment to favour populations of less desirables that thrive in unhealthy soil. In a dead soil, mineral breakdown and humus-building doesn't occur, so the farmer has to replace the missing inputs with more fertiliser. This leads to over-fertilising, to rising farmer indebtedness, and to more profits for the behemoth transnational chemical, pharmaceutical, energy and other agriculture-related companies.

Elaine Ingham explains that a healthy soil food web requires at least ten crucial biological elements. The base foundation is bacteria, which must be present, as these perform several important functions, including competing with disease-causing organisms, acting as essential decomposers, enabling nutrient retention and constructing micro-aggregates to improve soil structure.[14]

As important as the microbial-bacterial world is, there are also other key players in a healthy rhizosphere: the soil zone of perhaps two millimetres that directly interacts with plant roots. The next element involves the presence of a healthy balance of fungi. This often overlooked but hugely important component of healthy soils plays multiple roles, such as disease prevention, retention and transfer of crucial nutrients, making macro-aggregates, which form air passages and hallways to allow air and water to move into and out of the soil, and much more. So a dead soil nuked by fungicides (equivalent in effect to Agent Orange on a rainforest) and other industrial products, and/or compressed by numerous heavy wheel passages (which crush the fungal hyphae) only further accelerates the vicious cycle of dying soil and disease-causing, nutrient-bereft food produced from it.

This is not the place to outline even a short course on Ingham's soil food web – suffice to say, however, that the remaining eight elements of the web are equally amazing and diverse. They include other groups of soil life and their important role in maintaining and building healthy soils. The key point is that all this life – from the micro to the macro – provides food to feed the web: energy, carbon, minerals, nitrogen, amino acids and so on.

At the same time, the activity of the various organisms creates a healthy soil pH and structure for oxygen and water absorption by plants, which then further feed a healthy cycle and web. A diverse, healthy soil population also means effective disease control of pathogens and disease-causing bugs by other creatures.

The key message here for regenerative farmers is that all this amazing biodiversity needs to be well fed and 'watered' just like our livestock and crops require – and this is the farmer's job. Conversely, overgrazing and over-cultivation does the opposite and collapses the system and simplifies it down to a few single species, such as hardy bacteria.

Elaine Ingham is not alone in her influence on Australian regenerative farmers gaining knowledge about soil health and its determinants. Through various organizations bringing him to Australia, American Dr. Arden Anderson has had a significant influence, as have soil or nutrition evangelists like Jeb Gates, Gary Zimmer, Neil Kinsey, Hugh Lovel and Edwin Bosser, plus Australian Graeme Sait of Nutritech Solutions. Two other leading catalysts were effectively hunted from industrial-agriculture-committed organisations, and this led to them working with people who were more open-minded. One is Dr Maarten Stapper (whose work I briefly touch on in Chapter 21), and the other is Dr Christine Jones.

Maarten Stapper is widely active in Australia in both adapting industrial practices and scientific knowledge to a more biological approach, and in thus propagating a biological regenerative agriculture approach. As he stated in 2013: 'Biological agriculture gives industrial farmers tools for a profitable, successful transition to healthy soils and landscapes while they are learning to see and experience the living soil-plant system by "doing". This facilitates understanding and improves the capacity for biosensitivity. Practice is way ahead of science.' He concludes that, unlike industrial agriculture, farming should work with, not against, nature – and because 'nature confronts us with complex system, with intricate foodwebs, and with a myriad of dynamic, visible and invisible interdependencies ... People on Earth need to regain respect for their food and the land.'[15]

We have seen repeatedly in this book that many of the breakthroughs in regenerative agriculture have been made by brave individuals going in the opposite direction to others around them. Christine Jones is another such innovator. Her work relates to all key landscape functions, even though, in her case, it may have begun in grasslands before moving to soil carbon, water, soil biology and biodiversity. I could easily have told Jones's story in any number of chapters, from holistic grazing regimes to water, but I felt it best fits here in this most basic of all functions: a healthy living soil.

Many a maverick change agent experiences attacks and rejection from entrenched commercial and scientific interests. Jones is no exception, and this has inculcated a certain wariness when first encountered.

After an academic career that began in the wool industry via the CSIRO Division of Textile Technology, in 1981 Jones moved to Armidale in the New England region to teach agronomy, soil science and botany at the University of New England. Here she was strongly influenced by one of Australia's leading scientists and thinkers concerning native plants, the courageous Professor Wal Whalley.

As often happens in life, timing is critical, and Jones found herself in the New England area at the genesis of holistic grazing management in Australia. She was influenced early on by the thinking of both Stan Parsons and Allan Savory. And then, at a critical moment, Jones and fellow native plant botanist Judi Earl were able to conduct a three-property study. This compared a holistic grazing regime (then called 'cell grazing') with traditional set-stocking (and in an environment where traditional scientists were viciously attacking the new holistic management ideas). One of the three properties studied was Tim Wright's Lana, just as he had swung over to holistic grazing. The clear economic and ecological results of Jones and Earl's study found in favour of Wright's approach. This would have positive impacts for the ongoing thinking of not just the farming community but also both scientists involved.

In a seminal paper that pulled the rug from under set-stocking, a key point the two botanists made was that Australia (even without a history of cloven-hoofed animals) was no different from the other continents in terms of land degradation. The true causes of the degradation of landscape functions were not cloven hooves but the mouths of livestock: in other words overgrazing and/or over-resting.[16] That is, too many mouths set-stocked (i.e. grazing pastures too hard and too short for far too long) destroy too many green leaves (the solar panels), and thereby starves the energy production system while killing far too many roots. This in turn leads to a cascade of dysfunction, including degradation of the water cycle (infiltration and storage), associated destruction of soil life and ground cover, escalating erosion, the collapse of nutrient cycling and dynamic ecosystems, and so on.

Clearly this work in the early 1990s was a light-bulb moment for both PhD scientists, and each went on to work in the field of regenerative agriculture. It was soon after this that, on a farming field trip, Christine Jones visited central-west New South Wales farmers Darryl Cluff and Colin Seis when they were in the early throes of their evolution of pasture cropping (see next chapter). The two immediately began working together after this visit. Multiple pennies now dropped for Jones, and the huge significance of regenerative pastures and croplands not just for the basic landscape functions but also for addressing key issues such as salinity, climate change, and the entire water and other cycles became increasingly clear to her.

Jones now saw this as her life's work, and in 2003 she made the brave step of going freelance as a ground-cover, carbon and soils ecologist. She wanted to 'save the world'. Her first move was to found the 'Amazing Carbon' organisation and thence run a series of 'Managing the Carbon Cycle' forums. In 2007, she implemented the Australian Soil Carbon Accreditation Scheme.

Due to the tight interdependence of all the landscape functions, any entry point of regenerative agriculture leads to all the others. But given this, Jones has focused on the crucial role of longer-lived soil carbon. She says that the amount of vital carbon in the soil comes down to management – and thus can have profound consequences. This is because, she reiterates, 'carbon is the driver for every aspect of soil health and function – the MASTER KEY to every door'. For example, increased carbon in the soil, she says, can directly influence the capacity of the soil to store water': a capacity that is enormous.[17] That is, grassland, crop and pasture mismanagement in the first instance interferes in efficient photosynthesis: the source of living energy stored in molecular carbon bonds.

The secret, Jones argues, is to 'get life back into the soil' because living soils drive the whole process and this means more carbon – and especially long-lived, stable forms of carbon in the long-chain polymers called humus (given that humus is long-lived in the order of hundreds if not thousands of years, and by comparison to such organic carbon compounds as glomalin, which lasts around 40 years).[18]

Typifying the fact that much of the more life-empathic recent research in the area of soil health is led by women is Dr Kristine Nichols. She works on the Great Plains in North Dakota, USA, in the middle of ranching and some cropping country, and is a protégé of Sara Wright of the USDA, who in only 1996 discovered that magic substance glomalin.

We now know that glomalin comes from a particular class of mycorrhizal fungi that live inside plant roots (what are called arbuscular mycorrhizal fungi) in a symbiotic relationship. A number of researchers – Nichols at the forefront – now believe that glomalin is a key player in building healthy soil. In grasslands in particular, this is because, being a complex sugar-protein, it binds crucial elements such as iron and other ions. This means glomalin is essential in starting soil aggregate formation, which is a major part of what makes soil. In other words, says Nichols, 'Aggregates provide structures to soil for better water infiltration, water-holding capacity, and gas exchange, and increase soil fertility by providing organic carbon (that is, food) to soil organisms, which use this food as energy to release plant nutrients from the soil.' But glomalin is of yet greater relevance to regenerative farmers. This is because it helps protect fungal hyphae and enables safe transport of water and nutrients back to the plant host, while also protecting the host from decomposition and microbial attack. It also allows soil aggregates to store not just nitrogen but especially carbon deep inside the aggregate and so be immune from microbial access. This storage of long-lasting carbon as glomalin is a crucial pathway to combat global warming by locking up resilient carbon in the soil.[19]

Proof of the important role of glomalin is that Kristine Nichols is applying this work out in the tough grazing and cropping lands of North Dakota, where in winter, temperatures can drop to minus forty degrees Celsius. Here she has found that set-stocking decreases glomalin and that holistic grazing rotations increase it. Moreover, glomalin is especially high in concentration under native grass swards (with their long, coevolved histories). Indications are that this is mirrored in Australia. In cropping, Nichols has found that not just industrial fertiliser and chemical application but also fallowing and tilling decreases glomalin.[20]

Another type of fungi crucial to soil health (though not as widespread) is more associated with trees and shrubs and is known as ectomycorrhizal fungi. These fungi too form a symbiotic relationship with plants, but instead live on the plant roots, forming a net-like latticework around the roots. They also form dense hyphae sheaths whose cell walls comprise a highly resistant carbon compound called chitin (the same material comprising crustacean shells). This in turn is also a crucial anti-global warming compound, enabling (along with glomalin) the long-term storage of resilient carbon.

So, caring for all soil biology, and especially the under-recognised fungi, is crucial. Fungi have been coevolving with land plants for at least 460 million years, and are very smart at what they symbiotically do with plants in healthy soils. Not only do they improve plant nutrition and plant-soil biota feedback systems, they also source essential nutrients crucial for human and animal health (in exchange for plant root sugars). They also provide other multiple functions, including: influencing the phenotype and fitness of partner plants and thus their local adaptability; improving plant productivity and successional and competitive capacity; enhancing drought tolerance and plant survival and fitness; and stimulating of plant hormone production and the alleviation of stress.[21]

Today, at home, I am fungal-sensitive. I am careful not to use vehicles unless absolutely necessary and I look out for fungi now because I have learnt to actually 'see' them. After rain, I even try not to stock paddocks where large 'fairy circles' of mushrooms – fungal fruiting bodies – appear, and my family are sick of me exhorting them not to pick all the mushrooms in a fairy ring. My daughter Alison in a Christmas card (tongue in cheek) even christened me the 'protector of mushrooms'. But mushroom awareness is a clear illustration of how mind governs perception.

Through ongoing land degradation and modern management practices – via such activities as set-stocking, intensive cultivation, bare fallowing, and stubble, pasture and shrub burning – over the last 200 years more than seventy per cent of Australian agricultural land has been seriously degraded. This means soil organic carbon levels in Australia have dropped by at least eighty to ninety per cent of what they were before white settlement. If we had built carbon instead, this would have aided in drought-proofing farms and given us far longer periods of active plant growth and photosynthesis, with far more plant sugars being pumped in to feed the soil biology, and an overall huge stimulus to all landscape functions and resilience.

The message is clear: better grazing and cropping management that produces greater ground cover plus vastly more rainfall absorption and more water storage leads to enhanced photosynthesis, to greater root mass, and so to feeding a vast difference in active soil life between holistically managed and set-stocked farms or between biological cropping and industrial farming. For example, in a healthy soil there can be up to 5670 kilograms of living soil biological community biomass per acre (14,000 kilograms per hectare), all busily teeming with life. Conversely, in a dysfunctional soil there may only be 445 kilograms of bugs and creatures. This is the equivalent above the ground of between eleven yearling steers per acre on the one hand and less than one steer on the other. That is, one farm could be fourteen times more productive in the engine room that matters. This in turn can result in around an extra one million litres of water in the soil bank of healthier soils.

The father of modern biological soil science, Dr William Albrecht, once made the profound statement, 'The soil is the point at which the assembly line of life takes off.' How strikingly pertinent he was.

In 2010, following a research trip into the New England region of New South Wales, I cut across the head of the Liverpool Plains on a back road, heading for the locale of Spring Ridge. The road ran beside a railway line and before me lay mile after mile of rich, deep black soil. To the south, the west and behind me in the east, this plain was ringed by blue, eucalypt-clad hills, but to the north-west the valley of black went on and on into the distance as the plain widened out. I stopped my ute, not just to admire a belt of arguably Australia's richest soils, which were part of an elite group comprising around only five per cent of the Earth's farming soils. No, I also stopped because I noticed something offbeat.

It was a dry period at the time, and the night before (some eight hours previously) there had been a small rainstorm: perhaps eight to ten millimetres. What caught my attention were large puddles of water still lying on these deep and rich agricultural soils, plus evidence of tunnel erosion near the railway line.

Aghast, I took my camera and walked over to capture the scene. Normally in a dry period on such deep, rich soils – if they were healthy – that amount of rain would have been slurped up and disappeared in minutes. But here, some eight hours after only a moderate rain, large sheets of water glinted in the morning light. On closer inspection, I noticed the soils had thick, crusty capping and I realised there could only be one explanation. Here on Australia's best soils, there had been so much tillage and heavy traffic, and especially industrial fertiliser and chemical use, that not only was soil structure destroyed but also these magic soils were biologically dead and compacted: simply incapable, despite a dry time, of absorbing a mere eight to ten millimetres of rain.

In a sombre mood, I hopped back in the ute and headed towards the small locality of Coroona, near where I was to meet one farmer who definitely didn't treat his black soils in this manner. In the process, I came to understand what lay behind the erosion and the dead soils on the heart of the Liverpool Plains.

After a series of turn-offs and some intriguing back roads, I passed a group of silos, cattle yards and a large chook shed to arrive at Geoff Brown's house, located in the upper catchment of the Liverpool Plains (some thirty kilometres from its head). Of medium size, at a guess in his fifties, tanned face, strong handshake and open but searching gaze, Geoff answered the door with phone in hand, busily conversing as he directed me to a seat on the verandah. After a couple of hours (interrupted by further phone calls), I came to appreciate he was an overworked man. He was multitasking in a one-person show, with his workload temporarily inflated as he completed a swing into a different sort of agriculture. His head was already in the new space, his operations not far behind.

Geoff confirmed what I had seen crossing the Plains. 'The catchment zone is just above us here,' he said, 'and in front of us where my crops are is the flood plain. We're only twenty-four inches [600 millimetres] here but we have to be very careful how we handle it. But you know, when I was a kid I remember it used to take three days for the water to come down from the hills. Now it only takes from four to six hours … If there's a big rain, I've only got a few hours now to get my cattle off and onto the higher country. That's what's happened to the country up there, and down here. They've lost soil structure and humus. Here, I'm busy flat-out building humus; that's how you hold water on this country. You don't drain it.'

I came to realise that there were a number of reasons for this disastrous change. The first was over-clearing and overgrazing on the steeper lands surrounding the Plains. 'This place originally had a sawmill on it,' explained Geoff . 'Now, we're trying to replant trees and corridors and those sorts of things. Plus planting a diversity of grasses, crops, more perennials; and all this diversity gives us insect control for our crop pests. And the birds can now come in and out and look after both predators and good guys, because the bad guys come back with the good guys when you stop spraying.'

But a second key reason why the soils of the Liverpool Plains were in trouble was the new chemical agriculture. 'Farming agriculture is going backwards,' said Geoff. 'When I first started, taking over at the age of twenty-one when my father died, I followed the traditional line. I did everything like everybody else does. You do what you're told, basically. The local sales agronomist comes around. He says, "Right, we've got to do this, do that, put this in; put more product on more product," but it was "to hell with the consequences" … that was the primary driver, get tonnage out there. So I followed what everyone else did. But every time he came out here, he was trying to spend our money. And people around me, that's what they still do. Like they spread out a heap of nitrogen over their land. But it's like a steroid to an athlete: gives you a big hit, but as soon as it runs out, they fall in a heap, burnt out, had it.' Geoff paused, collecting his thoughts about the industrial approach. 'Yeah, we're driven by a quantity goal for production and we've lost our direction. And then came zero tillage and all its chemicals, which I initially adopted too.'

In short, Geoff began to realise that his high chemical, fertiliser, pesticide and weedicide use was killing off healthy soil life. The humus disappeared, to be increasingly replaced by more industrial inputs. 'Those cropping systems burn up carbon,' he explained. 'If they only knew humus holds four times the amount of water as anything else, but they've lost it. Basically, they're not looking at the symptoms. They're not looking after the soil.'

To emphasise his point, Geoff explained what he called 'downhill farming'. This was an approach designed to treat the symptoms of loss of soil structure and humus, where small furrows were created to direct water off the country. 'But the problem,' said Geoff, 'is that their soil is so glazed up she runs off anyway and causes damage. But that suits them. They can get back on to it quicker to do more work again, spread more stuff. But in two weeks they're looking for rain again. Basically, they want an inch a week just to keep going rather than use the inch and make it last a month or more.' To finish his point, Geoff was able to show me cattle-grid ramps on the flats that used to be level with the soil. Now the soil was a metre below them. 'That's all in half a lifetime,' he explained. 'In my lifetime.'

Now I understood what had so shocked me when I first saw the unabsorbed water as I crossed Australia's greatest stretch of soils. I was also beginning to see why Geoff Brown had begun to change his thinking.

'My father died when I was twenty and he was fifty,' recalled Geoff. 'We later put it down to a chemical issue. Back in those days, chemicals were all meant to be safe, but nobody knew anything about them. So that, and all these sorts of issues we've discussed, got me thinking ... From my point of view, I'm a farmer and my responsibility is to feed our fellow man a good product that he's happy to eat. So I just started from there, around fifteen years ago, looking at how I can learn to produce better stuff. Because I'd come to realise we've lost our direction.'

Then, as Geoff got more into regenerative agriculture and, as he expressed it, 'began weaning myself off all the nasties in agriculture' (first shifting to non-residual chemicals and finally to all natural products), he began to encounter more and more regenerative practices. He also consulted widely with many leading regenerative agriculture advisors and agreed he was an ecological cherry-picker. 'Where I'm coming from,' he said, 'is I'm gradually moving from a disabled system – from the traditional zero-till, chemical-type farming – into a minimum-till system which incorporates things like putting green manures in [i.e. ploughing back in growing grass or crops so as to feed the soil energy and not just minerals and nutrients] and refining key levels of limiting factors in minerals, like borons, silicas, that sort of thing.' Geoff now makes compost from his chook manure and uses additives such as fish and kelp, while in his pastures and crops he increasingly focuses on creating more diversity of type, roots, foliage, seasonality and so on. 'Basically, I'm working on the health of the country,' he said. And as for 'establishment' sources of knowledge? 'Product agronomists, Department of Ag. people and CSIRO? No, they're just so far behind,' concluded Geoff.

And the proof was in the eating. Due to his high humus levels, healthy soils and healthy plants and root systems, Geoff found that in a good season, while his industrial high-input/high-cost neighbours got higher yields, he was still in front on a gross margin basis. In a dry year, he was way in front on all counts, with better yields and no pinched grain (that is, partly unfilled seed, which is devalued on sale) because his soils captured and stored water rather than wasting precious rainfall that evaporated off the surface of energy-starved soils.

A pleasing part of the benefits of a healthy ecology for Geoff was pest control and fewer costs. By this, Geoff meant that in industrial agriculture when farmers spray insecticide to control, for example, sorghum midges and other pests, they also kill off biological control agents such as ladybirds, parasitic wasps and even birds. Without these natural predators, farmers become increasingly committed to the only pest-control method left open to them: repeated spraying of insecticide. So Geoff had eliminated chemicals and now relies on nature. He could then harvest mung beans without chemicals, while his neighbours were constantly spraying for 'pests'. His sorghum results were even more staggering. 'When the sorghum midges are on,' he said, 'the neighbours have their spray planes zipping down either side of me and around two and three times a week. But they're wiping out everything, including ladybugs and so on. But here, for example, the other day I watched a big swarm of midges come in, hover over the crop, but not land. Something about a healthy crop deters them. They took out a plant probably one in every hundred metres or so – but that's because that plant wasn't up to scratch, so they took him out. But everything else? Fine, hadn't been touched. So grow healthy plants, accept you're going to lose a few, that's part of nature. And it gives you a boost.'

At the time of my visit, Geoff was involved with neighbours and the community in the fight to stop destructive coal mining nearby on the crown jewel of the Liverpool Plains. He genuinely wanted to deliver healthier food to consumers but hadn't yet figured out a pathway for direct marketing. He also farmed now to be in control of his own destiny, out of the clutches of banks and multinationals. Pensively he concluded, 'Family is important. And you can't just exist on thin air. We've got to produce a good product for people – we've lost sight of that as farmers. But if we want to sustain our society, then we can't kill them off with bad food. It's as simple as that.'

CHAPTER 10

Farming without Farming

The main characteristics of Nature's farming can ... be summed up in a few words. Mother Earth never attempts to farm without live stock; she always raises mixed crops; great pains are taken to preserve the soil and to prevent erosion; the mixed vegetable and animal wastes are converted into humus; there is no waste; the processes of growth and the processes of decay balance one another; ample provision is made to maintain large reserves of fertility; the greatest care is taken to store rainfall; both plants and animals are left to protect themselves against disease.

Sir Albert Howard, *An Agricultural Testament*[1]

O ur farm resides on a high tableland, two hours south of Canberra in southern New South Wales. It nestles into an ancient landscape. The Aboriginal Ngarigo people call this country Narrawallee, or 'the big grass country'. The wider landscape is known as the Monaro: another Ngarigo word, for camp fire–like volcanic plugs, which we can see from our farm, sometimes capped by nipples of once extruded lava. Hence this Monaro country comprises a mix of basalt plains and khaki, straw-gold native grasses and poa tussocks. Mixed with this on the margins are over-cleared, rolling hills of granite soils and prehistoric, megafauna-sized boulders. The entire landscape is then interspersed by prominent darker blue islands of thick, eucalypt-clad hills composed of metamorphic rock half a billion years old. Hidden between these hills to the west are the serpentine valleys of the

Snowy River and its tributaries. Tough, lean, yet startlingly beautiful, our home country is irregularly drip-fed by an unreliable twenty-two inches (550 millimetres) of precipitation: mostly rain, plus a rare, big snowfall.

To the east, we have a clear view of forty kilometres across rolling, treeless basalt plains and hills to the coastal range. In summer, when the easterlies kick in, you can smell the sea salt, but you know the local trout won't rise and it is time to pack your rod and go home. We are in big sky country, and the sun is its principal player. I watch the eastern sky a lot: for a purple sun shrouded by bushfire smoke; a golden sun in a frigid winter sky; a crimson sun-curtain that beckons unsettled weather; and a sky on a winter evening with five degrees of rich scarlet-crimson then a distinctly deep-blue layer on top – sure sign of a coming frost.

But when it comes down to it, we sky-watching farmers are invariably looking for rain. Moisture 'in the soil bank' determines economic well-being and contented livestock, vibrant bird, mammal, reptile, insect and soil life. Conversely, when dry, one is surrounded by an increasingly muted world. Management is then an anxious time of constant decision-making and the study of holistic grazing, market and weather charts. When rain then arrives, it constitutes a release of a universal living spirit.

However, I learnt early in my management career that a long-awaited big rain or storm can also be enormously destructive if one has not carefully managed a desiccating landscape. Yet again, being a slow learner, I was educated the hard way.

By the late 1970s, I was starting to hit my straps as to what I thought was a young, progressive farmer on the Monaro. I was making an annual profit and busily cutting hay in November at the most hectic time of year. I regularly and successfully ploughed, sowed and established pasture, and all this in a landscape resource I regarded as an inert substrate upon which I mechanically intervened and grazed my livestock. The fact I was a keen naturalist and enjoyed my birds and other creatures seemed to me at the time to have no connection to my farming life. I believed I was in control of this inert agricultural environment, and it was an intellectual game of monetary lever-pulling. Even the seasons appeared predictable: that is, until the 1979–83 drought came along.

I can still vividly recall one scouring experience that finally began cracking open my mind.

Locally our property is known as 'easterly fall' country, as it is tucked into the lee of a high, timbered hill to the west, which forms the shoulder of a National Park reserve. This western ridge merges with our own country and runs as a long spine, some five kilometres to the north. Thus, it provides a wonderful remnant vegetation reserve comprising more than 1500 acres (600 hectares). This landform constitutes a natural watershed, part-protecting our land from prevailing westerly winds.

Beneath the hills and the National Park fence line, we have a lambing paddock. It is open country that blends into timber and then the shelter of the forest behind. Here the soil is sandy and fragile and, being derived from the weathering of ancient, metamorphosed sedimentary rock, is leached of many nutrients. We call this paddock the South Shelter.

This day was in early spring in the late 1970s before the big drought. Brimming with confidence at some recent pasture-establishment success, I sallied forth with a tractor and disc-plough to 'prepare' the paddock for sowing with 'improved' grass and legume species. I had determined the paddock sward was 'running out' and, because it was a crucial lambing paddock, that it needed 'renovating'. However, with the confidence of a young industrial farmer came arrogance, and so I ignored vital warning signs revealed by the landscape. Down the centre of the paddock was a small erosion creek, and, though its base had grassed up, it nevertheless dissected the valley-shaped paddock. There was also a silted-up dam halfway down this creek that the Department of Soil Conservation had put in years before. Like radiating veins, there were a number of secondary creeks that fed into the main creek. At the time, I just thought this was a normal landscape.

So, I ploughed the good earth over two days. Driving the tractor home late at the end of the second day, I recall noting how the soil had ploughed so easily into a fine, powdered tilth (tilled land). And I distinctly remember thinking, 'Gee, I hope we don't get a big storm before I have sown this down.' Within a month, I lost my gamble. One evening, a giant thunderstorm rocked and rolled, looming over Wullwye Hill as dark clouds ballooned and swirled, and then a white sheet of heavy rain and thunder swept down and across all our high country.

I drove up the next morning to find a scene of destruction. Mini gullies, like a network of fine crevasses, dissected the paddock; whole areas of fine, sandy topsoil were gone, filling further the already silted dam and piled in banks against the fence at the bottom. It took me half a day of shovelling sticks and compressed mud and sand to make the fence sheep-proof.

My heart was stricken, and I felt like a vandal as I comprehended the enormity of what I had done. I had cost the landscape and our family perhaps a thousand years of topsoil in only twenty minutes, preceded by two days of a thoughtless gambler's punt. I resolved then and there to never again plough any country with even the slightest slope. It would take another decade or more for me – the ponderous learner – to appreciate the full and unseen dangers of the plough in our landscape and to renounce such a mechanical object forever. But to this day, some thirty-six years later and even in a good season, I can still see the indent of the gutters that dissect that paddock.

A few years before I ploughed the South Shelter, I had read a book by English writer Edward Hyams titled *Soil and Civilisation*. Hyams methodically revealed how almost all great human civilisations in history had died within a millennia or two at most – and largely because of mistreatment of their most precious resource: a healthy soil that fed ever-burgeoning populations. Moreover, once healthy soils collapsed, so too did their interconnected landscape systems and functions. Clearly – and epitomised by my actions in the late 1970s – our species seemed to have some inbuilt but inexplicable inability to learn from history. Little wonder that Justin Isherwood so aptly observed that 'ploughed ground smells of earthworms and empires'.[2]

However, my ploughing folly wasn't just the result of inherited practice inculcated over 140 years of Australian land use. It went far deeper and was part of a long and powerful tradition of human agriculture. The belief systems and mental constructs accrued over many millennia were key to our journey into the Anthropocene. For it appears indivisible from our ancient agricultural heritage that there is a perceived need to plough the soil – to dominate and stamp our authority over Mother Earth. It therefore took another time and a uniquely different continent for more radical solutions to emerge to just 'ploughing the ground' – as we shall see directly.

Historians have written volumes on the history of agriculture since early domestication of plants and animals over ten millennia ago. A feature of early agriculture in its many sites across the globe was soil disturbance so as to plant seeds of newly domesticated plant varieties in what was subsistence agriculture. Early tools (not yet ploughs) were hardened sticks. Then came the first simple one-furrow plough, or 'ard' (scratch plough), initially pulled by humans and later by animals.

Through a series of technical and cultural revolutions, therefore, progress in agricultural technology can be traced to the present day. At its heart are, first, changes in tillage technology and the introduction of animal power, and then in time use of grazing animals to spread and enhance fertility.

By the medieval agricultural revolution of the eleventh to thirteenth centuries in north-western Europe, the use of grazing animals in a rotation system was integral to transferring organic nutrients to the arable lands so as to renew fertility. However, at this time there also arrived the heavy, iron curved-disc plough and all its attendant paraphernalia. In many ways, this was a powerful symbol of a further severing of our psychic umbilical bond with nature – a step begun upon the first domestication.

From the sixteenth to the nineteenth centuries, in temperate regions of north-west Europe, there occurred three modern agricultural revolutions that devolved from the medieval one. The first still involved rain-fed, culti-vated ecosystems based on the plough and was led by cereals. An important development, though, was the closer integration of domestic herding ani-mals in a pasture and crop rotation system. This was enabled by superior performing animal and plant breeds. These gave rise to increased animal densities and better organic fertilising, which in turn allowed the elimina-tion of fallowing. So this first modern agricultural revolution – what has been dubbed the 'triumph of horn over corn' – was in some respects eco-logically enhancing.

However, this first major modern agricultural revolution was the last where animals were integral to a self-sustaining healthy agro-ecosystem. This is because what followed were the second and then third modern agricultural revolutions from the late nineteenth century onwards. Critical to both of these was the elimination of the integration of animals in a crop-ping system and also of their organic fertiliser for building soil fertility.

The second modern agricultural revolution was characterised by the introduction of synthetic fertilisers, no fallowing, and the mechanisation of animal-drawn cultivation and transport. The next revolution continued these things, but this time it added a reliance on fossil fuels and the introduction of motorisation; intensive seed selection and specialisation; then in the twentieth century the massive use of chemical weedicides and pesticides; and the injection of genetically modified plant varieties. Therefore, modern practices had evolved to such an extent that it seemed virtually impossible to reintroduce animals to the landscape for regenerative and organic fertilising purposes. This is because the entire landscape had been modified to suit modern agriculture's giant machines and technologies.

On the flyleaf of his landmark book *Tree Crops*, J. Russell Smith (professor of economic geography at Columbia University) placed a fascinating drawing of our Earth but dominated in front by the North and South American continents. Circling the globe were eighteen lines of railway freight cars: the loops covering all of the USA and down to part of Brazil. Smith's caption to the diagram reads:

> The U.S. Soil Conservation Service reports that the soil washed out and blown out of the fields of the United States each year would load a modern freight train long enough to reach around the world eighteen times. If it ran twenty miles an hour continuously, it would take nearly three years to pass your station. We began with the richest of continents, but ...[3]

His point is well made: destroying and ploughing soil so as to prepare a seed-bed for another annual 'weedy' crop while killing so-called weeds at the same time is a short-term fix with disastrous long-term consequences. Plenty of history books tell us how many civilisations built around this annual cropping approach have fallen not to the sword but to the plough.[4]

Early ecological writer Edward Faulkner, in his prescient book *Ploughman's Folly*, wrote in 1945, 'evidence that trouble stops where the plough stops, has been almost universally overlooked'.[5]

So, when direct-drill technology and weed control via herbicides came in from the 1980s, it appeared a win–win situation, ostensibly delivering less soil disturbance and more residual vegetable matter, and thus reduced erosion. Since then, however, evidence is mounting that there are increasingly huge risks associated with this high-tech approach.

Undeterred, the big end of town undertook some diabolically clever reframing of the entire issue so as to hold and expand their territory. Backed by the Food and Agriculture Organization (FAO) of the United Nations from 2002, by other leading international bodies and the large multinational corporations involved in industrial agriculture, and by leading research and development (R&D) organisations across the developed world (including bodies such as the CSIRO and state departments of agriculture in Australia), the dominant paradigm power-base has relabelled the new approach 'Conservation Agriculture'. This clever con-job focuses on the capacity of the new direct-drill and herbicide system to conserve soil, but it conveniently ignores the considerable and increasingly apparent downsides.[6]

Clearly, therefore, entirely new and transformative solutions to addressing the unsustainable practices of modern industrial agriculture won't come from within the same old box. They need to come from outside; they need to be radical; and they need to include what we have learnt over the centuries while at the same time returning to first principles.

What is exciting is that a number of the revolutionary advances in regenerative agriculture involving cropping systems are derived from a mimicking of natural function. Three of the most radical of these have evolved in Australia, and all are dependent on the reintroduction of grazing animals into the cropping landscape. A fourth – based in the prairies of Kansas, USA – has sought to go right back to ecological first principles.

For nearly fifty years, one of the USA's leading agrarian and ecological farming thinkers, Wes Jackson of The Land Institute in Kansas, has sought to begin again at that take-off point of agriculture ten or so millennia ago. This involves an attempt to breed productive, deep-rooted perennial grass grain crops: as opposed to the direction the first human agriculture took – of utilising food from annual cereal grasses.

Now we know that in Europe humans have been opportunistically harvesting natural grasses when in seed for at least 23,000 years. But in Australia this began ten millennia or more before that. The work of Bruce Pascoe, Rupert Gerritsen and others reveals that not only did Australian Aboriginals manage and harvest a giant, continent-wide grain belt composed of diverse grasses and forbs (and most were gluten-free), but that this also involved seed selection, propagation and spreading (i.e. key elements of plant domestication), and resultant harvesting, storage and baking technologies. Moreover, this was based around perennial plants.[7] As Professor Ian Chivers of Southern Cross University recently observed, 'In Australia we have stunning examples of very long-term grain-food production that had no degrading impact on the environment, that did not require expensive fertilizers or pesticides, and grew without the need for irrigation water to be diverted from river systems.'[8]

The modern-day problem, however, is that such native grasslands don't yield the heavy crops required to feed the massive concentrations of contemporary humans. That is where Wes Jackson and his Land Institute are trying to restart or rewind the entire domestication process, but this time beginning with perennial and not annual plants (i.e. plants that live for more than two years rather than complete their life cycle and die within a year).

We know when it comes to the building of human civilisations there is no free lunch: that whenever we traditionally convert land to cropping, raising animals, extracting timber or building towns and cities, we grossly interfere with landscape function. However, all the regenerative agricultural solutions discussed in this book succeed in varying degrees to work with, rather than against, nature in their interventions. Trying to breed a productive perennial grass grain is another such attempt, but after decades of effort, delivering a viable solution is proving difficult.

This is because there is a reason annual grass-cum-cereals were domesticated: they have a simple survival strategy. Using shallow, opportunistic root systems through a winter growing season, they produce as much seed as possible. They then drop this at the end of summer and die – relying on their prolific and vigorously competitive seed to perpetuate the species. With human breeding skills, such cereal grain species can be manipulated for greater production and the enhancement of other traits. That is, such plants were pre-adapted for domestication because their functional traits suited human needs. Besides their unusually large or numerous edible parts, they could also self-pollinate and were easy to propagate and so repeatedly grown annually. They had lower seed dormancy, were easier to harvest and de-hull, and so on. And they had one other trait: their 'weediness'. That is, they thrived in disturbed fertile soils – like those found near human habitation.

What were not domesticated were perennial plants, simply because it is difficult to do. And this is the great pity, for domesticated perennial grains would seem the ideal ecological solution. It would mean no annual tillage and management cycle, with all its inputs and ecological damage. All that would be required would be seasonal management, say with a holistic grazing regime, and then harvest – after which the plant is still there. Crucially, such perennials maintain ecosystem function: functions that annual crops tend to degrade. For example, in healthy, well-managed grasslands, the plants are hardier and deeper-rooted, and so are more drought-resistant. They establish symbiotic or mutualistic relationships with a diverse soil microbiota, have longer growing seasons, are more efficient at intercepting and using rainfall, and are superior at scavenging for soil nutrients while being better adapted to their specific environments.[9]

All this means they are not as reliant on seed for species survival and so have inbuilt genetic resistance to producing vast quantities of it. In short, therefore, at first glance it would appear you can't have it both ways in plant breeding and can't do the impossible without major compromise. But Wes Jackson disagrees. He argues there need not be a trade-off between deep roots and lots of seed. This is because he sees a plant's deep roots as an investment and not a trade-off, because once established such a plant then has the potential to put more energy into seed production.

So we should never underestimate the capacity of the human mind. The many examples in this book reveal that it is possible to have a self-sustaining and regenerative agricultural system. A key reason for this is that such systems attempt to mimic the functioning of the ecosystems best suited to that environment. And this is what Wes Jackson and his team are attempting to do at The Land Institute near Salina, Kansas.

With a PhD in genetics, Jackson – like his wife, Dana – came from a Kansas rural and religious background. Each grew up in that time of change in the 1960s, witnessing the horrific soil erosion and other costs of modern industrial agriculture. As Wendell Berry described, this experience sensitised them to the 'political, social and ecological deterioration' around them, which 'they thought, mirrored an inward condition; the outward problems connected in spiritual blight'. Berry wrote, for that reason the Jacksons concluded 'technological solutions alone would solve nothing'. So they returned to their homeland and formed The Land Institute in 1967.[10]

The Jacksons believed the answer lay in the unploughed Kansas prairies, where diversity and healthy landscape function reigned; where energy was free via photosynthesis; where soil was being built; where the system provided its own nitrogen; where the water and mineral cycles provided all needs; and where the system maintained health through its own pest- and disease-controlling mechanisms. The same (with variations) was also true for Australia's native grasslands. So here was the healthy template.

As The Land Institute grew into a seminal research and education body, it became clear that it might be possible to grow grain off a diverse prairie ecosystem that was healthily self-sustaining. In the process, this would disempower the large multinationals whose mega-profits were dependent on substituting destructive inputs for natural function. That is, a new agriculture suited to smaller-farm agriculture would also entail a social revolution. Along the way, a brittle and unjustly imposed economic system could become more resilient.

The key to the Jacksons' thinking was biomimicry: of replicating a natural, complex, adaptive ecosystem and overthrowing the industrial paradigm for a natural one. After much analysis, Wes Jackson concluded that the key to the 'ten-thousand-year-old problem of agriculture' was perennial grain crops: what he and colleagues called 'natural systems agriculture', or NSA. This comprised 'the development of herbaceous perennial seed-producing polycultures in which native prairie is used as an analogy'. The Jacksons and team rightly claim that this is 'an entirely new paradigm of food production rather than a type of sustainable agriculture'. Therefore, NSA poses fundamental questions about the entire philosophical, moral, ethical and spiritual base on which modern human existence and conduct exists.[11]

Those at The Land Institute claim they have already demonstrated the scientific feasibility of NSA, even if, as Jackson says, 'we are at about the same stage as the Wright brothers were at Kitty Hawk'. This work, at its basic level, is paralleled by efforts in breeding perennial grains around the world in grain-growing nations such as Russia, China and Australia.[12]

However, the final verdict on NSA is likely to take decades to come in, and for some the jury is still out as to whether large volumes of grain that are required to feed humanity can be delivered off this system. This doesn't seem to worry Jackson, who once told ecological chef and thinker Dan Barber that 'If you're working on a problem you can solve in your own lifetime, you're not thinking big enough.'[13] Nevertheless, in conjunction with other regenerative-based cropping systems discussed in this chapter and Chapter 15, I am confident these combined approaches (including NSA) can fill the food gap at the same time as addressing – and not exacerbating – our Anthropocene dilemma.

In the meantime, a number of solutions parallel to NSA have emerged. These arose out of the heads of four separate but similarly battling Australian farmers in the dusty paddocks and dry, marginal cropping country of central-west New South Wales and semi-arid Western Australian wheat-land: the very antithesis of the rich soils and moist climate of north-west Europe or of deep-soiled American prairies. The Western Australian example (Di and Ian Haggerty) will be dealt with in Chapter 17. But concerning the New South Wales examples, what occurred would seem completely back to front, if not an improbability. This is because these farmers' innovations enabled them to annually produce productive grain crops off a perennial native grassland base while cleverly side-stepping the much longer and seemingly harder route of Jackson's and peers' breeding attempts. Their respective secret was the reintegration of grazing animals on native pastures into the annual cropping cycle, which enhanced natural function. These farmers had gone back to ecological basics of trying to work with, and not against, the key landscape functions – but in this case via a different route to those at The Land Institute. That is, they too had returned to 'thinking like an ecosystem'.

In historical terms, these developments – known respectively as pasture cropping and no-kill cropping – undoubtedly constitute twin breakthroughs of millennial proportions. And what actually occurred was a turning back to the basic principles that Sir Albert Howard had learnt in India in the early twentieth century from practitioners of ancient 'organic' systems that had developed over many millennia. Foremost of these was his precept that 'Mother Earth never attempts to farm without livestock'.

As remote as the chances may seem for a new livestock–crop integrated system to evolve in the hot, dry, harsh climate and lighter (more nutrient-deficient) soils of Australia, I thought it even more improbable that it collaboratively evolved, of all places, on two beaten-up farms in an area just north of the old gold-mining town of Gulgong in the central-west of New South Wales. This thought was reinforced on an early autumn day in 2010 as I drove north out of Gulgong along the twenty or so kilometres to interview Colin Seis on his 2000-acre (800-hectare) farm Winona. Seis is the best known of a trio of farmers in this region of New South Wales who have broken through to integrate the hitherto-separated components of cropping and livestock.

The reason for my surprise was the nature of the country I was driving through: what could only be described as light, marginal country. Judging by the unappealing adjoining paddocks on the drive out, this country was also easily degraded and eroded.

However, as I drove along absorbing the landscape, I came to realise that, after all these centuries of agriculture, a new, farmer-driven innovation was least likely to occur in the safest, richest farming environments. It was not comfort but adversity that was the stimulus of innovation: an observation that was proven time and again as I continued this remarkable and mind-challenging journey.

Colin Seis appeared to be your archetypal laid-back, laconic Australian 'bushie'. His speech was full of colloquialisms, his open, sun-blotched face frequently creased by a smile that betrayed his dry sense of humour. Until only thirty or so years ago, this hitherto apparently typical farmer – who is still in love with his stud merino sheep and kelpie working-dog stud, plus a regular beer at the pub with mates – was a traditional cropper and grazer. Though on tough country, Colin's farm nevertheless lay in an attractive landscape. Behind his house were high, timbered hills, with arable country in between. To the north-east was a drainage valley of mixed timber, grass and cropland: his rainfall around twenty-four inches (600 millimetres).

Colin's long-held family farm was selected in 1868 by his great-grandfather: an immigrant from Prussia who began growing wheat and merino sheep for wool. Since then, the place had been mercilessly ploughed and tilled for crops every year over four generations. During both his father's and Colin's tenure, Winona had been ploughed and fertilised: both to grow crops and to build a grazing system based on the widespread formula of annual sub clover and introduced grasses. Livestock were set-stocked or grazed on exposed earth in stubble paddocks. But this changed overnight when, in January 1979, the perennial spectre of an Australian bushfire descended on his farm. Most of Colin's fences, livestock (3000 out of 4000 sheep), farm buildings, woolshed and house were destroyed. 'I had to start from scratch with absolutely nothing,' Colin told me. 'I didn't even have the option of "low-input" agriculture. My only choice was "no-input" agriculture.'

Severely burnt in the fire himself, Colin lay in hospital for three weeks turning over in his mind how he could start again when he didn't even have the cash to plant a crop. But as sometimes occurs, out of extreme adversity combined with the miracle of timing, a lateral-thinking colleague, and a confluence of chance events and some luck, a unique – indeed revolutionary – agricultural development evolved.

The initial breakthrough came from extended discussions with fellow farmer and friend Darryl Cluff, who lived fifteen minutes away near the small town of Birriwa. The two regularly pondered Colin's cash dilemma and also the rising costs involved in modern cropping and pasture 'improvement' and their associated ecological damage. These conversations led to them posing the question, 'Why not try putting in "improved" pasture at the normal time in autumn/early winter but without any costly fertiliser?' Up to that point, Colin, in today's terms, had annually been spreading about $50,000 of fertiliser inputs. Over the next few years, both farmers (each at the time with no training in ecology) experimented with this new minimalist approach – beginning with Darryl Cluff.

'So why turn the tap off on the "super" and go against accepted practice?' I asked Colin.

'It was just a gut feeling,' he replied, 'because I thought that if we stopped putting out high rates of superphosphate, plants would grow that weren't dependent on it. There was no logic in this, no science behind it.

'What's more,' said Colin, 'the nay-sayers, the sceptics were right at first, because the wheels fell off everything, and our production crashed for seven or eight years. But then, we found that we evolved a natural grassland – a native grassland with some introduced species in it and from a better functioning ecosystem – through natural cycles, and which wasn't dependent on phosphorus. This particularly included perennials – many of which hadn't been seen for decades.' That is, and as with Tim Wright's experience, once the soil-mineral cycle and other functions began regenerating and as self-organising processes kicked in, the more valuable C_3 and C_4 grasses were enabled to re-emerge.

So with his pasture thinking now upended and beginning to favour a shift to a native grassland that was adapted to the Australian environment,

Colin turned to the next puzzle piece: cropping. Up to this stage, through the 1980s, Colin was still in traditional farming mode: still ploughing, or, as he put it, 'still working country'. By around 1990 or so, and after further yarns with his friend Darryl Cluff, or 'Cluffy', the two farmers swung over to the new industrial cropping technology then becoming widely adopted: that of 'zero tilling', or 'direct-drill', as it was also called. This eliminated ploughing, and instead weed control was managed through intensive herbicide use put out by increasingly big spray rigs (aided particularly by the widespread adoption of Monsanto's glyphosate, marketed as Roundup).

The main apparent advantage of this new approach that had begun sweeping the industrial farming world was elimination of the millennia-old destructive practice of repeated ploughing and over-turning of soils. This traditional practice killed off soil biology, created a hard-pan at shallow depths, required frequent 'workings' to control weeds with implements such as offset discs, and thus destroyed soil structure, dried out the soil and made the entire paddock more prone to erosion. It was also more time-consuming and expensive on fuel.

With the new industrial approach, after the weedicide application, initially tined combines were then used to sow cereal (and increasingly also pasture) directly into the un-tilled soil. In one sense, this zero-till approach was a step forward, as, besides destructive ploughing being eliminated, more organic matter and ground cover was preserved (especially if the farmer didn't burn or otherwise destroy his stubble from the previous year). But, in another sense, the intensive use of chemicals had opened an even larger can of worms (which we will deal with in detail in Chapter 21).

Either way, Colin and his mate Cluffy now had the second piece of their puzzle: direct or zero-tilling technology. Then, at almost the same time, Colin discovered the third and final key ingredient: holistic grazing management. Via newspapers and his personal grapevine, Colin heard about the new ecologically based grazing systems being propagated by Allan Savory and Stan Parsons around the world from the late 1980s. He quickly experimented with the Savory system, but on his terms. 'You see,' Colin recollected, 'I reckoned cell grazing was a load of bullshit, and I reckoned it wouldn't work. So I set out to prove it wouldn't work. I put a mob of about 1000 ewes together and started to move them around.' But, he continued, 'within about three or four months I'd bloody well proved myself wrong on that one'. Soon after this, Colin did a holistic grazing course and, like Tim Wright, began combining his sheep into large mobs of 2000 or so, with rapid rotations and long rests. As with Wright also, one of his aims was 'to have one hundred per cent ground cover one hundred per cent of the time'. Colin concluded, 'It was the way the grass recovered so well afterwards that convinced me I was wrong.'

So, puzzle piece number three was now in place. Then came the breakthrough, which rapidly evolved by the mid-1990s. This entailed the integration of all three puzzle pieces: (1) encouraging the regeneration of native grasslands by going cold-turkey on industrial fertilisers; (2) the implementation of no-till or direct-drilling with tines; and (3) combining the cropping with holistic grazing management at the same time.

Both Colin and Cluffy were thinking about their remaining dominant native grass at that time: a winter-dormant, summer-active C_4 species known as red grass (*Bothriochloa macra*). 'Up to that time,' recalled Colin, 'I thought red grass was bloody useless. But that was what we mainly had here in the early nineties, and only later did the even better grasses like Shot grass [*Paspalidium distans*] come in as we got the ecology functioning.'

As Colin says of Cluffy, 'he's a very innovative thinker, way out there sort of stuff, and the question came up: "Why can't we put autumn crops into this dormant red grass, and why do we have to kill the bloody stuff to put a crop in?"'

Perhaps a trigger was Colin remembering a particular comment of his father's after a good season years earlier. 'We had heaps of feed, especially clover, and we were getting the plough out to get ready for cropping,' recalled Colin. 'And this stuck in my head: my father said, "What a shame we've got to plough up all this stuff."'

Within a few weeks of their brainstorming session, both Colin and Cluffy tried the new approach of combining all three of their different systems in a cropping enterprise. Using minimum fertiliser and chemical inputs, Colin began growing crops by direct-drilling cereals into native pasture paddocks that were dominated by winter-dormant perennial species. Then he rotated his large mobs of animals so as to graze the edible winter cereal crops that had by now grown through the winter-dormant grasses. 'The first year, it worked surprisingly well,' said Colin, 'and then I guess from that time on I spent a lot of my time fine-tuning it.'

The next benefit in that first year after harvest was when Colin found that the summer-active red grass had woken up and that he had a healthy grass sward instead of bare ground and stubble as in traditional cropping. In effect, Darryl and Colin had turned the clock back to the first modern agricultural revolution of the sixteenth to nineteenth centuries: eliminating fallowing through better grass and crop species, which in turn allowed increased animal production and rotation, and thus the better recycling and application of organic fertilising.

Aiding this development was the emergence of new varieties of cereal crops (and later canola) that were suitable for grazing by animals. This was made to measure for a holistic grazing system, with the animals grazing the crops one or more times in winter, thus both stooling and thickening the crops and further recycling organic fertiliser. In his laconic way, Colin summed up what clearly constituted a major step in the progression of agriculture: 'The fixation on monocultures and controlling the landscape is because the thinking is that the only way you can grow a crop of wheat is to totally remove everything from the paddock to give that wheat every advantage you possibly can. What we're talking about here is the complete opposite.' As if to support this, Cluffy dubbed the revolutionary new ecological agriculture 'farming without farming'.[14]

At this critical juncture in the mid-1990s, Darryl and Colin met University of New England grassland ecologist Dr Christine Jones. As we saw in the previous chapter, Jones had begun challenging the deeply entrenched paradigm of industrial agriculture – galvanised by working on Tim Wright's property Lana. It should come as no surprise then, following visits to Darryl Cluff's farm and because of her scientific training in soils, botany and ecology, that when she visited Colin's Winona on a farm tour she was able to explain the property's remarkable functioning to a perplexed Colin. 'She saw it, and she instantly understood it,' said Colin, 'because up to that time, Cluffy and I were on our own, and everyone else was telling us we were lunatics.'

In short, what Jones explained to Colin concerning his unexpected recruitment of native plant species not seen before was that his new system was improving soil ecology. His minimal soil disturbance and dramatic reduction in harmful chemicals and fertilisers, said Colin, 'was somehow stimulating mycorrhizal fungi health and getting everything functioning

– you know, getting sugars and all that stuff along the row: an ideal situation for plants to germinate in.' Second, it seemed that grazing pre-drilling and while the pasture was still growing led to root-pruning, which put organic matter and carbon into the soil.[15] This fed the soil microbes (both bacteria and fungi). As Colin summed it up, these impacts, in combination with extra root sugars from the cereals, meant 'a whole heap of food was pumped into the soil. Plus you had the organic fertilising from 3000 or so sheep in a thirty to forty paddock rotation. That really fired things up.'

The result was that pasture cropping led to faster rates of soil-building than under pure perennial pastures, and that yields off an annual crop in a pasture-cropped system (when it is planted out of phase with perennial pastures) could sometimes equal those of a traditional annual crop alone.

Since that time, Jones and a number of scientists from various universities have intensively studied Colin's Winona – using his brother's traditionally cropped farm across the fence as a comparative 'control'. Colin quickly found the new system got better and better, and on comparative terms his approach is more profitable while being ecologically regenerative. For example, the number of Colin's native plants has risen more than tenfold, and today his farm carries over 200 diverse plant species. Evidence of true regeneration (due to self-organisation) is the reappearance of highly palatable but long-lost warm season (C_4) grasses: always the first to disappear under traditional set-stocking. In addition, this grass and cropping harbours a huge escalation (125 per cent) in insect biodiversity, and with 600 per cent greater biomass. This has meant that insect infestation and damage to his crops is now negligible.[16]

Some of the physical and ecological changes in Colin's farm are startling. Colin now has no soil erosion; remaining patches of salinity are disappearing; his soil moisture levels are double those of his brother's, and his country is more resilient in terms of withstanding dry times, while at the same time there is dramatic soil-building to considerable depth. Accompanying this is an amassing of more carbon in the soil (measured to have doubled already). Moreover, despite no superphosphate application on his pastures for over thirty-five years, Colin's phosphorus and other trace element and mineral levels have risen substantially, while at the same time there has been a dramatic shift in other major soil health indicators: the soil pH, for example, having jumped from high acidity levels to nearly neutral.

In short, due to pioneering work with Darryl Cluff, what Colin Seis had achieved – via an initial focus on enhancing his soil-mineral landscape function – was to get all his landscape functions operating healthily again. Through greater plant diversity, different root types and depths, and aiming for one hundred per cent ground cover, he was targeting the water cycle: to enable his healthier, more porous soils to absorb as much water as possible. By aiming for growing plants all year round, with greater diversity, he was enhancing the prime driver: his solar-energy cycle. Clearly, his dynamic-ecosystems function had then also responded, with a huge explosion in insect and other diversity.

Finally, Colin acknowledged that the biggest change was that he now thought ecologically (part of the human–social functional shift). This in turn set him on an open-ended journey. 'I find myself addicted to learning,' he said.

Colin now farms with his son Nick, who has also taken on the kelpie stud. This has allowed Colin to spend much time as an educator, both nationally and internationally. In recent years, he has begun refining the next stage in his innovative system: of what he calls 'multi-species pasture cropping', where he increases the diversity and different plant functions in his cropping with up to ten or more different cover-crop species. He also sells up to eight varieties of native pasture seed from a harvester he specially adapted.

At the end of my visit, we drank another cup of coffee at Colin's kitchen table and gazed out the window to the valley beyond. The paddock below my ute had tall, ripened native seed heads bending in the breeze. Colin pondered for a while on our discussion. 'You know,' he said reflectively, 'the closer that I work to nature or the function of nature, the easier it becomes, and the more profitable it becomes, and there's less costs, a lot less risk, and certainly a lot less work.'

He then concluded with a statement I would hear again and again from leading regenerative farmers, and which referred to the inherent self-organising capacity of natural ecosystems and other complex dynamic systems (even though many of the farmers weren't aware of this propensity). 'Nature will just drive it for you,' said Colin. 'Most of the time, all we've got to do is get out of the bloody way and stop interfering and it'll fix itself.'

Driving away from my yarn with Colin Seis and heading north on a series of spiderweb dirt roads in quest of his mate Darryl Cluff, my head again felt scrambled – just as with my earlier visit to Tim Wright. There was a huge amount to ponder on. But little did I know that even further down the track awaited more mental torture in the form of another radical innovator. This was Bruce Maynard, who had made another major breakthrough serendipitously around the same time as Colin's and Cluffy's. He had also used grazing animals to do this, but with significant differences from pasture cropping. Maynard was to dub his new system 'no-kill cropping'.

Now, while pasture cropping undoubtedly comprises an agricultural revolution of global proportions and application, in two key respects it still evinces traits of its industrial lineage. The first is that its adopters often use reduced amounts of herbicide and fertiliser. The second is that direct-drilling with tines can still create an aggressive degree of soil disturbance. Sometimes this can be constructive, but, misapplied or done too regularly, it can undo regenerative impacts, and paddocks then need a few years in ley pasture or rest to recover.

However, what is remarkable is that, at the same time as Seis and Cluff were evolving pasture cropping, some 140 kilometres to the north-west, between the towns of Narromine and Trangie, a remarkably free-thinking farming innovator called Bruce Maynard was also developing a new regenerative agricultural tillage system. This approach was even more ecologically radical than, and equally profound as, that of pasture cropping.

The significant difference is that Bruce Maynard's system is far less interventionist – first in the key area of soil disturbance, and second through the total elimination of both chemicals and fertilisers (hence the 'no-kill' title). In addition, unlike pasture cropping, which uses all the conventional tools in the modern cropping toolbox (tillage and spray equipment and sometimes modified chemical and fertiliser use), no-kill cropping uses a vastly less disruptive sowing technology while eliminating the chemical and fertiliser tools. In short, it is more ecologically purist and more radical. Thus, it comprises a synthesis of the old and new – as we shall see.

Bruce Maynard was the youngest of five children growing up on the family's 3700-acre (1500-hectare) mixed farming property of Willydah – part of the fourth generation on a farm that produced sheep, cattle, pigs and crops. I first visited him there in 2010. He has an open and friendly face, his bald pate somehow emphasising a seemingly perpetual friendly smile. But his intelligent eyes smiled also, and I realised there was a compassionate, ethical and razor-sharp mind constantly at work.

As we drove around in his ute to inspect his remarkable property, he told me that some early experiences had triggered hard-hitting realities that later became important. The Maynard farm had healthy soils, but it sat in tough climatic country that pushed out towards the New South Wales semi-arid Western Division. Hot summers, no defined seasonal rainfall pattern and regular droughts were the norm. When cropping was superimposed, it could be a harsh landscape. 'Yeah,' Bruce recalled, 'it was standard to see ploughed or bare and flogged-out dusty paddocks as a kid. That sticks in your mind, and I always hated the droughts because I've got a feeling for animals ... there was always lots of death and suffering. That had an emotional impact.'

He told me that he had an early environmental sensibility: 'The folks built this house. All the wood was milled off the place, and I remember asking Dad, "Oh, do you think I'll be able to do the same when I'm your age?" Typical child question. But he said, "Oh no, the trees are all gone now." And sort of my innate question was, "Oh, well why aren't we planting?" And because there were no answers, those questions just stuck there. But later in life I acted on those impulses. That was the emotional side.'

This aspect of Bruce was also why, when later in life as irrigation water became available for the rich soils out his way, Bruce made a conscious ethical decision not to go into cotton. 'I had a good grasp on what cotton had done in the US and elsewhere in the world – you know, chemicals, water use, general rape and pillage. Didn't want a bar of it,' he said firmly.

At the end of his schooling, Bruce had what he called a 'life-changing' experience: he scored a year in the USA on a Rotary Exchange – matched to live in a country town in southern Minnesota in 1984. 'There were metres and metres of this black alluvial soil,' he recalled. 'Highly fertile, lots of corn, big yields. But you know what the main agricultural lesson for me was? Despite the rich soils and the huge subsidies, there was a lot of farm rationalisation going on. Those blokes were going broke at a much greater rate than over here, and the community was under stress.'

This experience had two significant consequences. The first was a recognition by Bruce of the human capacity not to confront the truth. 'They were just like you find here,' he recalled, 'always blaming prices or the weather.' This fed into his second realisation. 'It's not about our soil, stock or grain prices or our rainfall,' he concluded. 'It's about our management ... about the human propensity to do things.'

Like many Australian kids who did exchanges in American high schools, Bruce returned a confident person, at ease with public speaking. He was also full of questions, but in 1987 he encountered strong adversity: the midst of the 1980s extended drought period and high interest rates, plus disastrous commodity prices. 'We'd over-extended in the previous decade and so had to rationalise. Our equity was negative fifteen per cent. So, Dad sold half the place, back to 3700 acres,' he recalled. 'And I now had to learn the trade of how to swing a shifting spanner.'

Bruce reflected on this and how they clawed their way back. 'That was also life-changing stuff,' he said. 'And, being the youngest, I just went along with the flow. But I know I didn't want to be "just a farmer".' Eventually Bruce remained to farm at Willydah, leaving just him and his father after three staff were released, but with no spare money for a tertiary education.

To adapt, they then moved into direct-drilling when it was introduced, or 'chemical farming' as Bruce calls it. But livestock (mainly sheep) were their mainstay. And so began a slow process towards radical change. I came to realise that it was exactly the same process that had occurred with Darryl Cluff and Colin Seis's innovation of pasture cropping: that a world breakthrough didn't occur on the rich, fat soils of the Darling Downs or the American corn belt, but in tough country and even tougher circumstances, where people are forced to survive on their wits.

As the tough times rolled on and costs kept rising while commodity returns continued falling, Bruce Maynard found his thinking and practice evolving. The next progression was an adaptation of the accepted direct-drilling approach. They wound back the expenses and dropped off a summer chemical fallow and just drilled straight in after one chemical treatment following rain. That is, they retained their grass over the summer (which enhanced soil health – evidenced by 'no clods after tillage' – and fed more stock). Bruce filed this lesson away: of being freed of the traditional rigidity of a summer fallow and bare, cooked and often stock-pounded soil. 'But,' he said, 'though a cheap cropping operation, all the same it was still full ground disturbance with tines in a one-pass operation.'

At the same time, they shifted towards growing more oats for their stock in addition to a single-use wheat crop. And so the dual-purpose oats gelled with the 'no clods' as more food for thought.

The next big jump in Bruce's thinking came in 1990 when his family did a whole-farm plan with the New South Wales Soil Conservation Service. 'That initiated a whole cascade of things,' said Bruce, 'because with that planning came a heap of videos of the Potter Farm Planning Scheme [see Chapter 16]. There was about eight hours of viewing, but I must have watched them a dozen times over six weeks.' Many things now began to gel for Bruce: from the soil-mapping to 'whole-farm' strategic planning thinking; to the use of land classes and significance of slope and other factors; to the importance of trees and vegetation corridors and their implications for productivity; and much more. What particularly hit him was the exhortation to 'take a hundred-year view'.

By now, Bruce realised he was a holistic and not reductionist thinker. 'The Potter Farm videos allowed me to dream,' he said. 'It's all about timing, you know. All these factors fitted together like teeth in cogs. It meshed, and it was a case of "when the student's ready the master appears".' So, through the early 1990s, Bruce began implementing a whole-farm plan, and it led to him viewing the farm as an integrated unit where key landscape functions were the basic drivers of farm and economic health. Nevertheless, he eventually realised the Potter approach was still limited, and he was still frustrated. 'We weren't stepping that fast towards our long-term plan,' he recalled. 'There was something missing.'

Then, at this very moment of frustration, came the next big jump. Bruce attended a life-changing holistic grazing management course run by Stan Parsons for RCS (Resource Consulting Services) at Dubbo in 1994. 'The light bulbs went on because I was ready for it,' said Bruce. 'It was a tipping point, because Parsons was particularly good on the business side.' So, in now understanding how a holistic grazing regime using ruminant animals could profitably regenerate landscape function, Bruce had a tool that would enable the world breakthrough of no-kill cropping. Aiding his evolving thinking was his wide reading, including of Allan Savory's *Holistic Resource Management*.

Surprisingly, Bruce recalled the next light-bulb event as positive. 'The best thing to happen to me after that course was we marched into the two horrible drought years of '94 and '95,' he said. 'At one stage, I destocked to where we only had thirty cows on the place.' That forced him to think through the principles he'd learnt on the RCS course. He recalled, 'That taught me to make decisions and make things happen, and to be incredibly flexible and responsive. And that's because you can't plan too much. Nothing's predictable out here.' In short, Bruce was shedding his 'European DNA', with its four safe, predictable seasons, for an Australian-adapted mind.

In the year after the ending of the drought in 1996, without realising it, Bruce was on the threshold of finding the last key to the cropping puzzle. 'It all came from the positive response to my new grazing approach,' he recalled. 'We were getting results, even in a tough year: more diversity, our stocking rate improved, and yet by direct-drilling into those paddocks we were taking them back down to ground zero.' So, like Cortés when he landed in Mexico, Bruce (who was now taking over from his father) decided to scuttle his boats and commit the crew (i.e. himself, family and a reluctant father) to a holistic grazing regime as their key farming strategy. 'So I stepped off the merry-go-round,' he said, 'and we sold all our farming gear, leaving us without an answer to go to.'

In the meantime, Bruce rebuilt an old combine [seed drill], this time with fine coulter-disc blades on it (blade-like small discs that slice through soil like a knife through butter), effecting a minimum amount of soil disturbance compared with the brutal intrusion and disruption of a tine, let alone a disc-plough. Then came the key moment of breakthrough in January 1996 – at the peak of summer, with dry soil and everything dormant.

'So we put together this machine,' recalled Bruce, 'and we hooked it up and gave it a couple of runs around the paddock to see how this new-fangled unit would go. And there was a father-and-son conversation. I threw a bag of oats in the seed-box and said to Dad, "It's putting the seed in, why don't we start sowing the crop?" It was a quiet conversation, as traditionally everyone only sowed after rain. Dad walked off. "Don't be so damned stupid," he muttered. But that was the light-bulb moment for me. It was in a native grass paddock, nothing special, near one of the sheds. And it worked. It left the surface undisturbed. Everything was still there. We weren't turning over the soil or anything – just popping the seed gently in the ground when it was dry, and the paddock was still available to graze.'

Other benefits soon became obvious, such as being able to sow dry and not frenetically race around spraying and sowing after rain, and also only using simple machinery. So while it then took a few years of tinkering, refinement and further thinking, that January 1996 was when the milestone breakthrough occurred – when no-kill cropping was born.

'As I gained some experience,' said Bruce, 'so came a lot of understanding and further refinement.' He simplified his farming system to five basic principles: principles that ensured minimum soil disturbance but enhanced the natural function of diverse grasslands. These were: (1) sowing is done dry; (2) use of coulter-disc sowing equipment (no tines) in pursuit of zero disturbance; (3) no fertiliser; (4) no herbicides or pesticides (both (3) and (4) meaning no simplification of soil and other biology); and (5) use of an ecologically enhancing, long-term holistic grazing regime. Cropping had thus been simplified to its barest essentials so as to enable nature and her self-organising functions to be the key player in landscape regeneration.

Oats was Bruce's preferred crop, and he, Colin Seis and a number of scientists suspect the extra sugars that oats release in the soil may have a synergistic effect on soil microbes in enhancing ecological function and plant recruitment. The advantages of no-kill cropping were manifold. First, vastly reduced risk due to much lower input costs (including of fuel). Less risk also accrued because if the crop failed because of seasonal factors, Bruce had still gained grazing off the oats while preserving his grassland, not to mention retained moisture at the start because the oats had been planted dry before the first rains. In addition, the crop–grassland–holistic grazing animal synergy increased dry matter production and thus, in the long run, also his returns and operating profit.

Furthermore, as Australia seems likely to enter increasingly unreliable climatic times in the face of global warming, a low-cost flexible system that enhances natural grasslands adapted to such a climate thus offers an increasingly viable alternative cropping and grazing option. But overall, and for similar reasons to that of pasture cropping, the non-disturbance nature of cropping with a dormant native grassland had huge ecological benefits in terms of enhancing the key landscape functions: of the solar-energy, water and soil-mineral cycles, and thus of dynamic-ecosystems and biodiversity functions. A key point of difference between Maynard's no-kill cropping and pasture cropping is that Maynard's system does not rely on dormant C_4 grasses nor any mix of C_3/C_4 grasses, just on working with whatever grassland components are there in variable seasons.

In a good year, Bruce Maynard and others using no-kill cropping have been able to gain acceptable crop yields. On average, however, in any given district these are one-quarter to one-third of high-input, conventional, industrial-agriculture yields. Given their low input costs, though, this makes them competitive with traditional cropping but with far less risk and much greater resilience in dry or drought seasons. In this respect, Bruce's flexible and opportunistic no-kill cropping approach is more akin to that of Australia's Indigenous grassland grain harvesting and semi-cultivation pre-1788 than to modern, aggressive monocultural cropping.

As Bruce had explained earlier, out on the edge of the rangelands (and as this ecological zone most likely spreads east with increasing climate change), 'you need to be flexible and responsive, as nothing's predictable out here'. That is why, with his holistic grazing, he trades stock and doesn't focus on the investment of a breeding operation. His cropping approach matches this. Rangeland ecologist David Freudenberger, who has studied Bruce's cropping operation, calls it 'Truly Australian'. In discussions with Bruce, they have likened no-kill cropping to the adaptive reproductive biology of the kangaroo: 'Low investment (a ten-gram joey at birth); flexible (mother can eject joey, cut off milk supply or pause pregnancy at any time); and opportunistic – she is already pregnant the day it rains!'[17]

Moreover, this new, ecologically based cropping approach has wider application beyond the fifty million acres (twenty million hectares) currently used for annual cropping in Australia. Farmers like myself in cold, dry winter climates, or grassland pastoralists seeking specific winter forage and cover crops, or to plant a greater diversity of grasses, now have a tool that could increase enterprise options and production without doing ecological damage. Worldwide, no-kill cropping has huge implications given that currently a massive amount of the world's food is grown on marginal country with a mix of grazing and cropping: and especially in places such as Africa and parts of Asia.

Bruce sees no-kill cropping and pasture cropping as complementary and not competitive. First, Bruce pointed out this is because 'they are the only two cropping methods that sow into grasslands and shrublands. Every other cropping method replaces grasslands and shrublands. And this is a huge turnaround. We're going against 7000 years of practice.'

Bruce also sees wider human health and marketing benefits in his approach. This resulted from his recognition of a massive shortfall (upwards of forty per cent) between the demand for pure, 'organic' grains such as oats and the supply. Bruce realised that no-kill cropping was the only truly organic cropping method around, apart from biodynamic agriculture. So he was setting up a business model and company for the training and extension of no-kill-cropping farmers in different regions, and so as to build consistent volume of organic grains for an increasing higher-value market. This approach and its brand is being called Grassland Grains, and it has enormous positive global implications. Meanwhile, Bruce farms with his wife, Roz, and their three children, while he consults, speaks and travels widely – both nationally and internationally.

In the end, as I finished my yarn with Bruce over the kitchen table, and after inspecting a farm of over 200,000 carbon-credit planted trees and more than 300,000 old-man saltbush shrubs (another 'buffer' approach for a tough climate, and which deliver a 'protein haystack') in alleyway and spiral configurations, plus mobs of 1500 cattle in grazing cells, I decided it was time to hit the road and let my head un-scramble. In his summation as we finished our coffees, Bruce said, 'Yeah, all this was a big change. And I guess we were ready to snowball. I liken our change to the snowball coming down the hill. Little things first, then as we went along, we're more able to be radical or to contemplate radical things and gather them up.'

As we shook hands beside my ute, Bruce paused. 'But you know,' he said, 'I call myself the "lazy farmer". I think it's a good aim to be lazy. To do things simpler, with nature. That frees up thinking time too.'

As I drove off and headed for home, I came to the conclusion that Darryl Cluff's description of this radical new agriculture was sufficiently oxymoronic to sum up the scrambled state of my brain: these new methods and thinking did involve a type of 'farming without farming'.

What is clear is that Cluff, Seis and Maynard brought about two of the most significant advances in broad-scale agriculture since domestication many millennia ago. What they have separately delivered is a universal approach to growing fibre, food and other crops that combines cropping and grazing into one land-management system where each benefits the other synergistically and symbiotically – as well as ecologically, financially and socially. Significantly, elements of these respective ecological farming approaches have now spread to Europe, Africa, and North and South America, as well as to most states in Australia.

On reflection, I realised that these innovating farmers had achieved for broad-acre grain production what Wes Jackson and his team were attempting at The Land Institute. While they had not evolved a system to consistently harvest grain off perennial native plants, they were producing grain off perennial grasslands, where the breakthrough tool was a holistic grazing regime.

I therefore don't believe it is stretching a long bow to conclude that this group of three farmers in the central-west of New South Wales have triggered the next major agricultural tillage revolution, following the inception of agriculture. But this time there is a key difference: that of positive, regenerative agroecological transformation. And the reason is these two new regenerative agricultural tillage practices are realigned to nature and her self-organising principles.

In some respects, therefore, these three innovators (and indeed a host of others whom I investigate in this book) have returned us to the Eden of the Fertile Crescent at the first stages of the Holocene after the last ice age. The challenge they are throwing down to the rest of humanity (as is also the case with Wes Jackson) is that they are daring us to turn back the clock so as to go forward, to dare to have managed ecosystems that attempt to mimic the natural, self-organising properties of a healthy dynamic ecosystem whose landscape functions are enabled to be expressed. And as we will see in Chapter 17 with Di and Ian Haggerty's extraordinary revolution on the other side of the Australian continent, this thinking back to first ecological principles is ongoing.

As I finished the final draft of this chapter, Bruce Maynard sent me a photo of his sixteen-year-old son Liam standing in a head-high crop of ripening grain. There is no bare ground in sight – just a diverse profusion of crop and native grassland. Bruce explained that Liam, entirely on his own, had planted, grown and harvested the crop 'without harming the ecology, using no fertilizers or sprays of any kind and with a tenth of the fuel use of conventional methods'.

Studying the photo I suddenly realised the scene symbolically represented much more than a new ecological cropping approach. Here was a sixteen-year-old youth, using self-modified, second-hand machinery, successfully raising a crop without any industrial inputs except a small amount of diesel or petrol (as in the soft soils, the seeder can be pulled by a farm petrol ute). That is, in combination with their own seed selection of grains suited to their landscape, the Maynards and others like them (and as they also seek organic and health grain markets) have broken free of the multi-national corporate industrial domination of the global grain-growing industry. Like the hundreds of millions of small farmers who have joined La Via Campesina, they have radically cut themselves loose of the behemoth that is taking Earth further into the Anthropocene.

That picture of a youth and his grain crop encapsulates a profound promise for our future.

CHAPTER 11

Dancing under the Moon

For more than 99 per cent of human history, the world was enchanted and man saw himself as an integral part of it. The complete reversal of this perception in a mere four hundred years or so has destroyed the continuity of the human experience and the integrity of the human psyche. It has very nearly wrecked the planet as well. The only hope ... lies in a re enchantment of the world.

MORRIS BERMAN, *The Reenchantment of the World*[1]

It is early March and the season is turning. This morning, I take the sheepdogs for a dawn walk. Below me, a long bank of fog fills the Bobundara valley, and an apricot cloud layer merges with soft azure on the eastern horizon. The air is cool on my hands and face, the grassland netted by hundreds of thousands of cobwebs strung across grass and fence lines, and laden with dew. Capping it all, a quarter way up the sky is the thinnest sliver of a crescent moon.

Besides the early golden and crimson tinges on exotic tree leaves, it is the native creatures and vegetation who are the harbingers of this seasonal change. The raucous, galloping songs of white-eared honeyeaters, down from the bush, echo in the garden. Later in the morning, currawongs will arrive to forage with mournful calling. And stolid as always are resident white-backed magpies – five pairs in our garden and surrounding trees and cottage gardens, generations holding the same territory in joined association with five generations of my family. Soon, however, their peace will be disturbed by bands of wandering white-winged choughs, who annually extend their home range down from the bush at this time. The resulting strafing battles are a repeat of internecine warfare that has gone on long before Cain and Abel were ever thought of.

Soon I turn for home, pausing to admire the gorgeous rose, crimsons and scarlets of candlebark trunks as they shed their bark covering to dress up for the coming winter. And I stoop to examine a group of gruggly bushes, now laden with creamy white–capped purple berries: food for insect, bird, small reptile and mammal.

Finally, I pass the now disused milking shed with its two cow bales either side of the milkers' calf pen. And memories come flooding back of battles of wits with rogue milker-calves dodging the evening pen-up; of my struggles to learn the craft under a German master; and of me later running Jersey cows so our children were raised on rich, creamy milk and large bowls of thick yoghurt that had condensed overnight beside a slow-combustion wood stove. Inevitably my thoughts dwell on my childhood.

When I was aged three, my father engaged as station hands a family who had recently emigrated from war-torn Germany: Willem and Annie Buggermann. Still in the Bonegilla transit camp near Albury, and in broken English, they had answered my father's advertisement in *The Land* newspaper. They would become an integral pattern in the fabric of my life. My mother died eighteen months after their arrival, and so I spent a lot of time with them, immersed in the German language, customs and habits. Luckily for me, their two children at the time, though older, became playmates. The older boy, Holger, still wore leather Bavarian short pants, and the girl, Rosita, beautifully patterned German wool cardigans. Their kitchen was full of strange foods and aromas that I came to love: such as pickled gherkins, sauerkraut and variously shaped, bitingly flavoured salamis and other cured sausages and cheeses.

Moreover, Willem had 'green thumbs', having brought the skills, knowledge and empathy for vegetable and food growing from old Saxony. His vegie garden – in which he quietly and contentedly worked of an evening and on weekends – was a delight of varied and constantly rotating produce. In addition, when the Snowy Mountains Scheme came to Cooma later in the 1950s, I would go to town with the Buggermanns and meet another social group: German-speaking people in the Austrian delicatessen, with shelves stocked high with exotic-tasting pickles, cheeses, sausages, salamis and delicate sweets.

Therefore, the key influences on my early food experiences were not just self-sufficient practices typical of rural Australia in the 1950s but also strongly Germanic. Only later would I come to appreciate how significant the German heritage and influence was in reawakening an organic and regenerative farming tradition both globally and in Australia.

We saw in Chapter 2 that an ancient, Earth-empathic and nurturing Organic world view was overthrown by the Mechanical mind. However, we also saw that this Mechanical mind always had its challengers. Early on, this was epitomised by the eighteenth- and nineteenth-century Romantic movement. It was out of this 'challenge', via Rudolf Steiner, that there arose in the early 1920s the theory and practices of biodynamics: the first modern retort to a mechanistic and reductionist industrial agriculture.

Biodynamics is the practice of viewing soil health, vegetation and livestock management holistically, where the farm is seen as a living, breathing organism. There is particular emphasis on generating healthy soils, including through harnessing and channelling what might be called spiritual and mystical energies emanating from the cosmos. Fundamentally, however, biodynamics involves a prime focus on enhancing soil health – particularly its microbiology. This is achieved by various methods, most notably through specially made, farm-specific biodynamic – or 'BD' – 'preparations'.

Biodynamics has its origins deep in the ancient soils, forests and mythology of old Europe – and specifically of Germany. Its founder, the Austrian philosopher Rudolf Steiner (1861–1925), was strongly influenced by Germany's leading Romantic, Johann Wolfgang von Goethe (1749–1832). And Goethe himself had been strongly influenced by the lost world of ancient Germany: a world steeped in ancient, peasant forest lore and pagan animism.[2]

Lawyer, writer, scientist, artist and statesman, Goethe was a true polymath and intellectual giant whose influence and creative works live on into the present. He referred to nature as 'The Great Mother',[3] and his intellectual approach – which he called a 'Phenomenology of Nature' – was to strive to offer a new way of 'seeing'. As Goethe said, modern humans had lost the capacity to perceive nature but could regain such skills. Therefore,

'Each phenomenon in nature, rightly observed, wakens in us a new organ of inner understanding.' In the words of Goethe-specialist David Seamon, 'Goethian science provides a rich, intuitive approach to meeting nature and discovering patterns and relationships that are not only stimulating intellectually but also satisfying emotionally and spiritually.'[4]

Rudolf Steiner did his doctorate on Goethe's scientific writings, and he immersed himself in Goethe's work for many years. It is no surprise, therefore, that Steiner was strongly influenced by the German genius and was himself open to non-scientific, non-reductionist phenomena. These included such things as cosmic energy forces impacting humans, plants, animals and other organisms (soil microbes, for instance). It was this openness that led Steiner to pioneer the first modern challenge to industrial agriculture. This comprised an attempt to develop a modern version of an ancient and truly 'organic', Earth- and nature-empathic approach that re-engaged with 'Mother Earth' and thus challenged 'mechanism' at the same time.

The son of an Austrian railway official, Steiner grew up among peasants in a rural society and so was acutely aware of an 'out there' spiritual world. Following his study of Goethe's work, by the early 1900s, after involvement in the German Theosophical Society, he set himself up as the leader of a movement of spiritual renewal called Anthroposophy (meaning 'wisdom of the human being').[5] Through this, Steiner aimed to try to reform life in the key areas of art and architecture, education, politics, economics, medicine, agriculture, and indeed Christianity.[6]

No matter how fantastic some of Steiner's theory sounds, and however dubious some of the foundations of his theory body, the Steiner philosophy has been found to have some effect and success in such areas as dance, art, architecture, education (e.g. worldwide growth in Steiner schools), medicine and, of course, agriculture. In the latter, the worldwide practitioners of biodynamics (whether they subscribe or not to some of the more extreme elements of Anthroposophy) attain some outstanding results in biological, regenerative, organic agriculture. This is evidenced in a range of areas (as we will see directly), including the healthy quality of their biologically alive soils and resultant food, and in areas such as yield, ongoing fertility, and disease and pest resistance.

What seems to have occurred with Steiner's development of biodynamics is that, as well as anticipating the modern science of healthy soil microbiology, he had combined elements of cosmic energy fields and positive thought energy into a provenly effective 'organic-minded' agriculture. That is, some of this success is clearly due to Steiner's phenomenological approach to tapping into the ancient German organic peasant knowledge, including of that relating to cosmic, planetary and other Earth and stellar energies.

However, it wasn't until 1922 – three years before his death – that Steiner began to unroll his thinking on agriculture and to test his biodynamic 'preparations'. The foundation text of biodynamics is built around a series of lectures Steiner delivered over nine days in June 1924 at the large Silesian estate of Koberwitz in today's Poland.[7] Much of his thinking grew out of the dangers he saw in the escalating use of industrial synthetic fertilisers and other inputs, especially to a healthy soil biology. He knew also that this would have devastating implications for food nutrient integrity and the health of human and animal consumers of plants.

In its essence, biodynamic agriculture, in the words of modern practitioner Hugh Lovel, 'views each farm as a living individual within the living earth and universe'.[8] As a result, it is focused on producing healthy, living soils and thus healthy, nutrient-rich food for both humans and their domestic animals, along with a thriving, biodiverse overall farm and human ecology. The name 'biodynamic' (derived from the Greek *bios* (life) and *dynamis* (energy)) was invented by and applied to the first group of German farmers whom Steiner worked with and who applied his principles. 'Biodynamics' thus means 'farmers and gardeners working with the energies which create and maintain life'.[9] Russian immigrant and biodynamics propagator in Australia Alex Podolinsky described the essence and context of biodynamics as 'farming within a vast cosmological and earthy environment … one huge cosmic ecology'.[10]

Steiner saw a healthy soil and the delivery of healthy food as the keys to a new agriculture. To enhance healthy soil biology, he laid out a system of farming techniques and biological 'preparations' (or what later came to be called biodynamic 'preps'). Both were designed to restore soil organic matter and improve soil structure. Ehrenfried Pfeiffer (a scientist and one of Steiner's trusted lieutenants, who carried on and spread his work) stated in later years that the aim of biodynamic practices was 'to restore to the soil a balanced system of functions', where soil was seen 'as a living system … a living soil' and its 'microlife'.[11]

Though Steiner died decades before the modern science of microbiology had revealed the vital role of microorganisms, subsequent leading practitioners placed a major focus on building the richest microbial and nutrient-dense humus possible (especially made from cow manures). These biodynamic preps were powerful 'biocatalysts'. They weren't just microbial starters but also contained crucial enzymes, scarce trace elements and other chemicals and organic compounds. Alex Podolinsky called these biodynamic preps 'soil-activator' sprays or 'microbial activators'.[12]

Such work and thinking has been verified today by chemical and microbiological research into healthy soil carbon structures and microbial function and diversity. Presciently, in his third 1924 lecture, Steiner stated that, in modern industrial agriculture carbon, was 'the corpse of what it should have been in "nature's household"', and this was because 'carbon is the carrier of all of nature's formative processes'.[13] But other techniques, such as correct tillage, crop/pasture rotations, cover crops and use of green manures, were also part of the integrated system of biodynamic farming practices. Remarkably, Steiner pre-empted modern microbiological research by four decades in recognising in his seventh lecture of 1924 that soil fungi play a crucial role in accessing desirable soil nutrients and screening out harmful toxins, factors and agents.

Both Steiner's recommended farming practices and biodynamic preparations were rigorously tested by a group of scientists and farmers beginning before, and completed after, his death in 1925 – in Germany, Switzerland, the UK and the USA. Key to many biodynamic techniques was Steiner's recognition of the crucial role of trace elements – often in the most diluted quantities – decades before modern research showed that there was more to agricultural productivity (let alone health) than just massive NPK applications.

Biodynamics, as Steiner conceived it, had wider dimensions. He saw it as a truly holistic approach to agriculture. The 'task of spiritual science with respect to agriculture,' said Steiner in 1924, 'is to observe the macrocosm, the broadest dimensions of the workings of nature, and to understand these workings.'[14] For Steiner, it was not just solar energy that was involved. Both his biodynamic preps and his annual calendar of farming activities were designed to capture the available but unrealised cosmic energy forces (moon, planets and other cosmic or electromagnetic energies) that modern farmers had lost knowledge of.

Around the world today, therefore, biodynamic farmers have moon and planetary charts (based originally on old peasant almanacs) by which they do their planting, harvesting and other farming activities. We know from broad areas of science that cosmic and lunar periodicities and cycles (including ocean tides) have significant and long coevolved effects on living organisms: ranging from plants (for example, on germination, flowering, growth and pollination) to insect navigations and behaviour, coral spawning behaviour, diurnal crab behaviour, and on humans (for example, in such areas as reproductive physiology and menstruation). The same appears to apply to other cosmic energies.[15] As Pfeiffer described, once these energies were captured, the plant could then 'transmit both substance and energy as food'.[16]

Research into use of Steiner's nine biodynamic preps (conducted after his death in the 1920s, '30s and later by such scientists as Ehrenfried Pfeiffer, Guenther Wachsmuth, Eugen and Lili Kolisko, and others) has also revealed significant increases in desirable microbial populations and in trace elements and other factors. There was also a corresponding decrease in harmful or undesirable microorganisms (such as disproportionate amounts of anaerobic bacteria).[17]

In such work and practice, there is a stark contrast between the minute concentrations of microorganisms in the biodynamic preps and the blunt bludgeoning of industrial fertilising practices. In the latter, massive amounts of fertilising chemical agents are known to rapidly swamp or poison the plant's microbial populations and the soil's capacity to handle the overload. The result is that large amounts of industrial nutrients have been found to be either leached out of the soil or rendered inert.

Industrial nitrogen fertilisers are another example of such industrial overkill. We know from Chapter 2 and increasing worldwide evidence that such nitrogen fertilisers (made by the Haber–Bosch process, and now with global applications exceeding one hundred million tonnes) are a key factor in destabilising the global nitrogen cycle in the Anthropocene – through greenhouse gas emissions. Because of the unnatural origin, nature and application of such nitrogenous fertilisers, it is increasingly recognised that in many cases at least thirty per cent of industrial nitrogen applications can be lost (volatised) to the atmosphere or leached, when in the form of nitrates, by rain from the soil. Other forms of denitrification also occur.[18]

A key focus of biodynamic preps, as Pfeiffer concluded, was to free the farmer from a drug-addict-like dependence on industrial fertiliser and chemical companies. That is, said Pfeiffer, 'Here is capital which the farmer has on hand and does not need to buy.' Another aim was for farmers to harvest their own vegetable and grain seeds and retain the original vigour of these. Even more now than in Pfeiffer's time, this flies in the face of both growing multinational control via genetic engineering and the hybrid seed business, and also of a drastic decline in seed quality, vigour, variety, germination rates and nutritional value, and concomitant rising levels of disease susceptibility.[19]

A further challenging element of biodynamic preps, as envisaged by Steiner and practised today, is the positive thought element, or mental 'intent', that should accompany their use and which Steiner believed enhances their efficacy.

It is undeniable that, to a Western-trained scientific mind, some of Steiner's ideas are challenging. Nevertheless, in its pragmatic application (as we will see directly), there are concrete agroecological benefits from this first modern form of regenerative agriculture.

Encapsulating his thinking, Steiner stated in 1924 that not only 'did the soil function as an organ in the wider area of nature's household' (a particularly modern ecological term), but also that 'the whole starry heaven is involved in the growth of plants. We have to know this and start to take it seriously.'[20]

It is not until late July that I return to my office to complete this chapter on biodynamics. Moreover, my location seems a hundred worlds away from the semi-opaque Anthroposophy of Rudolf Steiner and its roots in the dark forests of ancient, animistic Germany. On our farm, we are now in what I call 'silver wattle and mountain duck time', which always coincides with early lambing. Those talismans, the proud mountain ducks, have arrived from who knows where to familiarise themselves with their ancient territory before nesting, while golden blobs of emerging silver wattle flowers stipple our landscape.

By now, I am beginning to understand some of the thinking behind biodynamics. This wasn't always the case. I remember when I was about eight or nine coming home from a family Christmas party. One of my older cousins had married a biodynamic farmer (the practice being quite new in Australia then). I had noticed how he stood out as rather 'shy' and different at the gathering, even in the homespun clothes he wore. On the drive back home, the way my stepmother discussed this bloke you would think he consorted with the devil and cavorted with nymphs and goblins as they danced around a fire under the moon.

The truth, of course, is quite different because not every biodynamic farmer I know subscribes to the deeper, Steiner-centric Anthroposophical world view or 'religion', though they certainly apply the Steiner agricultural principles. And biodynamic practitioners, far from being satanic, were in fact the pioneers of Australia's ecological-agriculture movement.

While a few innovative farmers dabbled in biodynamics in Australia soon after its birth in Europe (Italian immigrant Ernesto Genori, for example, making biodynamic preps in Melbourne as early as 1927), the real fillip to the growth of biodynamics in this country began in 1938 on a small, light-soiled, two-acre farm and bush lot in Castle Cove in Sydney. Owned by Bob and Louise Williams, the farm was called 'Kawana' – a local Aboriginal word for flowers. Louise Williams had grown up on the black soils of the Liverpool Plains, Bob in inner Sydney (though he had spent much time on farms). Bob's daughter Dianna Outhred described her father to me as being a 'seeker' who was passionate about 'care of the land'.[21]

Before marrying, the Williamses had been introduced to Steiner's thinking by 1936. Walter Burley Griffin and Marion Griffin (the American architects who designed Canberra) were particularly influential later – with Marion reading Steiner's *Agricultural Course* (colloquially known as his Agricultural lectures, but contained in his reprinted work under 'Spiritual Foundations for the renewal of agriculture – as in my bibliography') to the Williamses when they were first married. The Williamses then joined the Anthroposophical Society. By 1938 Bob Williams, who had also joined Steiner's Agricultural Experimental Circle at Dornach in Switzerland, had begun making his BD preps with cow manure sourced from a BD farm in the Blue Mountains. After buying Kawana in 1940, the Williamses founded the Biodynamic Information Centre and then became the first suppliers on a regular basis of BD preps to other farmers across Australia. It is clear that the Williamses – from their micro-farm with goats, vegetables, an orchard and reserved native bush section – were the first pioneers of BD in Australia. Thus in 1947 Bob Williams registered the trademark 'biodynamics' in Australia.

Bob Williams continued regularly supplying BD preps to scores of Australian farmers well into the late 1960s, and also provided initial start-up preps to farmers in South Africa and New Zealand. Whilst regularly corresponding with the likes of Ehrenfried Pfeiffer in the USA, Bob Williams mentored many farmers (including large landholders such as Ruby McPherson), and he conducted a huge correspondence while also giving lectures around Australia. Ironically, it was Bob's pursuit of grinding quartz crystal (a key Steiner element) that led him to getting 'miner's disease' or silicosis. But his tradition was carried on by people like Terry Forman and Hugh Williams. Forman went on to found the influential BD newsletter *News Leaf* in the 1980s, and also founded the Biodynamic Farming and Gardening Association.

In time the BD movement in Australia (as with many farming movements) developed its factions and sub-brands. But a man credited with giving it further impetus arrived in Australia in 1952 – his name was Alex Podolinsky.

Of Russian-Ukrainian parents who fled Russia in 1917, Podolinsky was allegedly first tutored in biodynamics in Switzerland by some of the participants of Steiner's 1924 lectures from the late 1930s. He and his family survived the war, and then he escaped to Australia, where, in the early 1950s, he began biodynamic farming near Melbourne on an early BD farm. He subsequently went on to become one of the main global propagators of the modern biodynamic approach, as he developed an extensive and influential client base in Australia and internationally – effectively initiating phase two of BD in Australia on the platform the Williamses had built. He too corresponded with Ehrenfried Pfeiffer in the USA, developed a range of his own biodynamic preps, and helped articulate the integrated practices of biodynamics. On his farms, he also linked biodynamics to other regenerative agriculture methods, such as the use of deep-ripping, chisel ploughs and different forms of rotational grazing.

Known for his deep knowledge but also irascible nature, Podolinsky was one of the co-founders of the Biodynamic Agricultural Association of Australia (BAAA) in the early 1960s, along with people like Allan Earle and Andrew Sargood. By the 1970s and '80s he was training urban gardeners and smallholders, who in turn formed biodynamic gardening associations. By 1980, Podolinsky's BAAA members were allegedly farming nearly nine million acres (3.6 million hectares) in Australia, producing a wide variety of agricultural products, and on farms ranging from small market gardens to 9880-acre (4000-hectare) wheat and sheep properties. By this time, he was also teaching internationally.[22]

Due to the early work of the Williamses and their protégés – and later Podolinsky and those who became leaders and teachers after him, such as Terry Forman and Hamish McKay (of Biodynamics 2024) – Australia today is the world leader in recognised organic agriculture, with thirty million certified organic agriculture acres (twelve million hectares) – or thirty-two per cent of the world's total organic acreage.

And the proof is in the eating concerning the practical use of biodynamics. My experience is that the soils of biodynamic farmers are beautifully healthy, vibrant and water absorbent, and as a consequence their crops, vegetables and meat are extraordinarily tasty and healthy: packed with nutrients, trace elements and other factors. Many are certified 'organic' growers because their farming system is ecologically robust and healthy enough to function without weedicides and pesticides. But besides the 'eating proof', there is also a considerable body of research into chemical, biological, soil, food quality and other parameters that now shows biodynamics has major physical and health benefits.

Some research trials have been ongoing for many decades. In such trials in the UK, Europe, North America, Australia and elsewhere, and from a range of results published in over thirty peer-reviewed scientific papers, biodynamics has often proven superior in diverse ways when compared with conventional industrial agriculture. Such areas include: improved soil health activity and structure (more soil organic matter, greater enzyme activity, more efficient soil microbial mass, greater fixation of organic soil carbon, greater soil macro-porosity and structure, low soil bulk density); enhanced soil microbial diversity and function; enhanced crop, pasture and viticulture yields, and superior produce quality and taste (including higher plant sugar or brix levels); and other areas of desirable chemical levels and utilisation, enhanced hormonal impact and suppression of fungal and bacterial plant pathogens, and so on.

One area where biodynamics (like most regenerative agriculture practices) is thus playing a crucial role is in addressing key Anthropocene boundary levels (such as enhancing phosphorus cycling, greater sequestration of carbon dioxide and atmospheric nitrogen, and the fixing of more long-lived soil carbon – thereby reducing greenhouse gas emissions, not to mention lower fossil-fuel emissions).[23]

The reason for such results is that well-managed biodynamic practices enhance all five landscape functions, starting with the soil-mineral cycle (as we will see directly with a farming example).

It was with thoughts of the rich and distinctive taste of potatoes I had bought at our local organic shop in Cooma (so distinctive that they could have been a different vegetable compared with the industrially grown produce labelled as 'potatoes') that one particular morning I set off towards Canberra. I was seeking the producers of those startling potatoes: friends of mine and an intriguing couple – Tobias and Beatrice Koenig. An hour away from us, they are successfully practising biodynamic farming in our generally challenging Monaro environment: an environment that is a far cry from the rich, lush, gentle light and moist climes of Germany where the 'organic' practice evolved. It also seemed even further removed from Germany's ancient contiguous farming history and organic–animist lore and traditions – and from the land and society there where both had grown up.

I soon arrived at the gate of the Koenigs' farm, Ingelara, which represented a green oasis in the midst of frosted, grey-white and love-grass-infested neighbouring paddocks. Their house rested on a hill and was surrounded by small paddocks. Below sat a large, willow-thronged dam and a cluster of other houses and sheds. The backdrop was a semicircle of steep, eucalypt-clad hills. Further to the west was the high, blue-grey western tier of the Murrumbidgee escarpment – the famous river buried somewhere deep below in the shadows.

The Koenigs emerged from a patch of sunflowers and rhubarb. Both were fit, lean as whipcord and tanned by the Australian sun. They exuded health and seemed at peace with each other and the world. But while there might appear a neat circularity in two more Germans bringing Steiner and biodynamic beliefs to Australia, their journey had nevertheless involved a tough apprenticeship.

Tobias was born in Düsseldorf, in the Neder Rhine (the lower Rhine region), while Beatrice was born further south in the Black Forest. Both were trained in biodynamics in Germany and in Switzerland, and then gained extensive experience on farms in Germany. Therefore their cultural milieu was entirely foreign to that in Australia, because Anthroposophy and biodynamics comprised a holistic approach to society, including an integration between farming and philosophy. It wasn't just about growing healthy food. Education, architecture, music, medicine, caring for the handicapped, spirituality, philosophy and overall human health were all integrated in a society that already had a large biodynamic movement.

Nevertheless, following a visit to Australia in 1990, Tobias and Beatrice then moved here in 1994: leasing a farm near Narrandera. It proved a disaster. The farm was giant in size by German standards, and they walked into an Australian drought. The Koenigs still carry vivid memories of regular dust-storms sweeping through. On termination of their lease, a tough period followed, until eventually the Koenigs met regenerative farmer Michael Coughlan, who had a farm at Narrandera. Things then slowly began to look up.

Valuing the Koenigs' skills, Coughlan employed them to grow biodynamic wheat. In the process, while gaining other experience, Tobias also worked with Alex Podolinsky. Then, after ten years at Narrandera, the Koenigs finally moved to their current farm south of Canberra. Their decade of hard yards had prepared their minds for both the Australian environment and the dearth of Earth-empathic thinking that rules under the economic-rationalist world of Australian farming.

It was now 2004, and the Koenigs had teamed up with fellow biodynamic farmer Peter Bottemly to buy that parcel of land near Canberra. By this stage, the Australian biodynamic fraternity had realised that in the Koenigs they had gained serious depth and knowledge for their small community. Beatrice began teaching at a Steiner school in Canberra. Tobias, as he began biodynamic farming in his own right again, also became a biodynamic inspector and certifier for Australian Certified Organic. Today, the Koenigs sell all of their healthy produce either to organic outlets like ours in Cooma, or to the Canberra Farmers' Market, where they and their produce are well known and popular.

In between a simple but beautifully cooked meal made from their intensely flavoured and healthy produce, and shared with rich, freshly brewed coffee, we talked in a sunny living room. Here, a warm stove crackled with wood, baskets of apples and other produce lay neatly arranged, and Beatrice's own attractive artworks adorned the walls. Later, we would walk the farm.

Their farm was divided into four sections. A mob of up to a hundred mixed, largely black cattle rotationally grazed the larger paddocks. The resultant intensity of animal impact and adequate rest was slowly regenerating some love-grass-infested areas. A second section comprised thickly and attractively timbered bush areas, and finally there were the more intense sections. A few sows and their litters were raised in rested sections here, while chooks and geese wandered about – the geese loudly complaining at us interrupting their business.

When not helping Tobias and Laurence (their son who had just come home) at key times with the larger volume cash crops, and since giving up her teaching, Beatrice mainly worked in a small market-garden section: their third land component. Here, she grew beautiful hilled perennial rhubarb, pumpkins and other vegetables. Lost in a natural world and the rhythm of work, and tuned to the sounds of bees, other insects and birdsong, Beatrice had planted alternate rows of flowering plants and crops such as millet and sunflowers for their aesthetics but also as nectar for the bees and the birds. Later in the season, larger areas of mustard, ryecorn, pumpkins and sunflowers would be left for the pigs, bulls, geese, chooks and other creatures. There were also rows of beet, kale, shallots, lettuce and other foods.

Finally, there was the fourth and main cash-crop section, which Tobias, and now Laurence, largely worked on. This section involved a carefully planned, seven-year rotating cycle of crops and pasture, which were designed for food production, soil enrichment, pest-control breaks and fallows. All were tended by the application of various biodynamic preps specific to the timings of the calendar, sun, moon and planets. Overall, it was a highly planned annual process, adjusted by season, acute observation and resultant reaction, and an intuitive feel for the health of animals, humans and plants. I quickly realised that their entire operation was based upon the regeneration of the four biogeochemical landscape functions, with the entire process, of course, under the impact of a deeply thought and felt world view: the fifth or human–social function.

The process for this fourth cash-crop section started with a green manure crop of diverse species, planted in the autumn, then turned-in in October.

Depending on the season, the green manure crop could be grazed by pigs and/or cows. Then potatoes were planted – to begin finishing eleven or so months later, in mid-September.

All planting was carefully aligned to the phases of the moon, though Tobias was enough of a pragmatist to adjust this to Australian conditions and the degree of soil moisture. Mental intent injected into all operations was crucial.

For the green manure, the Koenigs made their own biodynamic preps – what they called 'Cow Pat Pit'. The cow pats for this were collected from lactating cows, as they were rich in calcium: the recipe being based on that of famous German biodynamic innovator Maria Thun. Then ground-up eggshells, rock dust and other manure preps were mixed in. Cow Pat Pit was placed not in a cow's horn, as with many biodynamic preps (a practice said to harvest cosmic energy in the soil), but just in the ground, and it emerged after eight to twelve weeks as very rich soil full of microorganisms. Out of this was made a compost prep, which was hugely diluted and then sprayed onto the crops: what was called a 'compost tea'.

In addition to Cow Pat Pit, the Koenigs used the other typical biodynamic preps. Cowhorn 500 they made themselves, and others they accessed within the local biodynamic community. The Cowhorn 500 was vigorously stirred in two directions to create a vortex, as biodynamic farmers were confident that, in addition to adding oxygen, this imprinted the memory of cosmic energies mixed into the water. As biodynamic 500 impelled energy, it needed to be sprayed in the afternoon when the earth was believed to be inhaling. All the different preps were sprayed out regularly – sometimes every eight to twelve weeks.

Once the spuds were harvested, the paddock was ready to be worked again, helped by the rooting of pigs. Then came a summer green crop, perhaps a mixture of millet and legumes, but including soybeans, clover, peas, sunflowers, buckwheat and turnips.

Finally, after two to three years of cash crops came pasture: a mix of a deep-rooted legume such as lucerne, plus perennial coxfoot and so on. In the autumn, ryecorn, oats or ryegrass could be drilled into the lucerne. The pasture remained under grazing for four to five years, building soil, before the cash-crop cycle began again. With potatoes and other crops involved in the cycle, the pasture also provided an excellent disease-break.

With mixed success, the Koenigs had also tried 'peppering': an ancient German and biodynamic approach where the skins or shells of pest creatures (such as kangaroos, snails, grasshoppers) and weed seeds are burnt and then the ashes placed in strategic positions around the farm. This was known in some instances to deter pests. It could work for animals, especially if an energy broadcaster was used and powerful intent projected, but despite Steiner claiming it took four years for weed vitality to decrease, the jury was still out in Australia on perennial grass weeds such as love-grass and serrated tussock.

The integration of the four main land-use sections on the Koenigs' farm, with their diverse practices that all work in synergy over a seven-year rotational cycle, is a clear demonstration of the clever utilisation of all five landscape functions.

The regeneration of active soil biology and diversity encourages all sorts of organisms to inhabit and thus condition the soil, leading to deeper, more biologically active, water-absorbent soils. The micro-life – bacteria and fungi particularly – in a self-organising process plays an active role in accessing required and scarce soil nutrients (the mineral-cycling component), while screening toxins and preventing disease. In addition, through fostering healthy soil biology, the natural system also looks after and regenerates the soil's chemistry and physics – which modern institutional soil lecturers are so hung up on. Each of the other landscape functions are thereby enhanced – influenced by a philosophical commitment (the 'human–social') to the principles of biodynamics.

In short, at Ingelara, the Koenigs are clearly enhancing each landscape function: healthy soils rich in microbial and other life (the soil-mineral function); lots of green leaves (solar); heaps of diverse, deep roots (water cycle); perennial plant cover, with nearby native bush to enhance biodiversity and ecosystem dynamics; the integration of diversity of plants and well-loved animals; and the fact they are part of a human community of supporters (also a component of the human–social). Their story also illustrates that the human–social function needs to be *long*: many generations of living with the land. It also takes time to build up a trusted and solid reputation for biodynamic produce, as the Koenigs have now achieved: yet another element of the human–social function.

By now, after twenty-two years in Australia, the Koenigs had begun to fully appreciate the stark climatic, physical and social differences between Australia and Germany. This included the way biodynamics was practised. They had found that Australian farmers operated less collaboratively and more in isolation. Perhaps this was due to Australia's vast distances, but it seems to me that a contributing factor was Australia's young, non-Indigenous and European history. An arguable result of this was an absence of deep physical, emotional, cultural and spiritual embeddedness in the land among post-1788 immigrant people (combined with their prevailing exploitative, hard, mercantile approach to farming). By contrast, in the biodynamics fraternity in Germany, farmers collaborated more in groups, meeting each other every month, sharing and making preps at the right calendar times, plus exchanging tools and so on.

Beatrice – a deep-thinking, sensitive and spiritual soul – still ached for the deeper connection to earth, land and cosmos that she found lacking in Australia. 'We are missing the "middle way",' she said, 'which also involves the spiritual and the spiritual in nature.' Beatrice paused. This was deep territory, rarely articulated but fundamental to their farm. 'In the old days in Germany,' she continued, 'the farmer walked his fields. They were humble; they prayed with humility. It was a Christian culture. Farming then was called "husbandry": all about how you walked and lived with your land and animals. Today's production agriculture is totally different.'

After walking through paddocks and crops close to the house, we were back in their living room for more coffee, freshly broken walnuts and crisp, sweet apple slices. Ruminating on our walking and talking, and to the sound of galahs squabbling on the millet stand outside, Tobias explained how they were still constantly learning: via field days, books and journals, and the internet. I noticed an issue of the journal *PLENTY: Biodynamic Living in Australia* lying on the table. The American cropping innovator Gabe Brown from the American prairies interested them with his remarkable yields from healthy soils and smart cover crops and enterprise variety stacking. They wished to travel there, the Koenigs mused, but had just used any spare cash on a visit to Switzerland, Germany and France, studying organic and biodynamic garlic production. 'You know,' said Tobias suddenly, 'we've done BD farming for forty years and we're just beginning.

We're not at the end of our learning curve: give us another ten years and we'll know more.'

Finally it was time for me to go, and we casually strolled through apricot and apple trees in the house garden, commenting on how our respective fruit trees behaved and performed last season. The galahs were still squabble-chatting in the millet. Pigs snuffled in the corn stalks, and in the midst of a stand of sunflowers a group of three contented bulls poked about, rustling the drying stalks and leaves. Tobias explained how he watched a bull the other day take five minutes to work out how to eat a sunflower head as big as our hats, by wrapping his large tongue around it. 'Notice how they focus on their stomach when concentrating on eating,' he mused. 'They go internal, become self-inward focused. It's a sort of meditation. And you can see it in their eyes.' Bovine they might be, but that didn't make them stubborn or dumb, which was a human interpretation.

The incident was a nice conclusion to my visit, for Steiner saw early on the importance of bovine digestion. He was aware of their inward, stomach-focused nature, and the crucial role that bovine microorganisms and digestion played in enhancing soil fertility. That is, grazing animals – especially ruminants, but including kangaroos and others – are walking composters. Their guts, as my friend and scientific colleague Dr David Freudenberger says, comprise supercharged garbage bins that daily convert tough, fibrous plant material into rich compost that is immediately colonised by dung beetles, worms, micro-bugs and other organisms to be incorporated back into the good earth. Out of this process came Steiner's preps through using buried cow horns, and his understanding of how humans can play a crucial role working with animals and natural forces to regenerate the Earth and its people.

As I drove home in the late afternoon, I reflected on how the Koenigs were adapting the first modern regenerative agriculture techniques to Australian conditions. I pondered their hard work, upright characters, humour and visible health. And I concluded that Steiner was right in respect of his concept that a healthy biodynamic farm, via the thinking, acts, beliefs and character of the people working and living on it, develops its own personality: that it truly becomes a unique living organism.

CHAPTER 12

Keep a Green Bough in Your Heart

Keep a green bough in your heart and God will send a singing bird.

It was last summer at the end of a day's shearing, as I was following a mob of shorn merino hoggets up a lane from the shearing shed, when I was reminded of how powerful and fundamental the processes are within natural systems. At the top of the lane in an adjoining paddock of mixed bush, there are a couple of lovely candlebarks, surrounded by blackthorns and native grasses. As I passed these, mooching along behind the sheep with my dogs, I suddenly heard an awful scream coming from the vicinity of a candlebark. I had heard rabbits screaming when caught, but this was different: a piercing sound of sheer terror by some poor creature.

I climbed the fence to investigate, but then stopped. Two metres in front of me was a three-quarter-sized rabbit in the entrance to a burrow. But around it were three coils of a large brown snake, whose head rested on the rabbit as it gazed at me with those dark, glittering and implacable eyes that send a thrill through one's backbone. Clearly it was waiting for its venom to take effect, while the rabbit continued screaming.

I left prey and predator alone, and as I walked back to my dogs and sheep, the words of Tennyson sprang instantly to mind: that nature was 'red in tooth and claw'. However, as I continued slowly following the sheep,

I thought about the incident. Yes, Australia had its share of fierce predators, yet this 'jungle law' (what Herbert Spencer in the nineteenth century coined the 'survival of the fittest') wasn't the only mechanism operating Down Under.

In fact, for life on Earth in general and the process of evolution in particular, collaborative symbiosis and forms of mutualisms[1] are fundamentally integral mechanisms. In their seminal book *Microcosmos*, Lynn Margulis and Dorion Sagan powerfully reveal that, 'in contrast to the usual view of neo-Darwinian evolution as an unmitigated conflict in which only the strong survive', evolution is as much 'symbiotic and interactive'. There is, they concluded, a very 'thin line between evolutionary competition and cooperation'.[2]

Today, as a result of their and others' work, we now know that much of the history and evolution of life is symbiotic: that a lot of the major evolutionary jumps in life on Earth have come from symbiosis. Much of this is via bacteria, given, as Margulis and Sagan reveal, that life on Earth 'has been largely a bacterial phenomenon'. This has involved different forms of life being ingested or captured by others but them then working together to form new organisms or organelles with new, enhanced functions. That is, all present and past life forms (including we humans) are the result of these extraordinary collaborative arrangements.[3]

As Margulis and Sagan challengingly put it, the very fact our own cells maintain the carbon- and hydrogen-rich environment that existed when life began billions of years ago ('a medium of water and salts like the composition of the early seas') reveals that 'we coexist with present-day microbes and harbor remnants of others, symbiotically subsumed within our cells. In this way, the microcosm lives on in us and we in it.' Thus we and all living creatures are nothing but evolutionary mosaics of microscopic life. And it all began with bacteria.[4]

Crucially, due to Australia's ancient history, our leached soils and scarce nutrients have stimulated a whole raft of unique collaborative arrangements to better share such scarce resources. This includes arrangements between birds and plants; bacteria and mycorrhizal fungi and plants; insects, bacteria and fungi; placental mammals, marsupials, microbes and fungi; honey possums or honeyeaters and flowering plants; cooperatively breeding birds, and so on. That is, Australia can boast – via cooperation, symbioses and mutualisms – an extraordinary and long-coevolved variety of organisms working together for mutual survival. This Gondwanan ark truly was and is like no other place on Earth.

It was while driving away after a head-rattling morning with Bruce Maynard that I realised he, Colin Seis, Tim Wright, Ron Watkins and many other regenerative farmers I knew shared a number of key similarities. For a start, all spoke a language laden with ecological terms. When compared to the mainstream, they also thought quite differently about their land. Furthermore, not only did they hold a long-term view as to what they wished to create on their land but also their vision involved a great deal of ecological complexity across their landscapes.

Yes, they were using new forms of regenerative agriculture to make a living, but their management quiver seemed full of many arrows. They each preached and practised diversity in both their landscapes and what they did to and in those landscapes. The first question they asked seemed to be, 'What is best for this landscape?' This had strong echoes with an Australian Indigenous question I frequently hear: 'What can I do for country?' Indeed, I was beginning to realise these questions were now coming together from different ends of the spectrum.

I also recognised one more similarity shared by these leading innovators in a new ecological agriculture: a natural consequence of their approach to farming was an enormous enrichment of biodiversity and thus of a functioning and healthy ecosystem and its community dynamics in their landscape. In turn, this enhanced the resilience of their farming systems: an ability to withstand seasonal fire, pest, economic and other shocks. All such innovators clearly prized this and aimed, by various methods, to strengthen it further.

On thinking about this some more, the reason why became obvious. While Tim Wright had begun focusing on his grass-growing and management (that is, focusing initially on his solar-energy flow), he found that regenerating this aspect had huge and positive benefits on the water, soil-mineral and dynamic-ecosystem aspects of his farm. Likewise Ron Watkins. In helping to restore water function in his landscape, he had found significant knock-on benefits to solar-energy flow, production, soil health and ecological function on his farm. And the same applied to farmers such as the Koenigs, Geoff Brown, Colin Seis and Bruce Maynard. In addressing the soil-mineral aspects of their farms, they likewise reaped enormous benefits in the other three areas of landscape function and processes.

In this and the next two chapters, I will be examining the fourth functional component of the landscape: that of dynamic-ecosystem communities. This is often visible as a rich farm biodiversity (communities of living organisms and life-webs). But it was becoming increasingly clear to me that all of the landscape functions were intimately related, one to the other and indivisibly entwined. That was why I had constantly heard from many of these leading innovators such terms as 'complexity', 'landscape function', 'diversity' and 'biodiversity'.

Any farmer who has done courses relating to holistic planned grazing and read Allan Savory's defining texts on this, or even read more recent textbooks or articles on ecology, will have become aware that the landscape functions discussed above are the foundation to ecosystem health. The converse is that if just one of these aspects is rendered dysfunctional, then landscape degradation and ill health follows. As Allan Savory put it, if a landscape manager modifies any one of these processes, then he/she will 'automatically change all of them in some way because in reality they are only different aspects of the same thing'.[5] Savory uses the analogy that the four landscape functions he articulates are four different windows that allow you to observe the same room: the latter being the broader ecosystem environment. As we have seen, to Savory's model I have added one more landscape functional element: that of the human–social. Importantly, however, as Savory summed up, 'Ultimately these … processes are the foundation that undergirds all human endeavor, all economies, all civilizations, and all life.'[6]

A dynamic ecosystem can be described as a complex community of organisms in any locality that, while being in a state of relative 'balance', is nevertheless in constant flux and interaction – thus dynamic. Many ecologists, including Allan Savory, regard this function – the dynamics of healthy biological communities – as the most vital of all. The way diverse and dynamic healthy biological communities sustain themselves varies with each different environment, characterised by differently marvellous and complex processes.

The detailed attributes of dynamic healthy biological communities will not be dealt with here as they already fill many textbooks. But as we continue on this book journey, I will be touching on a number of the most crucial and general principles that modern systems ecology is coming to understand. These include the fact that all living organisms are adapted to their own specific environment through a process of coevolution, and when they become established in a community they alter their surrounding microenvironment. In turn, they are altered by their environment again.

The result is a community that consequently develops over time as a dynamic whole. To paraphrase John Donne, no organism is an island, entire unto itself. If such a complex, interconnected and interdependent environment is then altered – for example, by inappropriate farming practices – then such actions can have huge and unforeseen consequences. Conversely, when regenerative farming practices enhance dynamic, interacting ecosystems, then they become more resilient as complexity increases.

Linked to this in turn is the fact that most biological activity occurs underground: a crucial point when we consider the role of the key landscape functions. Further connected to this is the importance of another observed principle: that in a healthy, evolving or regenerative landscape or ecosystem the process of change tends to be one of 'succession'. This means both a gradual and often staggered build-up of species diversity and biomass, in combination with changes in the microenvironment. As we have seen with Tim Wright, Colin Seis and Bruce Maynard, due to revived landscape health, long-buried grass seed from valuable grasses that disappeared after white settlement have been brought to the surface by 'ecosystem engineers' (insects, earthworms, reptiles, burrowing mammals et cetera) to germinate and begin to restore diversity and functional integrity in modified landscapes.

Such complex, indivisible ecosystem communities have been described as the 'web of life'. The first to begin fully articulating this was the extraordinary polymath and 'organic' thinker Alexander von Humboldt, who, back before Darwin, realised that 'nature is a living whole' and not a 'dead aggregate'. In the words of his biographer Andrea Wulf, Humboldt believed nature was 'a web of life and a global force', and he was thus 'the first to understand that everything was interwoven as with "a thousand threads"'. In other words, he saw the unity in variety, and that nature was composed of patterns.[7]

Since the days of Humboldt, Darwin and others who forged a pathway in our thinking on the complexities of nature, coevolved ecosystems and such, we now know – as will be demonstrated in this and the next two chapters – just how vital the diversity or biodiversity component of ecosystems is (including species richness). Moreover, more and more work investigating complex levels of ecosystems support this. Pertinent to this book, in the words of a group of forty scientists involved in a famous long-running experiment examining ecosystem processes in Germany that is focussed on grasslands (and known as The Jena Experiment), is mounting evidence 'that biodiversity is a significant driver of ecosystem functioning' – and that 'high diversity communities' tend to provide 'higher and more stable levels of ecosystem functioning'.[8] This conclusion is increasingly supported by not just mounting research but also practical experience in farmers' paddocks.[9]

The problem with conventional industrial agriculture – indeed with agriculture going back ten millennia and more – is that it is about the simplification of ecosystems. This is still the dominant mode of global and Australian agriculture, notwithstanding the fact the examples in this book reveal that a profitable, productive and sustainable agriculture is possible by working *with* and not *against* ecosystem functions.

Hitherto, in exploring a range of remarkable families who have dramatically restored landscape function to their farms, we have seen they all began by identifying the landscape function most out of whack in their landscape and addressing it. In doing so, they found that their actions triggered positive change in the other three biogeochemical functions.

I now turn to examine an intriguing and innovative pair of landscape managers, Richard and Jenny Weatherly, who focused initially on the fourth landscape function: dynamic-ecosystem communities. This focus was triggered by a recognition of the massive post-1830s loss of biodiversity in their region of Victoria's Western District, where they run their 4200-acre (1680-hectare) property Connewarran.

From where I live across the Snowy Mountains, it is a good nine- or ten-hour drive to the Weatherly property near Mortlake. I had known Richard Weatherly for a number of years, because we shared a common interest in merino breeding, ornithology and the corrupted state of wool agri-politics. For the last two hours of my drive from home, after negotiating the famous gold city of Ballarat, I headed towards Skipton (meaning 'sheep town' in Old English) and then generally south-west. From the roadside and across small pieces of remnant vegetation, I could glimpse the privacy of ornate Victorian homesteads and secluded gardens: the equivalent of British aristocracy's grand estates but with names like Mount Elephant, Larra, Mount Fyans, Wooriwyrite, Tandarook, Warrambine and Ercildoun.

This was country that, between 1850 and 1870, had been one of the great sheep walks of the world. At that time, it was also the leading merino sheep stud region, which served as the engine room that helped Australia to become the global giant of a redesigned industrial wool fibre. But then the sheep ate the heart out of the country and its vast swathes of diverse, long-adapted natural grasslands. Consequently, this breeding centre moved to regions further out and the next newly opening frontiers of unexploited natural herbage and grass.

So as I headed south-west and into the catchment of the Hopkins River this day in 2010, I found it hard to visualise what the landscape had looked like pre–white settlement. This was because not only had most of the original vegetation – trees, understorey and grasses – been cleared, ploughed, poisoned or eaten, but also the region was in the early throes of the latest phase of industrial agriculture: of a gigantism where fences, trees and other impediments were making way for massive machines. I particularly noted the elongated spray rigs, their jointed arms like giant science-fiction insects, bouncing across monocultural land as they excreted pink blobs of spray-marker – until I had to wind up the window as the acrid odour of glyphosate dried my palate.

On his third expedition as surveyor general of New South Wales, explorer Major Thomas Mitchell in 1836 renamed the great plains of western Victoria, thus instantly erasing for posterity a long Aboriginal heritage. The word 'explorer' is culturally contextual, as Aboriginal people had walked and fire-stick farmed these lands for tens of millennia until Mitchell (with his cumbersome party of twenty-five men dressed in military garb, plus five bullock drays, fifty bullocks, twelve horses, a flock of sheep and a large cart that carried two whaleboats) came on the scene to popularise the new country. He told his audience 'back home' that the region was so rich 'flocks might be turned out upon its hills or the plough set to work on the plains'. Mitchell later recorded that the land was 'open and available in its present state, for all the purposes of civilized men. We traversed it in two directions with heavy carts, meeting no other obstruction than the softness of the rich soil; and in returning over flowery plains and green hills, fanned by the breezes of early spring, I named this region Australia Felix.'[10]

The 'rich' soil, besides much of it being of volcanic origin, was also the result of extraordinary grassy woodlands, where early settlers such as Niel Black spoke of the ground being spongy and soft, the grass heads brushing his saddle flaps, and the grass 'so rich and close that the horses went down in it like as if they had been treading over a lump of hay'. Indeed so soft was this spongy grassland that Mitchell's cartwheels left such deep ruts they became the track, known as 'The Major's Line', which the early overlanders and settlers then followed, and which remained visible into the twentieth century.[11]

But when Richard and Jenny Weatherly returned in February 1985 to take up and manage half of the family property in this region, they initially took possession of over 4000 acres (1618 hectares) of an almost totally cleared and denuded landscape. This was divided into only eleven paddocks and contained much sour, swampy country and eaten-out and eroded watercourses. There were also the first signs of salt in the Hopkins River that traversed their land. They found no soft soil, no rich grasslands, and indeed virtually no remnant vegetation bar a sick population of aged river red gums that were in the throes of an accelerating die-off. 'It was pretty bleak. We had no shelter on the place and so nowhere to lamb,' Richard recalled. Moreover, unless one ferreted in old station diaries and letter collections or stumbled on a private sketch or painting, Richard Weatherly said it was almost impossible to get an accurate impression of what their country had looked like in the early nineteenth century. So they had to begin from ground zero.

Connewarran was originally taken up in 1839, but, said Richard, 'My great-grandfather William got this place in 1895 and he really struck it lucky and made a lot of money.' However, the next two generations 'enjoyed a pleasurable lifestyle and were very generous', he said, with the result that he discovered on coming home that his father 'had run down the capital'. So, he and Jenny returned to '4000 bare acres, no capital, no sheep, no shearing shed, no sheep yards, and a whacking debt'. And the only sheep he could buy in the midst of a wool boom were a hotchpotch of everybody else's culls. 'The place was unviable, and we were effectively stony broke,' concluded Richard.

But Richard knew he wanted to live on and conduct agriculture on this land. 'I had a very strong recognition of the history of the place,' he said, having grown up on the farm. He pointed out to me an oil painting by family friend Jessie Trail of himself and his sister, when little bigger than toddlers, tadpoling in a creek. Their jars were full of tadpoles of the now rare and endangered growling grass frog. 'And around us was all bare ground,' he said. Reflecting further, he added, 'In fact, I've never seen any portion of this place with its natural vegetation on. Never. And I'm probably one of the very few landowners who has ever sat down and mapped the place and drafted the probable vegetation groups over his place and looked at how they're associated with soil types and things.' This came about after decades of intensive research, plus checking small vegetation clues on roadsides and other remnant patches.

The Weatherlys' house was near the banks of the Hopkins River, and, with two other major creeks that followed fault lines, their watercourses carved through tertiary sedimentation. Such country was once all open woodland, built around magnificent river red gums. Across the Hopkins, the Weatherlys had country that was part of a rich and vast basalt plain. Contemporaries alleged these were only ever grassy plains, but Richard now knew the basalt plains once had a rich diversity of not just grasses but also shrubs and other plants dominated by wattles, casuarinas, allocasuarinas and banksias: of what he called 'Dripping she-oak and blackwood'.

Richard and Jenny Weatherly thus set out on a hard journey – not to recreate the arcadia of the original Australia Felix but at least to bring back to their land vegetation diversity and healthier functioning ecosystems. The result that I saw was an extraordinarily transformed landscape of rich riparian corridors, connected vegetation patches, tree-breaks and plantations across the entire property, with a marvellously diverse flora and fauna: from massive bird numbers, with over 200 species, to repopulated growling grass frogs and platypi, insects of all sorts, and right down to a highly valued mix of landscape-renewing earthworms.

On this journey, Richard and Jenny had been pilloried for their alleged 'green' and radical ideas, and Richard for his enthusiastic range of innovations. To understand why they swam against the strong current of reigning land-management attitudes, and why they persisted in following a hard but ultimately rewarding road of restoring landscape function to

Connewarran, we need to trace the path in particular of Richard's remarkable career: as someone who could be described as a contemporary ecological-naturalist polymath farmer and artist.

From a childhood spent surrounded by plants and animals on his family's farm and beyond, Richard developed first as a brilliant naturalist and bushman, and later as an active farm manager and merino stud-breeder. Besides tadpoling in creeks and showing a keen affinity with nature, he grew up in a unique environment. 'I lived on the banks of probably one of the great duck shoots in the world,' he said. Moreover, Connewarran and district was the location of serious zoological collectors in the early twentieth century, including from Melbourne Zoo, Melbourne Museum and Melbourne University. In addition to keen birders visiting the farm, researcher Charles Barrett did his pioneering work on the platypi at Connewarran, while other leading naturalists worked in the area. Almost inevitably, Richard became interested in nature very early on. He said that as a child he just imbibed a 'great sense of natural history'.

I too had become entranced by ornithology at a young age. My favourite book was a battered, green, hard-cover 1932 third edition of Neville William Cayley's *What Bird Is That?*, with an embossed kookaburra motif inside a question mark on the front cover. I constantly pored over it until I knew every bird's name and was able to tick off the sightings. It turned out that Richard's early favourite bird book was also the Cayley, though he eventually gathered five editions, from his grandfather's stripped-apart first edition on through. 'I can remember at the age of six being really perturbed if I saw something and couldn't put a name to it,' Richard recalled. In putting a name to things (including in his case their Latin names), he noticed them at home, and he 'knew' birds in Cayley 'without having seen any'. In later life, he would rectify this. So, over the next decade he grew as a naturalist. About the same time, when aged eight or nine, Richard also became intensely interested in bird drawings and art.

At the end of his rather idiosyncratic school years – where an interest in art history and nature continued to blossom – Richard left Australia in 1966 for England. He would not return for six formative years. Like a lot of Western District and South Australian squatters' sons, Richard was sent for an obligatory experience to Cambridge University. 'I read history because it was a good mind-trainer and because I had no bloody clue of what I wanted to do,' he said. But he soon lost interest in history, for by now art, and particularly its ornithological aspects, attracted him more. By chance, his wooden carvings and drawings caught the eye of intrigued art galleries, and he realised he could make a quid through artistic pursuits. In both England and then Rhodesia, Richard trained with leading bird artists and naturalists, while learning to hunt and track: to begin thinking like the animals he stalked and studied.

After six years away, Richard returned to Australia. He immediately threw himself into the Australian bird artistic world, painting and organising exhibitions, and was soon invited on scientific, surveying and collecting expeditions to remote parts of the country. Here his bush skills came to the fore. In the process of illustrating a groundbreaking and award-winning book *The Fairy-Wrens*, Richard collaborated with leading taxonomists and zoologists. 'The upshot,' he said, 'was that I spent eight years on and off working with the CSIRO survey and travelling through Australia. I worked with the scientific groups in reptiles, amphibians, with top botanists and of course mammologists and ornithologists. It was phenomenal training.'

This work continued, and Richard further advanced his zoological and ecological knowledge. In later years, as his reputation as a bird artist spread, he fulfilled overseas commissions and thus travelled for field work extensively overseas. Summing up this journey, Richard said, 'I had very good experience of birds and zoology on four or five continents [including at Mawson base in Antarctica, where he studied Adélie penguins], and that is why I tend to think taxonomically in my ornithology [that is, classification of birds], because of my good practical experience in those areas.'

Within a few years, following a trip into the Simpson Desert that was being filmed by the natural history unit of the ABC, Richard met Jenny, who was helping in production. Like Richard, she had travelled widely to interesting places, and they shared a passion about the natural world. They married in 1976 and lived on Connewarran, though it wasn't until 1985 that Richard began farming, having leased the place for twenty years since he bought it from his father in 1966.

Despite not actually farming the land yet, Richard and Jenny turned their eyes to the degraded state of the property. They began planting trees and implementing a whole-of-farm plan to turn things around. As the key driver and enthusiast, Richard brought something to his property rarely seen in farming landscapes in Australia: an artistic sensibility and deep knowledge of his landscape and its biota. A key part of this was that Richard had the mind of a true artist, along with retaining a child-like wonder and enthusiasm about the world.

This was quickly demonstrated on my first visit to Connewarran when Richard got down on the kitchen floor to mimic the jerky, startled and nervous behaviour of the marsupial mouse *Antechinus*. He and Jenny found these precious creatures in the living room of their bush retreat in the Grampians National Park each time they visited. Associated with this was what some have labelled Richard's 'grasshopper mind': an endless fascination with multiple ideas and hence a propensity to energetically pursue too many projects at once.

It was soon after returning home from overseas that Richard received a jolt. 'I was riding a mare one day and noticed our remaining red gums were dying,' he said. 'It was just obvious that the landscape was in real trouble.' 'Furthermore,' added Jenny, 'besides the red gums, we were becoming aware that this landscape probably had a hundred or so species in it originally, but we were lucky to only now have five – and five in a very, very unhealthy state. And there was nowhere we could go for material that replicated what was originally here.' Clearly, some different thinking was required.

This began an ongoing and evolving quest to restore landscape function to their farm, with ever more bifurcations along the way on the Weatherlys' path to greater ecological literacy. For a start, by the late 1970s, Richard and Jenny realised they couldn't save their 400-year-old red gums, and that other strategies were needed. The result, over a thirty-year period, was that Richard and Jenny transformed their once bared land. But unlike some regenerative farmers, the Weatherlys were very pragmatic about natural versus exotic grasses: an approach inculcated by the fact that Richard started with nothing except a denuded landscape, a large debt, and the 'arse' effectively out of his trousers. So his farming operation was to an extent traditional, with the addition of industrial fertilisers where a need was identified by soil tests, and the regular establishment of exotic grasses. He admitted he had neglected his native grasses, but said he would have gone broke if he had tried to regenerate these alone and ignored modern exotic pasture production. Now, however, he and Jenny were increasingly focusing on these native grass species.

What was not traditional was the extraordinary regeneration and encouragement of biodiversity on Connewarran. This was integral to his farming, and so Richard's farming philosophy was simple: 'We are managing an open-plan ecosystem to harvest the excess production.' For starters, Richard and Jenny were some of the first to instigate a long-term, whole-farm ecological plan, starting fifty years out and done in overlay sheets of decreasing time horizons: twenty-five years, ten, five and then working forward again in three-year projects. This immediately changed the traditional short-term-farming perspective to a much longer time horizon. He also did a map of what he thought the original vegetation was pre–white settlement (and matched to soil types). On the original whole-farm plan, what was mapped early on were wetlands, tree-break corridors and larger bush blocks.

It didn't take the Weatherlys long to realise that hand-planting expensive tube stock (plants bought in small cylindrical pots that are ready to be put in the ground) would hardly make a dint in the required task. This was because, said Richard, 'By the late seventies, Jen and I had begun seriously looking at how to revegetate this whole area.' The result was that Richard became the prime mover in developing Australia's first practical native plant direct-seeding machine. With a small team involved, including a local engineer, they self-funded with a syndicate of ten (from three states) as the Western Tree Sowing Syndicate. They were trialling the machine by 1981. Much research into seed treatments, herbicide use for weed control and so on accompanied the engineering R&D. In time, the finished machine became known as the Eco Tree-Seeder, and was copied and used widely throughout eastern Australia.

Besides these technical aspects, the biggest breakthrough was the realisation that a new type of ecologically diverse planting was needed. 'Up to this time,' said Richard, 'when we first started planting it was pretty monocultural – mainly red gums and a few others. Other people just had one or two species. But we realised there were ecological consequences of this approach: that pest predation of any species planted would occur in what was an effective monoculture.' This led to the Weatherlys appreciating the need for ecological diversity: for year-round food sources such as nectar and pollen for birds and other creatures such as parasitic and predatory insects and spiders, derived from a great diversity of all levels of vegetation. 'So in our planting,' Richard continued, 'we quickly moved from seventeen or so species to forty-five and up to 115 species of vegetation, and then let Mother Nature sort them out.' Richard also sought provenance diversity within the same species from other regions, which broadened the timing range both of flowering between species and flowering within species.

The Weatherlys' appreciation of the need for a functioning, balanced ecosystem was based on their extensive reading and their increasing work with scientists. They came to realise there were many pests out there ready to eat any exposed plantings, such as sawfly ('spitfire') larvae; defoliating caterpillars; scarab larvae (curl grubs) and their adult form of Christmas beetles; various other leaf-eating beetles; a range of moths and their caterpillars; and psyllids,[12] among many others. But, they reasoned, if a healthy functioning ecosystem was in place, then dozens of control mechanisms would be too. These controls included a wide range of birds, possums and gliders, and particularly other insect species (including various types of parasitic and predatory wasps, spiders, flies, ladybird beetles [*Coccinellidae*], hover flies, lacewings [*Neuroptera*], dragonflies and many others). However, the Weatherlys also realised that a number of desirable predators could not fly further than 200 metres from their nearest nectar, pollen or food source if they were to attack predatory grubs or adults out on the landscape.

Richard and Jenny copped a lot of criticism from 'green' purists for introducing native plant species from other regions, and even for bringing in different provenances of the local species. But they were pragmatists and knew by now in their quest to get food and nectar from their vegetation to provide twelve months' nourishment for insects and other species that the key was a functioning ecosystem – not some impossible dream of returning to a theoretical pre-1830s biota.

In a little over two decades, the Weatherlys seeded and planted over half a million native plants, and soon were reaping the benefits in insect control of crop pests. This included some extraordinary but unanticipated insect arrivals and synergistic behaviours within their developing ecosystem. 'The joy of it was that we were learning as we went,' said Richard. 'We had to, because at the start we knew nothing at all.' This learning was corroborated in a stand-out incident in 1991, when they found that their predatory insects controlled an outbreak of diamondback moths in a grazing canola crop, thus eliminating what everyone else was using: heavy-duty insecticides.

Many other regenerative farmers I studied, though primarily addressing other weak functional elements in their ecosystems, had experienced similar pest-control results arising from their more diverse and better functioning ecosystems. A number in cropping (such as Ron Watkins) soon found that increased insect biodiversity enabled them to eliminate insecticides. Leaving this control to nature has yielded large financial rewards.

The other breakthrough in his direct-seeding pioneering was Richard's discovery of the importance of 'scalping'. This technique, based on Melbourne University work, involved using a station grader to take off around one-and-a-half inches (forty millimetres) of excess and degraded topsoil before then direct-seeding. This topsoil, in country as severely modified as Connewarran following droughts and exotic seed blow-in, contained huge amounts of competing seeds, and especially unwanted exotic species. Removing this also eliminated the need for herbicides. The result? 'We found we could actually revegetate the landscape with things that were originally here,' explained Richard. 'We could revive plants that we never knew were on the place at all. They all unexpectedly came up in profusion along with the planted seed.' So an amazing variety of interesting natives – from wattles to all sorts of understorey shrubs, forbs, grasses, *Glycines*, *Hardenbergias*, lilies and orchids – appeared, plus others from the seed they had also drilled in.

In the end, the Weatherlys could proudly boast over forty-five kilometres of plantings and connected tree-breaks. Within this was a huge variety of flowering shrubs, trees and other species, which, along with other elements such as undisturbed litter, provided food and safe living environments for insect- and other biodiverse-controlling species so critical to any ecosystem. That is, by the time scientists in New England were finally putting the horrendous eucalypt dieback story together (a disaster striking that region from the 1970s – see the next chapter) and had begun realising the importance of biodiversity and a functioning landscape, down at Connewarran the Weatherlys had also begun demonstrating that farmers needed to work with and not against nature.

As we will see later, one of the key principles in enhancing landscape function is the propensity of a natural system to self-organise. In his own way, Richard summed up the issue. 'Very ambitiously,' he said, 'we were trying to develop a 400-year-old climax forest, and had put the primary and secondary colonisers in … We had the design, and we put the whole bloody lot in there; now we were letting nature sort it out. And,' he continued, 'this was because from early on we were waking up to the fact that the real key to the health of the country was within six inches below the surface and six inches above the surface, and how this fed into the trees and the pastures. But all we can do is put in place a repairing system.'

However, the Weatherlys didn't stop with their grazing and farming land. Perhaps the jewel in the crown was their riparian system, focused around the sixteen kilometres of Hopkins River frontage that flowed through their landscape. So at the same time as they were fencing and direct-seeding, they were also addressing the denuded riparian areas. This devolved from their whole-of-farm ecological plan. 'Ultimately,' said Richard, 'we came back to a system of ecological niches: bush swamps, wetland reserves and things interconnected by corridors. The concept was to have shade and shelter around all four sides of each paddock and all these interconnections coming into these environmental niches and to try to get to the point where no portion of the place was unsheltered.'

Significantly, even by the early stages of their thinking and planning, the Weatherlys were using the phrase 'from an ecosystem view'. That is, like Aldo Leopold's exhortation to 'think like a mountain', more and more the Weatherlys were 'thinking like an ecosystem'. In discussion, Richard talks the language of an ecologist. He frequently refers to the 'edge effect' in their diverse system: that interface between, say, a grassland and a tree-break or patch of native bush. Such areas are often more biodiverse because of greater microclimate diversity and a wider mix of species and generalists favoured by the two environments.

Therefore, crucial to the Weatherlys building a better functioning ecosystem was their water-cycle function and its interrelationship with the solar-energy, soil-mineral and dynamic-ecosystem functions. The spur to fence off and better manage their river frontages was twofold. First was rising salinity and degradation of the riparian zones. The second was related to the first: the platypi were under threat. 'So we began fencing both sides,' said Richard. This fencing created wildlife corridors and a series of interlinked wetlands, which in turn were connected to all the tree-breaks and plantations.

Benefits immediately came from fencing off the river. There was much vegetation regeneration, including of river red gums. The platypus population was now thriving, as was a plethora of other forms of biodiversity. 'The result of all our work,' said Richard, 'is a very noticeable species flow back onto Connewarran: mammals, amphibians, reptiles, birds and beneficial insects.'

For example, the endangered growling grass frog was now in large numbers. And the birds – Richard's favourite birds – were astounding. Connewarran now boasted sightings of 205 species – which was extraordinarily rich for such a small area. Richard believed it was a combination of three eco-regions, which had now intersected with all their regeneration work. For an ornithologist and bird artist, he was a pig in mud. For example, when one drank a cup of tea in the Weatherly kitchen, which overlooked a big lagoon below the house, it was not unusual to be interrupted by Richard reaching for a well-used and battered pair of binoculars. He then checked the birds on a lagoon full of activity and redolent with the evocative cry of duck and waterhen. Behind the kitchen table was a rack of well-thumbed bird field-guide books.

The Weatherlys now had thirteen wetlands of various sizes – the biggest being a seventy-eight-acre (thirty-two-hectare), metre-deep, ephemeral wetland. 'It's just crazy when at its peak,' enthused Richard, 'like over a hundred pelicans at once, 5000 ducks. And at a time when there's a general outcry about where all the frogs have gone, well, we've now got an unprecedented nine different species.'

Richard had come to realise that the translocated European approach to agriculture in Australia would always fail. 'Post ice-age Europe inculcated a competitive ecology,' he said. 'But in Australia everything is very cooperative, like the complexity involved in breaking down a eucalypt leaf: you need cooperation among organisms to do that, and we see it in the cooperative breeding in Australian birds. The Aboriginals understood this. We're talking about two totally different approaches. The British never took a cooperative ecological approach. Their agriculture was based on mining the system. They sprang out of the country and said, "Hang on, we've brought in some barley, get that growing." It was an ecological arrogance.' Pausing briefly, he concluded, 'Basically, I've spent enough time in Africa to see the value in abandoning traditional British agricultural practices and looking for things which are more adaptive to a country's landscape.'

In short, as Richard and Jenny Weatherly demonstrated, the fourth landscape functional element of dynamic-ecosystem communities was no less important than the other three previously discussed (the solar-energy, water and soil-mineral landscape functions). The Weatherlys' lifetime of work is living proof of the importance of a dynamic-ecosystem function, because it involves other unseen and certainly unperceived nested and networked communities at all levels. A dynamic-ecosystem function is thus fundamental to a healthy landscape because the other landscape functions need diverse plants, microorganisms, ecosystem engineers, insect predators, different trophic levels, and so on – along with their complex interactions.

And, like all the other landscape functions, the fifth – the human–social – is crucial because we humans, through our world view and dominant mental constructs, can either enhance or else rapidly degrade all landscape functions, not the least of them ecosystem community dynamics.

It is therefore fitting and appropriate that, when speaking at a conference in October 2009 (his talk titled 'Ecosystem Decline in Isolated Habitats'), Richard ended by quoting the wise words of Aldo Leopold. Said Richard, 'Writing of his 1937–38 trip to the Gavilan River in Mexico, Aldo Leopold referred to "the standard paradox of the twentieth century: our tools are better than we are, and grow better faster than we do. They suffice to crack the atom, to command the tides. But they do not suffice for the oldest task in human history: to live on a piece of land without spoiling it."'[13]

'That paradox,' concluded Richard, 'remains at the beginning of another century.'[14]

CHAPTER 13

Blessed Are the Meek

All living beings are members of ecological communities bound together in a network of interdependencies. When this deep ecological perception becomes part of our daily awareness, a radically new system of ethics emerges.

FRITJOF CAPRA, *The Web of Life*[1]

Like many before me, I dearly wish I could be transported back in time to go for a long walk through the pastoral ecosystems of Australia prior to white settlement. Just once to walk across grasslands un-grazed by the cloven-hooved animals of white settlers; to see the plant diversity and different seed types and colours; to feel the soft ground underneath and assess the depth of layered mulch; to witness fully hydrated landscapes and see how the original chains of ponds looked and functioned; to hear the cry of bustard and reed-warblers, and on dusk of curlews and nightjars; to glimpse bettongs and bilbies busy about burrows beside inland streams; and to listen to insect and birdsong under river red gums and she-oaks. And, if possible, to watch Indigenous hunters and gatherers going about their tasks – perhaps stringing nets across streams for frightened duck – and to see how they burnt country. But that is only a dream, though every now and then we can snatch a brief glance into a mysteriously beautiful past.

One such glimpse was afforded us by one of Australia's early ornithologists, J. A. Leach. In 1926, he recorded that two naturalists 'came upon a flock of [straw-necked] ibises breeding in the Riverina'. Their estimate was

240,000 birds in the colony. Within each bird they shot for investigation, there was an average of 2000 young grasshoppers. Leach observed, 'Think on it; 480,000,000 grasshoppers a day! Where are those birds now, when needed to stem a locust plague?'[2]

For the ancient human psyche, trees – those giant superstructures of entire ecosystems – are buried deep and carry great mythical power.

Whether in the dark, mysterious arboreal forests of ancient Europe or North America, the savannah-forest margins of Africa, the giant eucalypt forests of Australia, or in the organic profusion of rainforests on most continents, trees loom large in our ancient memories.

Our most powerful myths and beliefs reveal this. In ancient animistic cultures, where the world was often dominated by immense primeval forests, trees were – and still are – believed to be inhabited by tree and nature spirits that give rain, sunshine and fertility. This sparked myths and tree worship: including of sacred groves, oak trees (by Druids), and wood-spirits and 'shades'.

Our very origins are connected to the collective subconscious replication of the 'Tree of Life' – found in so many different and separated cultures. This is exemplified by the tree with forbidden fruit in the Garden of Eden. Here, primordial 'man' was placed in the 'divine garden' to guard the 'tree of life': a concept buried deep in our mytheme or archetype, and a mythical tree that connects all forms of creation while being redolent with religious symbolism. In some cultures, trees were believed to be the ancestors of all life, and quite often fruit from the tree of life conferred magical properties, including immortality. In the Christian world, the tree of life represents the perfect state of humanity and eternal life before 'the fall', when Adam and Eve ate the forbidden fruit off that second tree: the tree of knowledge.

So trees are integral to human conceptions of their place in the world and in a religious cosmos. However, humanity has a shameful record when it comes to our forests.

While estimates vary, since humans developed agriculture we have totally cleared somewhere between thirty-three per cent and fifty per cent of the world's forests (around five billion acres [two billion hectares]), at the same time degrading around eighty per cent of the original forest cover. This is notwithstanding the fact that forests still comprise one of the world's largest base ecosystems. Many ecologists regard forest ecosystems as the most important component of the biosphere. This is largely because of their role in maintaining landscape functions such as the solar-carbon, soil-mineral and water cycles, and acting as the principal repository of terrestrial biological diversity and dynamic-ecosystem communities (around ninety per cent of terrestrial species are located in forests).

Unfortunately, as a recently settled colonial country, Australia has fared even worse than the globe as a whole. This is because the post-1788 Australian track record on destruction of tree cover is frightening. For example, of Australia's intensive land-use zone (that forty per cent of the continent where vegetation can be removed and replaced for active agriculture), around fifty-two per cent has been cleared or significantly thinned. Nationally, forty per cent of forests and ninety per cent of temperate woodlands have been cleared. The same alarming situation applies to loss of grassy woodlands and understorey. Replanting such over-cleared areas with monocultures of foreign pines (or, heaven forbid, clearing virgin land to do so) most definitely doesn't restore landscape function. Instead, after harvest there is generally just a gutted, poisoned, dysfunctional wasteland.[3]

What is alarming post-1788 is that Australian settlers to the present day have no cultural nor historical perception nor record of what the landscape was previously like. This was and is exacerbated by being coupled with an exploitative as opposed to nurturing approach to the land. This has often had unexpected dire consequences: as graphically revealed in the spread of 'white cancer' or dryland salinity in large areas of Australia, and as will be directly discussed with the catastrophic New England eucalypt dieback syndrome.

By the twentieth century in the industrial world, the harvesting and management of trees had largely been taken under the industrial forestry umbrella. This had followed, from the eighteenth century, the fact that modern forest and woodland management had evolved into an economic system in Prussia and other German states. While this German tradition morphed into a fully scientific resource-management regime, the Scottish and French foresters in time would spin off different traditions that led to more multiple-use forestry. These eventually included the elements of recreation, aesthetics and conservation-biodiversity – and ultimately agroforestry: that branch connected to farmers.[4]

As with other aspects of regenerative agriculture, agroforestry consti-tutes the alter ego of industrial monocultural forestry. The latter is little different from industrial agriculture. The land is more generally regarded as an exploitable resource, where efficiencies in terms of increased unifor-mity (i.e. monocultures) and large scales for ease of management, along with industrial fertiliser inputs and chemicals, is more the norm. These forests can often cover entire landscapes irrespective of diversity of slope, aspect, soil, biotic and biological diversity existing there prior to industrial action. However, aside from the obvious negatives of interventionist indus-trial practices, such forestry exposes landscapes to increased threats: whether climatic, bushfires, pests and diseases or markets. Part of this higher risk exposure in Australia, therefore, is because biological, social and economic diversity has been compromised.

Agroforestry, however, is not about simplistic monocultural timber production. Instead, over the last few decades a richer, more diverse and ecologically regenerative concept of agroforestry has evolved in Australia, led by a few leading innovators – including farmers, academic-derived thinkers, and government employees in, for example, natural resource management areas.

Some of these leaders are now world famous, such as Tony Rinaudo (founder of the revolutionary worldwide use of regrowth trees in develop-ing nations, known as 'FMNR', or farmer-managed natural regeneration); Robert Vincin and his extraordinary work in revegetating Chinese deserts; and others such as Bob Wilson and Andrew Sippel – pioneers of edible shrubs for Australian farmers (and in particular the exotic shrub tree lucerne or tagasaste, and native shrubs such as saltbushes). Ian Dickenson and Peter Downey, both of Tasmania, are leaders in long-rotation native forestry harvesting.

But a particular example of outstanding agroforestry leadership has come through the Melbourne University School of Forestry and its leader Rowan Reid – and especially as a result of Reid linking to a group of farm-ers in the Victorian Otway Ranges who collectively formed the Otway Agroforestry Network.[5]

Farmers grow trees and shrubs for multiple reasons. These include: to prevent erosion; to restore landscape function in, for example, saline risk areas; for commercial timber harvest; for carbon or biodiversity credits or even eagle nest offsets; for native plant seeds; for aesthetics and/or to add capital value; for shade and shelter, or edible shrubs for livestock; for bush tucker or medicines or cut-flowers; for firewood; and for apiary, just to name the key uses.

Many of these reasons clearly are the same ones that motivate other regenerative farmers. But almost invariably the key common denominator is the improvement of landscape function.

Rowan Reid says, 'It is not the scale, the planting pattern, the species or the purpose of a forest that makes it a "farm forest" or "agroforest" – it is ownership ... ownership of the decision to do it and how it is done'. That is, agroforestry is the result 'of a farmer's decision to practice forestry' for all sorts of reasons, and often many combined. So agroforestry is not a predefined set of land-use practices, but instead it comes down to a farmer's choice to willingly engage in agroforestry for any of a large number of reasons.[6]

In its simplest form, as Reid and peers confirm, 'agroforestry ... is a useful all-encompassing term for the deliberate management of trees and shrubs on farm'. Thus, healthy agroforestry is about a process and not outcomes in regard to establishing and managing trees and shrubs (woody perennials). Crucially, therefore, agroforestry is about meeting farmers' needs. To this end, Reid and peers define agroforestry as 'the commitment of resources by farmers, alone or in partnerships, towards the establishment or management of forests on their land'.[7]

Invariably, good agroforestry practice restores landscape function and thus usually also confers concomitant economic and social benefits. That is why agroforestry can be a key practice in regenerative farming.

However, for one farming family in the New England Tableland of New South Wales, their unanticipated move into agroforestry arose because suddenly all their dominant over-storey eucalypt trees began to die. What then unfolded must still surprise them today.

The unexpected, drastic consequences of conventional industrial agriculture's approach to simplifying dynamic-ecosystem communities are many and varied. But the eucalypt dieback that occurred in the New England Tableland of northern New South Wales from the 1960s was especially dramatic. Moreover, one of its key lessons concerned the unpredictable and varying time it takes, following injudicious human interference, before an ecosystem collapses. For John Robertson's grasslands in western Victoria, it took only fifteen years; for dryland salinity to emerge in Western Australia, it took three to four decades. But for region-wide eucalypt dieback to suddenly strike the New England Tableland, the time lag was 130 years of progressive ecosystem simplification.

Unknown to me for a time when I first came home to take over management of the farm, this massive ecological disaster was unfolding to our north. This tragedy, along with other emerging problems such as dryland salinity, would serve to shine a spotlight on the consequences of land managers adhering to the rigid precepts of industrial agriculture: the belief that to successfully farm, one must grossly simplify and 'dominate' the natural landscape.

Like the Monaro, the New England is a high tableland, ranging from 2600 feet (800 metres) up to 5000 feet (1500 metres) and higher in the east. It has less naturally treeless country than the Monaro because most of it was once grassy woodland, tending as in the Monaro to more forested country in both the west and the east. And on its eastern edge also, the New England Tableland suddenly plummets through gorges and high rainfall escarpment forests down to the coast. Further like the Monaro, the New England is predominantly high-quality sheep and cattle grazing land, with minimal cropping. In covering over 7.5 million acres (three million hectares), it comprises one of Australia's major agricultural regions.

However, in some respects the New England became its own worst enemy after white settlement. It has relatively rich soils and a higher overall rainfall than the Monaro, and its rolling, downs-like country (which explorer John Oxley described as 'open woodland') was easily cleared of timber, beginning a few decades after white settlement in the 1830s. Until that time, this ancient coevolved landscape was one also adapted to fire, and this would have included frequent cooler-season Aboriginal fire-stick farming over many millennia.

Decade by decade, more country was cleared of timber and understorey so as to make way for pastoralism. From the early twentieth century, European intervention escalated with the aid of mechanised agriculture, leading to intensified clearing and ploughing of grazing land so as to plant non-native, introduced or 'improved' pastures, which also saw the beginning of industrial fertiliser use. Widespread planting of non-native, exotic European trees such as conifers was also well established by this time. From the 1960s, such industrial practices only escalated.

In short, in less than 150 years, European settlement had methodically removed many of the interlinked components of a series of dynamic-ecosystem communities across the New England. This involved gross destabilisation and simplification of the four main biogeochemical landscape functions and cycles. As the nation was now finding with regard to salt problems in agricultural zones, the New England was about to realise that humans (the fifth landscape function) can only interfere so much with complex dynamic systems before they reach a tipping point.

That tipping point was reached in New England in the 1960s and became known as the New England eucalypt dieback. What unfolded is widely regarded as one of the worst, whole-of-landscape mass die-offs of trees in the world. One by one, and then in massive swathes, vast areas of the grazing land of New England lost most of their eucalypt (gum) trees. What then slowly became clear was that this collapse of an ecosystem's base structure was multi-factorial. It arose because of the systematic degradation of all key landscape functions.

It was quickly apparent to farmers and scientists alike that the main cause of the dieback was insect defoliation of eucalypt leaves. This prevented trees from building sufficient energy reserves via photosynthesis. Most insect attack was in summer, when, for example, Christmas beetle larvae (scarab grubs or curl grubs) hatch into mature beetles. The savage defoliation by beetles in particular then led to succulent (and thus more appetising) foliage regrowth on the crown (as in the sum total of the tree's above-ground parts) and in epicormic shoots (those shoots growing from a bud that lies underneath the bark of a trunk, stem or tree branch). These were eaten until the whole canopy died and the trees ran out of energy, thus losing the ability to grow more crown leaves. Moreover, while the prime defoliating culprits were Christmas beetles, there were other insects that could also reach epidemic proportions, such as sawfly (spitfire) larvae.

However, there is a second problem with Christmas beetles, whose thirty-five endemic species are members of the scarab family

(*Anoplognathus*). While the adult Christmas beetles' favourite tucker is eucalypt leaves, their larvae – the scarab grubs – live underground. The problem is their favourite food is plant roots – originally native but, with pasture improvement since the 1950s, exotic species as well. This would have huge ramifications when native grasslands began to be modified by pasture *improvement* and superphosphate, as the scarab grubs thrived on exotic grass roots.

But while insect defoliation was the obvious final symptom of dieback, the underlying or primary causal factors were many and interrelated, and it took scientists some time to unravel this story.

Land clearing followed by ploughing, pasture *improvement* and super-phosphate application (which all resulted in a doubling of stocking rates) grossly changed the solar-energy cycle for trees, understorey and grassland communities. The original wide variety of plant species (adapted to functioning within an interdependent, coevolved ecosystem) now suddenly either disappeared (as with most of the understorey and many native grasses) or else, in the case of trees, were left like shags on a rock. This meant trees being isolated out in the open and exposed to increased stress and insect attack, with sheep and cattle camping in high density underneath them, which also served to further compact the soil.

Once much of the vegetation went, stocking pressure intensified on what remained. Besides the increased trampling and soil compaction, any native vegetation seedlings were eaten by stock, while hungry cattle would frequently ringbark trees also. These changes, in combination with the clearing and modification of other vegetation, completely disrupted the dynamics and functioning of ecological communities, both under and above the soil surface.

The planting of thirsty exotic grasses, increased stocking rates and associated increased soil compaction in turn affected the water cycle. Fertiliser, animals and exotic grasses also dramatically changed the soil structure and its traits (such as the pH, leading to greater acidity), and thus mineral cycling. And the loss of litter for such functions as nutrient recycling, thermal insulation and biodiversity fostering at the micro and macro levels was significant. Moreover, the application of massive doses of phosphorus in superphosphate, we now know, has dramatic negative impacts on soil mycorrhizal fungi and some bacterial populations and their role in nutrient recycling. This was accompanied by the disappearance of native plants that were adapted to low phosphorus soils.

Once the initial over-clearing and disruption of the key landscape functions occurred, then a whole range of secondary causal factors or symptoms came into play: insect defoliation being just the most obvious and severe. The altering of previous fire regimes (both natural and from many millennia of Aboriginal burning) would have led to changes in the soil-mineral cycle. Given the dependence of many native plant species on fire for regeneration and function, and in conjunction with other disturbance factors, a gross imbalance in age cohorts of vegetation developed. This was because (among other factors) fire-generated recruitment of regenerating eucalypts and understorey species virtually ceased from even before the 1960s. The result was a dominant population of increasingly geriatric trees that was rendered more vulnerable to increasing stresses.

The loss of diverse vegetation meant a reduction in food and habitat necessary for predators and parasites, including those that once controlled eucalypt defoliators such as Christmas beetles. For example, vegetation clearing (including not just food species but also general habitat and nesting hollows in dead trees) meant a drastic decline in species such as microbats, which catch some insect defoliators, but especially in marsupial gliders and possums. The sugar glider is the main predator of Christmas beetles in New England, as one of their favourite summer foods is the beetle, of which they can eat their own weight in a night. In addition to losing their tree habitats and living and breeding hollows, sugar gliders lost much of their non-summer essential foods due to over-clearing of understorey vegetation (especially of acacias and their energy-rich sap).[8]

There was also a drastic decline in both insectivorous birds and in natural biological insect control. Again, it was the creeping loss of vital understorey that had knock-on effects here, because the diverse species and litter in understorey provide essential food to such things as predatory and parasitic wasps (one such crucial shrub, for example, was blackthorn).

Added to this was an improved underground environment for scarab larvae. Their numbers exploded as they feasted on the more prolific succulent roots of new grass species boosted by chemical fertilisers. The ongoing dieback syndrome then triggered a vicious circle, whereby the extra Christmas beetle attack (derived from thriving larvae) accelerated tree decline, which thereby promoted pasture growth because of reduced competition from the trees, which in turn favoured the scarab larvae, and on and on.

By the 1970s, therefore, the totally unforeseen impacts of massive agricultural intervention meant that an ancient, coevolved, complex, dynamic series of interconnected ecosystems had begun coming apart at the seams. With the progressive simplification and destabilisation of landscape function and the dismantling and corrosion of complex feedback loops, checks and balances, other factors came into play. These rendered the eucalypts and their entire supporting system vulnerable to any shocks and disturbances. The weakened remaining trees were more vulnerable to stem-borer and sap-sucking insects and also fungal pathogen attack. A senescing population of eucalypts included a higher proportion of unhealthy and dying trees, upon which increasing numbers of defoliating insects now concentrated. Such a population, in an altered soil- and water-cycle environment, was more vulnerable to drought and increased soil acidity due to the pasture-fertiliser regime.

This latter industrial regime now exacerbated the eucalypts' plight because the 'richer' nutrient and moisture environment ironically rendered the trees even more vulnerable to insect attack as their leaves became more palatable to Christmas beetles and other defoliators. And these are just some of the factors that eroded the health and vigour of the trees.[9]

So this was how the gross simplification of a complex ecosystem through major disruption of all four landscape functions led to a tipping point where massive ecological collapse suddenly occurred. With no checks and buffers left, mass dieback occurred across tens of thousands of hectares.

Ever since a young age, I had been keen on the natural world. A self-sufficient only child, I relished exploring the bush on our farm, or hunting rabbits among granite rocks with accompanying dogs and poddy lambs. I was then packed off to boarding school at the age of eight. This small school of barely a hundred pupils proved a blessing, for it was surrounded by farmland and bush in the New South Wales Southern Highlands near Moss Vale. Besides companionship and sport, one could go camping and exploring in the broad surrounding farmlands and native forests. But the greatest bonus was a young teacher called John Hutchins, who was a keen ornithologist and who took under his wing a small number of boys interested in nature. This included me. With nearby rainforests and tall, wet sclerophyll forests on the coastal sandstone escarpment to explore, weekends became a joyful discovery time in pursuit of elusive lyrebirds, rainforest pittas and other mysterious denizens of these rich ecosystems.

Consequently, on leaving secondary school and after two years jackarooing on the family farm under my father, I enrolled at the Australian National University in a science degree, with the intent to major in zoology. My heroes of the time were scientists such as the founding ethologists (animal behaviourists) Niko Tinbergen and Konrad Lorenz. I wished to become an ethologist also, working in wild and beautiful regions of our planet.

However, the professor of zoology was a strict Oxford professor and reductionist ethologist to boot. One day, I remember him ripping up an essay I had lovingly laboured over. My piece was the result of hours of difficult and time-consuming trapping, colour-banding and then stalking a curious, social-nesting bird: the white-winged chough (which was plentiful in the bush on our farm). 'What I want,' he said, 'is rigorous quantitative measurement, not this opinionated, free-wheeling observation and supposition of yours!' To pass the unit (and ultimately zoology), I was required to count how many times a rat licked sugar off a lever as a reward in what was classic psychologist-behaviourist methodology.

The professor's view of science was not for me, and, though I passed in zoology, I soon gravitated across the campus to the first course in holistic thinking of any university in Australia: that of human ecology. Here, many of my fellow students and teachers were decidedly 'green', even radically so. Most of them dressed in colourful or 'hippie-type' loose-fitting clothes, while others smoked pungent roll-your-owns, many of which definitely didn't contain tobacco. On a few occasions, I joined them in protest meetings on campus or marches in the city, for it was a time of a rising green consciousness in the wake of the tragedy of Lake Pedder. These marches concerned slogan-led issues such as 'Save the Whales' and 'Keep Uranium in the Ground' – though one day I recall one larrikin student bringing to a march a memorable handmade sign that read 'Keep Whales in the Ground'.

Joining this new, challenging (to the university, at least) holistic-based course was to be a life-changing event. However, reality then intervened.

My father suffered a major heart attack and I made the decision at the age of twenty-two to return home, take over the farm and finish uni part-time. Thus began a forty-year land-management journey that is a minor thread in this book.

On returning home, I continued developing my interest in birds, mammals and other creatures. However, at the same time I seemed to live a schizoid life. I was quite ignorant about the rest of the insect world, not to mention the arthropod and underground microbial world. In fact, concerning the latter, invisible world, I lived as if it didn't exist. Yet this underground component constituted ninety per cent of life on the farm. Even more alarming was my ignorance about the vastly more obvious plant world: the very system on which our farming landscape was founded.

Yes, I knew most of the exotic trees and a gum tree or two, and I knew most of our introduced exotic pastures (what we termed the 'improved' species), but I could barely name more than one or two native grasses beyond the most obvious one called 'corkscrew' (*Austro stipa scabra*), and this mainly because corkscrew's seed penetrated both sheep's skin and woollen socks just before Christmas.

That is, when it came to farming, I had somehow compartmentalised my keen interest in nature to consider only a few obvious and attractive components. To the rest I was effectively blind.

Then, one day after a decade or so of management, I was confronted by this blindness. I remember setting off mustering on a horse, my two kelpie sheepdogs trotting behind as we meandered along a two-wheel track through open granite country. This country had once been grassy woodland and would then have comprised a dominant but scattered over-storey (canopy) of snow gum, candlebark and ribbon gum, plus large blackwood hickories; a scattering of understorey shrubs and forbs; and then diverse native grassland. After earlier phases of clearing, ringbarking, and then more recently set-stocked grazing, all that remained on this day as I mustered were a few dead grey stumps and the odd silver wattle or hickory surviving high up between granite tors and out of the reach of sheep and cattle.

So it was therefore puzzling that, as I rode along, I suddenly noticed the strong, sweet smell of what I was convinced was snow-gum blossom. I stopped the horse and looked around, but there wasn't a snow gum in sight.

Intrigued, I circled the horse in a wide arc. Still no snow gums, but the strong blossom scent seemed to come from near a clump of rocks. The only living thing, which partly covered these boulders, was a nondescript dark-green bush that I had barely taken account of previously.

Dismounting, I walked over to the bush. It was perhaps a metre high and shaped like a dome, while roughly circular and around two to three metres in diameter. It was entirely covered in rigid thorns, which were surrounded and interlaced by very small green leaves, perhaps the size of a small pea. Covering some of the thorns was some yellowy-green lichen. Later on, I would learn that the combination of thorns and small leaves was an adaptive defence by some plants to grazing in landscapes. Dating back possibly millions of years, the plants and grazing animals (in this case, mainly kangaroos, wallabies, potoroos and so on) had coevolved.

But at that moment, I was intrigued by this nondescript bush because when I bent down I discovered that it was the source of the overpowering honey-blossom scent. And yes, there, of the most delicate size and shape, were a myriad of little, bell-like, pale yellow flowers hanging upside down within their protective screen of thorns.

That wasn't the only thing I noticed. Clearly the strong, rich honey scent had a purpose, because surrounding the flowers were small native bees, equally small, nondescript flies, and bright-orange parasitic wasps. These must have been attracted from a long distance by the shrub's scent. Moreover, either the funny little bush had evolved a scent mimicking that of snow gums, or vice versa. Either way, the result was the same: a host of attracted insects to complete fertilisation of the flowers.

On looking up my reference books, I quickly discovered this nonde-script plant was called a tree violet (*Hymenanthera dentata*), otherwise known locally as a gruggly bush. But I did not recognise its importance to an ecosystem nor what its strong scent and attendant insects implied for eco-logical function. I had been blind to this bush, and it would take time for the blindfold to be fully removed. But I have never forgotten that day, and for me our gruggly bushes are emblematic both of what was lost and also of what could be.

As I travelled further on my journey of ecological literacy, I came to realise the significance of 'gruggly' and a myriad of species like it in the vast range of ecosystems across our continent. It is the 'unseen' elements such as gruggly bushes and blackthorn shrubs in New England that provide vital food for animals and insects in a functioning ecosystem. Such insects include parasitic wasps, spiders and the like, which prey on scarab grubs.

And, suddenly and unbidden, those radical words from the Gospel of Matthew came to mind:

Blessed are the meek, for they will inherit the earth.

It was the role of dynamic-ecosystem communities and life-webs that were integral to a family of landscape managers who were virtually catapulted into becoming national leaders in addressing the New England eucalypt dieback story. This family are the Taylors, led by husband and wife team of Jon and Vicki Taylor. They reside some fifty or so kilometres to the south-east of Tim Wright on the New England Tableland, and near the small settlement of Kentucky.

The Taylors were early settlers since the 1840s. Jon and Vicki are the fifth generation of Taylors on their 1606-acre (650-hectare) farm The Hill – a farm name that became rapidly recognised after the 1970s as a key site of agroforestry pioneering. This is because the Taylors were leading innovators in finding a solution to the impacts of New England's eucalypt dieback.

As with many such innovators, it was a life shock or trauma that triggered the Taylors' move to regenerative agriculture. In their case, the shock was the sudden and mass die-off of most of their beautiful eucalypts, and the loss of ecological benefits and the aesthetics of a lovely parkland landscape.

The Hill provides a typical case of progressive ecosystem simplification on the New England Tablelands, which culminated in the eucalypt dieback disaster. Beginning with woodland tree cover and understorey over much of the farm in the 1830s, bursts of clearing occurred between 1860 and 1890, and then again between 1910 and '30. By this time, the farm's tree cover had declined to around thirty per cent of what was there originally.

When Jon Taylor was growing up in the 1950s and working with his father, Walter, The Hill still had a nicely balanced cover of around twenty-five per cent, all of native timber. Then, from the late 1950s into the '70s, the Taylors aerially spread superphosphate and introduced pasture seed. But then, as Jon recounted, the trees progressively began to be defoliated two or three times a year by massive insect attacks, especially from Christmas beetles. Jon recalled that 'between 1967 and 1990 nothing came up anywhere, and the tree ecology all fell in a heap. Things got out of balance pretty badly.'[10] 'Out of balance' meant the Taylors' tree cover got as low as around one to two per cent of the farm before Jon and Vicki started turning it around.

Little wonder, in the midst of a skeletal landscape, that this triggered change in sensitive souls: souls somehow still linked to elements of our ancient Organic mind and its empathy with trees in our landscape. What made things worse in retrospect was that from the late 1960s Walter and Jon began 'cleaning up' the dead timber by burning it, and sowing fodder crops and pasture. Jon stated that this final act of removing dead trees and logs and sowing oats and pastures had 'a big impact on the environment. It increased exposure, removed the dead wood that was a source of nutrients cycling back into the pasture, reduced the habitat diversity and produced a monoculture with spans of hundreds of metres of oats or pasture without a log or dead tree for sheep, lizards or other creatures.' Further, and based on his experience, Jon believed that 'insufficient biodiversity causes problems. The beetles got the upper hand here because there were not enough counteracting parasitic wasps'.[11]

However, at the start of their journey, Jon and Vicki were no more ecologically literate than the next New England farmer focused on producing good fine wool and beef cattle.

I first met Jon and Vicki when I was an undergraduate at the Australian National University. A good friend of mine at college was Jon's sister Sally, and one day Jon and Vicki dropped in for lunch, where we swapped stories of our respective tableland experiences. At the time, and indeed on subsequent meetings after their extraordinary innovations, the Taylors just appeared normal, unaffected 'cockies': slightly wide-eyed and innocent in that disarming bush way, but also incredibly positive about their ongoing journey. On that first meeting, Jon had that endearing, slightly awkward, open-faced demeanour that many farmers convey when entering an urban or academic world: a mien that is the result of dealing with the basic natural elements of life. Ever present, however, was a rapier-like sense of humour, which his wife also shared. Vicki appeared somewhat quieter, with an obviously sharp and imaginative brain that would combine with Jon's to make them such a breakthrough team.

What triggered Jon and Vicki's innovation journey was initially the need for economic survival. Like most farms in New England, their land was high and exposed and had cold winters. Of their 5500 fine wool flock, half were ewes, and a key profit driver was lamb survival on the windswept paddocks. As more and more shelter disappeared, the Taylors realised that correcting this had become an economic imperative.

However, not only did the Taylors act when all around watched like stunned mullets, but they also spread their innovations to many others beyond themselves. In the process of planting huge numbers of initially *Radiata* pine, the Taylors went on to pioneer planting, propagation and nursery techniques, and then played a role in popularising on-farm agroforestry and vegetation regeneration management that would be of seminal national influence.

So in 1979, after they lost the last of their susceptible eucalypts to dieback (and aware that planting more eucalypts in the face of a huge imbalance of predatory insects was futile), the Taylors acted. They planted 20,000 *Radiata* pine seedlings in linear and narrow tree-breaks. (In retrospect, it was an approach I strongly identify with from my own early mistakes, and which has minimal ecological benefits.) Subsequently, as they became increasingly engaged in agroforestry techniques, and as they gained greater ecological knowledge, the Taylors evolved different strategies in what was a never-ending journey of trial and error and interaction with others in the field.

From 1982, the Taylors then also began planting mixed native (both trees and understorey) and exotic pine breaks so as to improve diversity and the robustness of their tree-breaks. A decade later, they began planting over whole paddocks along contour lines. This came about because 'birds of a feather flock together'. Not only had Jon and Vicki become hungry for any information on agroforestry via whatever medium, but they were inevitably drawn to interact with like-minded innovators and people who were invariably positive. These included Andrew Campbell and others involved in the Potter Farm Planning Scheme in Victoria, but also Richard Weatherly and many others. And in 1992, the Taylors interacted with Ron Watkins. He gave them the idea of designing their latter planting on a Keyline-type contour pattern using both pines and natives in double rows, and with their tree-breaks 200 feet (sixty metres) apart. It was a visit to

Ron's farm that convinced Jon of the veracity of the new approach: an approach that became known in New England as 'engineered woodlands'.

As I had also seen on Ron Watkins' farm and other farms in Western Australia (and enacted on our farm), the result was a replication of shelter provided by natural timber cover plus other increasing benefits from improved landscape function. Within twenty-five years from 1979, the Taylors planted 400,000 trees and shrubs and effectively transformed their sheep, beef and now multi-enterprise farm. Total plantings today are pushing half a million, and each year they continue to plant around three hectares of vegetation: a continuing mix of pine and varied native species aimed at providing insect, bird and other species' habitat and food sources.

In a result that I have found replicated in many different environments across all states of Australia, the Taylors found that overall production levels were not adversely affected by putting fifteen to twenty-five per cent of the farm back into vegetation. Indeed, in some cases elsewhere the increased vegetation cover actually increased production from the original baseline.

Nearly twenty per cent of the Taylors' farm is now revegetated, yet pre-1979 farm production levels have been retained and income increased due to greater drought resilience, better animal performance and agroforestry income, and eliminating the major cost of fertilisers (having ceased fertilising their pasture paddocks over twenty years ago). Regarding the greater drought resilience, Jon says the trees have paid for themselves 'two or three times over'. This came from the Taylors being able to judiciously use un-grazed feed in their tree-breaks, which led to significant returns in avoided drought fodder purchases, greater animal performance and the retention of key breeders and other productive animals.[12]

'The trees have given us an appreciation of how exposed our pastures and stock were,' Jon observed in 2004. 'We used to have harder frosts, the feed tended to burn off quickly on dry windy days, and the stock suffered – not just in winter. If we want to find our stock on a summer's day, we know we'll find them in a tree plantation,' he concluded.[13]

Equally spectacular has been the remarkable turnaround in landscape function – and not just in dynamic-ecosystem communities but also in all the others. Regarding obvious impacts on wildlife and biodiversity, Jon says that before they began replanting trees, 'we were down to mostly just a few magpies and crows, and very few of the smaller birds in particular. The other wildlife, the kangaroos, were hardly seen on the place, and my father lived his whole life here never seeing a koala on the place.' Fur-trappers during the 1860s and then loss of trees meant the koalas had long deserted the property.

But now, among many other spectacular benefits, echidnas and koalas of all ages are back on The Hill. The first koala returned in 1995, and today many live in the pine tree-breaks. Local scientists at the University of New England, such as biodiversity specialist Dr Nick Reid, are now quantifying a spectacular improvement in biodiversity. There are many more echidnas, and present for the first time are swamp wallabies (a tree and shrub browser) and black wallaroos. Nevertheless, due to the early stage of regeneration and the absence of nesting hollows, up to this point possums and gliders have continued to decline.

Jon and Vicki calculate that bird numbers are twenty to fifty times higher. This includes the return of a number of threatened woodland insectivore species. A study by Reid and his University of New England team has revealed eighty-two species of insects on the exotic trees and eighty-seven on the natives, but with only three of the fourteen insect orders being common to both tree types. The same results were found for spiders. As Jon concluded, 'this shows you the benefits of any trees and vegetation, both exotic and native'.

When I stayed with Jon and Vicki Taylor in 2010 and met their son Michael and his wife, Milly, who now also work the farm, I went for a walk at dawn just as the landscape was waking up. Traipsing past farm sheds and an on-farm saw-milling plant with cords of neatly stacked, freshly milled *Radiata* pine slabs, I climbed a hill and looked over a vibrant landscape. Below me, as I had experienced on Ron Watkins' farm, I could barely see open paddocks. This was because the lazy-S of contour-winding tree-breaks composed of mixed species at varying altitude levels seemed to merge into the one forest. Mist rose off the rehydrated lower country, but most impressive of all was a cacophony of birdsong. It was dominated by the loudest combined chorusing of magpies that I had ever heard: all seemingly competing with one another. As I drew closer to the vegetation, other insectivorous and nectar-eating species familiar to me at home then became audible: flycatchers, fantails, honeyeaters, wood-swallows, whistlers and others. Truly, here was a whole squadron of unpaid pest-controlling, polli-nating and fertilising helpers willingly on the job.

Another obviously impressive enhancement of landscape function could be seen in the Taylors' water cycle and repaired riparian zones. Before they began revegetating their farm, their main drainage line – the Terrible Vale Creek – was, under regular grazing, becoming an eroded drainage gutter of bare-sided waterholes and wounded banks. Trees were then planted along the banks, and in time the entire riparian area was fenced off. The result was spectacular. As Jon explained 'the creek is disappearing! The banks have grassed up, reeds have come back, and the vegetation retains sediment.' Thus in each flood, regenerating reeds and other vegetation retained deposited soil, while the creek was also taking on a chain-of-ponds look, with defined waterholes. Water was now clear and algal blooms vastly reduced. In addi-tion, there was a large increase in reptiles, amphibians and other species.[14]

These spectacular results, stimulated by the dieback disaster, were only achieved because, in addition to everything else, Jon and Vicki Taylor became leading on-farm pioneers in tree-establishment and silvicultural (forestry) techniques, and, more lately, in on-farm timber harvesting and milling. This pioneering of techniques, ideas and hard-won lessons by the Taylors came to play a major catalytic role at a vulnerable time for Austra-lian rural landscapes. It was also when programs such as Landcare were on the threshold of beginning (starting in 1990), which meant an urgent need for practical and viable revegetation techniques.

The Taylors' first innovation came in 1983 when costs for tree estab-lishment leapt from fifteen cents for bare-rooted stock to $2 for tube stock. Jon developed a mechanical tree-planter, and as the Taylors' example of replanting trees began to be emulated, he was able to cover costs with a contracting tree-planting business. 'We were so busy,' quipped Vicki, 'our kids, who were only little then, nearly divorced us!' The next step was when Jon and Vicki travelled to the USA and then on return introduced the superior Swedish Hiko seedling tray system to Australia in 1991. They quickly established a business selling these plus planting equipment, which also put them in contact with other agroforestry and biodiversity innova-tors such as John Weatherstone (see next chapter), who at the time was running his own nursery. The Hiko trays allowed the Taylors to develop their own nursery and experiment with a greater variety of tree species.

At the same time, they turned their pine thinnings into treated pine posts (they now tow a portable post-peeler behind their tractor into the paddocks), and then they moved on to milling their own pine timber. They budget to harvest around 3.7 acres (1.5 hectares) of trees a year, and thus calculate a return of $6,100 to $7,300 an acre, just for timber.

In the meantime, through their local Harnham Landcare Group but led by Vicki, in 1992 the Taylors hosted a national Treefest on a dedicated seven-and-a-half-acre (three-hectare) site on their farm. Six thousand people came from around Australia to share ideas and view the latest thinking in agroforestry and tree-planting technology. For a number of years, the Treefest became a biennial event on this site.

Two generations of tree-wise Taylors are continuing with their ongoing innovation. Fodder and human-food trees (including willows, apples and pears) are currently being trialled, as are long-lived exotics including many different oak species, which have the potential to recreate an Australian *Dehesa*[15] environment. Jon and Vicki travelled to Mexico and elsewhere to research these. The Taylors are now finding that remnant and new natives are beginning to self-generate again, and having begun and led a major tree-planting movement, their flexible approach for planting to suit different paddocks is being copied.

The synergies between the Taylors' work and that of the Otway Agroforestry Network (see Chapter 19) and individuals such as Richard Weatherly, John Fenton of Lanark in Victoria and John Weatherstone (whom we shall meet directly) are striking. Nevertheless, the Taylors believe they have much more to do. Next generation co-manager Michael Taylor is aware that other aspects of agroecology remain to be explored, such as regenerating diversity and function in native grasslands possibly via a holistic grazing regime, and fencing off other areas for greater vegetation diversity for dynamic-ecosystem communities.

The simple fact is that New England dieback suddenly hit because no one understood the complexities of Australian landscape function. Sadly, similar crashes are poised to occur in other major regions of Australia. The beautiful river red gum (*Eucalyptus camaldulensis*) country of western Victoria – the heart of Australia Felix – is dominated by tens of thousands of acres of geriatric trees, many hundreds of years old. One day, the 'old age home' will be emptied. Some scientists describe such landscapes where old trees are now continuing to die and not being replaced as landscapes of 'the living dead', as they are but relics of an original vegetation cover with no progeny following them.

This alarming situation applies across broad swathes in other areas, such as, for example, the temperate Lachlan catchment in central-west New South Wales. For those prepared to 'see', all such cases are canaries in the cage for impending ecosystem collapse, and especially as we increasingly slip into an era of man-made global warming. It is hoped that farmers can learn from such leaders as the Weatherlys and Taylors and rise up before it is too late.

Moreover, reflecting on the Taylor family's track record, I concluded it was unlikely that ongoing innovation wouldn't be constantly on the agenda. For example, recently Michael initiated an imaginative but idiosyncratic piece of macro landscape art through tree planting. With a mix of tree and shrub species, the Taylors designed and planted an area that reproduced a giant frog with spreadeagled legs. It is now visible from jets at 30,000 feet on the Sydney–Brisbane leg.

Jon Taylor is pragmatic when he says the particular focus on exotic trees doesn't necessarily apply in other regions. Pines were essential in their part of New England owing to the levels of insect attack, but not, for example, in Tim Wright's part of New England. 'It's a mix and match,' Jon said. Vicki concurred: 'If we lived somewhere else, we would have planted natives for sure.'

When I drew my visit to a close at the Taylors' in 2010, and as the extended family laughed and quipped around a kitchen table covered in maps, photos and also fine wool fabric from yet another enterprise they were engaged in, Jon paused, gazing out a window to gather his thoughts. 'You know,' he said to nods around the room, 'we probably wouldn't still be here if we hadn't gone on this journey. But even if we were still here, we'd probably be very miserable human beings and our kids wouldn't be here, I would say.' Vicki agreed. 'You see,' she said, 'we call it our garden. The whole property's a garden … It's a type of creativity. We've wanted to soften the landscape, to create open spaces and vistas and sheltered areas. We've done that and it's changed our lives.'

CHAPTER 14

Listen to the Land

When we try to pick anything by itself, we find it hitched to every-thing else in the universe.

JOHN MUIR, *My First Summer in the Sierra*[1]

January, mid-summer 2014. A furnace. Day after day of pulsating, searing heat that shimmered off the ground. Heat that blazed from a ferocious sun and bleached blue the waterless skies. A record run of days of thirty-five to forty degrees Celsius and more, pressuring the eardrums. Acrid smoke from distant bushfires clung to the hills, washing the valleys in a reddish-brown haze. All stock-work needed to be done by morning 'smoko' before they became immovable, locked down in the heat. This meant pre-dawn starts.

With sheepdogs jostling for position on the back of my quad bike, I would head off, forever in wonder at the fresh rejuvenation of a new day, but knowing a crimson-orange furnace would soon lift above the rounded, treeless basalt hills on the eastern horizon. Even the dried grassland reflected what was coming. Its serried ranks of native swards and seed heads glowed a golden-straw colour before any sunrays showed on the horizon, as if the whole landscape was lit by an internal energy source. Then came the sun, highlighting the purple tint of poa tussock seed heads and emphasising the contrast of dark butts and shining gold stalks. Beside them, kangaroo, wallaby, fox and quail prints lay embossed in talcum-like dust.

At times, even the best techniques can't retain soil moisture, so sensitive ecological management becomes especially paramount. I was in that window now. Around the district, naked, inert soil was being exposed in large areas, while poa tussock clumps stood pointed like sharpened pencils from overgrazing. I realised it was time to accelerate our destocking, to carefully plan grazing rotations, and to keep our trough waters clean and flowing for livestock. The latter had already left their camps for a morning feed. After this, they would come in for a drink and then camp the long day, idly flipping ears against flies as they dozed in the heat under shade, or locking heads in resigned stoicism in the open.

Such searing summer experiences immediately trigger in me sometimes painful memories of two of the archetypical experiences of Australian land management: bushfires and droughts. Each inflicts trauma in a different way. The former evokes memories of adrenalin and danger, smoke and confusion; of shooting fatally burnt sheep with blackened wool, feet and mouths; and of the speed with which a tranquil (albeit uncomfortable) hot summer's day changes to a black landscape with smouldering embers and tree-boles, burnt fence posts, yellow overalls and firefighters, vehicle tracks embossed on the soft grey-black of talcum-like ash off once diverse golden grasslands. The second archetypical experience – drought – is the opposite. Here, there is slow torture and helplessness in the face of mocking blue skies; the sights and sounds of endlessly wandering sheep and bellowing cattle; and the constancy of no-win decision-making.

Fires, like drought, are part of the nature of this continent while being part of the ancient patina of our minds. The irony is that not only do we live in a fiery continent where a high percentage of its many bio-systems and species are adapted to fire, but fire was one of humanity's first tools. As such, fire was undoubtedly a critical factor in our cognitive, cultural and even physical evolution. Fire thus defines us as a species.

Once early humans learnt to control and then make fire, the Earth changed forever. This was because the tool of fire (along with agriculture) would allow our species to colonise and dominate almost every region on Earth. The deliberate usage of fire became a turning point in human evolution as it allowed our ancestors (certainly *Homo erectus* and possibly earlier hominids) to warm and protect themselves from predators and insects; to cook food; to allow their species (through torches) to expand into dark caves and colder night hours; and thence to enter colder climates. While there is evidence of human use and control of fire 200,000 to 400,000 years ago, new evidence in East Africa and China is emerging that pushes back hominid use of fire to at least 1.7 million and possibly even 1.9 million years ago.

What cooking led to was improved nutrition from cooked proteins, which probably triggered knock-on evolutionary responses, both culturally and biologically. For example, aside from changes in dentition, we are now adapted to digesting cooked foods. Because cooking rendered complex carbohydrates more digestible, this would have allowed greater energy consumption and an increase in brain size. Also, camp fires would have enabled greater socialisation – a key part of hominid evolution. Culturally also, fire would have allowed refinement of weapons (wood and stone); the use of glues and pitch for weapons and tools; the making of pottery; and later for the production of metals.

In time, fire became integral to clearing fields, improving pastures, easing travel, establishing tenure, signalling during warfare and performing rituals. Fundamentally, the use of fire by foraging, horticultural and pastoralist groups would have led to increases in both productivity and the quantity and quality of food and fibre plants. Such 'fire management' could even be continent-wide – as occurred in Australia with Indigenous fire management in both forest and grasslands. However, when burning was practised over long periods (even millennia) it could alter pathways of vegetation succession, change nutrient cycling processes and thus, in time, alter the distribution and abundance of species. Such burning practices over centuries and millennia have left long-term impacts on landscapes.[2]

Nowhere is this more true than in Australia, which of all the continents is a land particularly exposed to fire in what, for vast reaches of time, has been a blazing continent. That is why, in any discussion of the rise of regenerative agriculture in Australia, we need to at least briefly address the use of fire.

Allan Savory lists fire as one of the eight management tools available to farmers, and especially pastoralists (the other seven tools comprising money and labour; human creativity; rest; grazing; animal impact; living organisms; and technology). Regarding landscape function, an inappropriate fire regime is clearly just as dangerous as an inappropriate grazing regime such as set-stocking (i.e. no rest and recovery of the grassland ecosystem).

Savory (whose views are much influenced by his extensive experience in Africa and the rangelands of the USA in particular) believes that increased frequency of fire in the last 10,000 years, when combined with a reduction in the disturbance to soil surfaces and vegetation caused by the inadequate herding of animals by predators in the last few centuries, 'is one of the prime factors leading to desertification in brittle environments'.[3]

Typical brittle environments are the grassland savannahs of the world, including much of the top half of Australia. These are tropical and sub-tropical regions that get most of their rain in a short period during the monsoon or 'wet' season, which stimulates rapid growth of tall vegetation if it is not properly grazed. This is then followed by a prolonged dry period. At the end of this 'dry', there is little of both green feed and readily digestible protein. In such environments, various herders and livestock or game managers tend to burn the dry grass before the rainy season so as to create short green feed for stock. In the 'top end' of Australia, the end of the dry season is also usually associated with storms and lightning, presaging the breaking of the season. The result either way is hot or holocaust conflagrations – often burning many millions of acres.

Ecological damage stems from multiple causes. First is the repeated regularity of burning at the same time of year, for nature abhors a lack of variance and diversity. Therefore a lack of variation in time of burning means that different grass, forb, shrub and tree species will be either preferentially penalised or else favoured in terms of reproduction and survival. As Savory nicely summarises:

> Fire that is not followed by any other soil disturbance [animal impact] tends to cause major changes in living communities. Any influence that creates essentially the same microenvironment over vast areas favours establishment of the few species of plants, insects, and other organisms adapted to it.[4]

That is, what is not created are healthy mosaics, patches and diversity to all range of conditions, including different fire regimes.

But second, the hot burning severely impacts landscape function as it bares the ground and kills soil microorganisms. Soil surface management and high ground cover and diversity are crucial to healthy landscape function. Degradation of these impacts all the other landscape functions. That is why we see an increasing lack of biodiversity and an increase in less digestible, seedier and less desired species in regularly burnt grasslands, and increased woody shrubs instead of desired grasses in the rangelands. But even in non-brittle environments, we need to be careful.

Consequently, while someone like Allan Savory generally slams the use of fire as a management tool (and most of the time he can be right), nevertheless we need to be careful not to throw the baby out with the bathwater. This is because in this Gondwanan ark of Australia various large chunks of its biota are extraordinarily adapted to regular fires – including extreme events. Therefore, when used judiciously and strategically, fire can also be a creative tool in regenerative landscape management. In short, as experienced desert and fire ecologist Peter Latz observed, 'I came to realise that the present can only be understood by consideration of the past.'[5]

As we saw in Chapter 1, much of Australia's dominant vegetation types (such as eucalypts, acacias and hakeas) have adapted to Australia's arid and variable climate, its strong sunlight and higher temperatures, along with its nutrient-poor soils. However, there is one big downside to these adaptations: they make such vegetation highly flammable. But to this they are also somewhat adapted. And here we can see what appears to be a circular process of coevolution, because Australia's scleromorphic vegetation actually promotes its own burning (even to the extent that some species would probably become extinct without periodic burning). Thus, by the beginning of the most recent geological period (the Holocene, from 10,000 years ago), fire tolerance had become standard for as much as an estimated seventy per cent of all Australian floral species.

From the point of view of landscape function, a crucial role of fires is the triggering of nutrient recycling. In their 'Nutrient-Poverty/Intense-Fire Theory', the rangeland systems ecologists Gordon Orians and Antoni Milewski claim that 'most anomalous features of organisms and ecosystems of Australia are the evolutionary consequence of adaptations to nutrient poverty, compounded by intense fire that tends to occur as a result of nutrient poverty'.[6]

We now know that, over long time spans, intense fires in Australia have been regular events. With this comes the final step in the circle: these intense fires burn up or volatise the scarce micronutrients that are so essential to animals (such as nitrogen, sulphur, iodine and selenium), thereby exacerbating the nutrient poverty and driving the whole cycle again. Quite simply, hotter fires destroy more scarce nutrients. As Orians and Milewski conclude, intense fires in turn complement the effects of nutrient poverty in building more fire fuel by 'perpetuating the cycle of accumulation and loss of energy in the form of fuel'.[7]

This unique, circular nutrient-poverty/intense-fire syndrome appears to have led to a large number of the anomalous or uniquely different traits seen in Australia's biota and ecosystems. The result has been to set Australia apart as both an entirely different 'Gondwanan ark' but also a blazing continent with a range of features seen nowhere else on Earth.

With this wide adaptation of so much of Australia's tree, shrub and grassland vegetation to frequent fires, the question arises: how hot and how frequent is too much? As we saw in Chapter 1, it appears that for possibly forty millennia or more, Australia's Indigenous people had largely worked this out. They had become adept fire-managers of the Australian landscape: skilled in what anthropologist Rhys Jones dubbed 'fire-stick farming'[8] on this 'Biggest Estate on Earth'. Such farming maintained and sustained viable ecosystems, mosaics and other landscape components while also benefiting humans. Importantly, Indigenous burning was conducted not just at appropriate times of the year to avoid holocaust crown fires but also at the right time of day. Such burning, usually involving low-intensity cool burns, not only favoured fire-tolerant species but also tended to maximise species diversity in each region through the creation of mosaic patches.[9]

Significantly, it is becoming clear to some scientists and skilled observers alike that both Aboriginal cool-season burning and judicious diurnal burning was and is, in the hands of skilled craftsmen, an extraordinarily sophisticated operation. Not only were the ecological impacts of cooler burns at different moisture levels and at dew point appreciated (in terms of encouraging ecological change or seed, propagule and other vegetative regeneration), but there was a sophisticated appreciation of local air moisture and cooler temperature factors involved in creating long-lasting biochar via pyrolysis (the latter being the irreversible thermal conversion or decomposition of materials at increased temperatures – producing products like biochar and charcoal).

That is, unique sets of knowledge were being combined – and which I have noticed when working with my friend, local Ngarigo elder, Rod Mason. Peter Marshall (see Chapter 8) has articulated them in the language of physics too. He says that the earth breathes slow, out in the evening and in at morning time. It is diurnal.

Healthy soil has its own atmosphere, higher in CO_2 and humidity than the air above ground. As the sun goes down and the air above ground cools and pressure drops, higher pressure air in the soil pores moves upwards, bathing the surface in cool, damp, fertilized air. In the morning the process reverses and dew, fog, frost and oxygen is drawn into the soil.

Aboriginal fire-stick farmers would have understood this process and used the near-ground blanket of humid air to regulate their burns. They would have also known to stratify their burns to manufacture pyrolysis chemicals as growth stimulants and fertilizers to be drawn back into and stored in the soil come morning.[10]

Experienced Indigenous burners thus seem to understand that the dynamics of flame, fluid flow and radiant heat can be manipulated. If the climatic conditions are right (as described above), and the fire temperatures not too high, a top-down fire will radiate to the fuel-bed below. Incomplete pyrolysis of the converted organic material results in charcoal or biochar too: that is, a solid compound rich in carbon, and a great soil conditioner and environmental stimulant (as well as long-term storage substrate of carbon).

Charcoal and ash produced by different woods and organic material at different temperatures have differing values as nutrient stores, fertilizer, medicines and pigments. The downside to this long history of Indigenous burning is that the use of broad-scale fire-stick farming probably meant that Indigenous hunting coupled with fire was a possible factor in the last great extinction event of Australia's megafauna: with implications of this for healthy landscape function and thus regenerative agriculture.

The relevance of this event to the story of regenerative agriculture is that, when Australia had marsupial megafauna – and as still occurs in Africa, for example, with elephants, rhinos and the like (and bearing in mind that a marsupial diprotodon was as large as an elephant [two to three tonnes] and likewise produced up to 100 kilograms of dung per day) – there would have been a higher degree of constructive nutrient recycling. That is, there is evidence that our megafauna likewise acted as constructive eco-system engineers in our grasslands, through modifying large vegetation, turning over large volumes of topsoil, burying seeds and generally incorporating litter and vegetable material into the soil.[11]

In a burst between around 50,000 and 46,000 years ago, nearly all (ninety-four per cent) of the remaining suite of megafauna rapidly died out or were severely depleted to non-viable levels. This die-off overlapped with Aboriginal arrival and was probably due to the conjunction of climate-change-induced drought and Aboriginal hunting and fire-stick farming.[12]

For long periods, the megafauna would have played a huge role in the ecology of Australian plants and thus in landscape function via dominant impacts in controlling and modifying vegetation ecologies. To then suddenly lose this significant group would have had further significant impacts on landscape function – due to changes in photosynthetic capacity and a decreased level of nutrient recycling.[13]

All of this (including loss of dung breakdown and recycling) would have led to a massive drop in energy and nutrient recycling in landscapes. Hence, the Australian landscape after the megafauna disappeared was ecologically changed, and some key elements of healthy landscape function would have decreased. Arguably, Aboriginal over-hunting and especially burning contributed to this.

On removal of the megafauna, many grassland and woodland habitats would have reverted to dense and uniform formations of woody shrubs or tall, non-diverse grasslands. These are far more fire-prone and would have encouraged more frequent, larger, hotter and more-damaging fires. These in turn would have reduced fire-sensitive plants to the advantage of those more fire-resistant or more fire-dependent.[14]

A number of people now believe that the increasing growth of holistic grazing regimes across Australia (especially when well executed) is in part leading to a modified reintroduction of the equivalence of megafauna grazing and resultant enhanced landscape functional components. For starters, sheep and cattle when holistically grazed are fulfilling the role of Africa's vast migratory herds in stimulating healthier grasslands and woodlands through reactivating healthy landscape function. Recycling of the more digestible dung of these ruminants (for example, by introduced dung beetles and other organisms) is also assisting better function.

But this isn't the only ecological impact of regenerative grazing on Australian grasslands and woodlands. After the first Australian edition of this book was published, I received a letter that pulled me up short. It came from a leading Australian soils scientist whom I had met previously, and who is now based at the University of Oregon, Professor Greg Retallack. He is a *paleobotanist* and *paleosol* (ancient soils) scientist who has worked around the world, including in Africa and Australia. Whilst being a strong supporter of holistic grazing management as a tool for both regeneration of grassland ecology and fixing of long-lasting carbon to address climate change, he disagreed with my analysis of Australian grassland succession. He told me: 'Your book implies that regenerative agriculture is recreating Australian grasslands, but I beg to differ. Agriculture is recreating African grasslands with ungulates, and works primarily because of that, even though a lot of Australian plants are recruited.' He then explained: 'There are no preagricultural *Mollisols* in Australia [very thick grassland over 18 centimetres and of organic clayey and crumb-structured surface horizons] nor any sod-forming grasses as far as I am aware.'[15]

I was slightly puzzled by this, as I had seen grasses in South Africa of the same genera and very similar to our *Themeda*, *Eragrostis* and other plants. Moreover, these were C_4 grasses, whose evolution had occurred since eight million years ago – long after Australia had separated from Gondwanan Africa. Also, scientists kept reiterating that such C_4 grasses hadn't crossed into Australia over Wallace's Line and the northern Asian corridor. So, what was going on?[16]

I turned to Terry McCosker, head of RCS and a man who had thought deeply on these issues. Terry had worked with Stan Parsons in Africa, and he told me he agreed with Greg Retallack that regenerative grazing in Australia is recreating African grasslands – though Terry believes it is only 'a small point but technically correct.'[17]

The reason for this, says Terry, is 'that we have many of the light-seeded grasses seen in southern Africa (e.g. *Themeda*, *Heteropogon* and *Aristida*) but none of the larger seeded grasses … My theory was that over millennia light seeds were lifted to the jet stream by wild fires,' (that is, to cross east to Australia). Terry concludes that grazing herbivores are therefore having the same effect on grassland regeneration here in Australia as in Africa 'but not recreating the forms of grasslands that were here at White settlement.' Terry concurs with Greg Retallack, he concludes that 'with good management we are in fact recreating an African savannah or grassland.'

This in turn brings us back to fire. Terry then finished by telling me that, 'One of the most common questions I have had over the last three decades is "We did not have grazing herds in Australia, so why would grazing management work in Australia?" The reason I provide', he continued, 'is that our grasslands at settlement were the result of a fire climax, but now that we have ruminants we will not recreate what was there but should strive to create a grazing climax, which will replicate what was created by the large herds in Africa and America.'[18]

Finally, the point needs to be made that in addition to this 'African grassland recreation' in Australia, it appears that such regenerative practices do encourage mycorrhizal fungi in particular and thus the development of healthier, deeper soils and the laying down of long-lived carbon in the form of glomalin and chitin. So perhaps one day we too can have deeper Mollisols!

And that inevitably brings me back to the point of my own landscape management and another of many questions as I look to the future: can the fire-stick – one of Savory's eight listed management tools – be of any use in our environment? And with the widespread loss of millennia-evolved fire-stick skills post-1788, can a revival and application of such skills allow farmers like me to work with and not against the Australian continent's coevolved landscape functions? For it is becoming clear to me that the issue of fire management once again assumes critical importance as we enter a period of increasing climate change and ongoing land degradation.

Indigenous fire-stick farming skills are not lost, as they reside with highly skilled Indigenous people in many regions. Furthermore, I believe that if we landscape managers wish to further understand the unique landscape and ecological functions and the Australian continent's unique self-organising capacities then we need to begin to learn from, and work with, our Indigenous people – given the huge role of fire in the coevolved history of our landscapes, let alone facing up to the past reality of colonisation and the frontier wars.

To this end, we ran a 'cool patch-burning' day on our farm in April 2016. Our teacher was my friend, local Aboriginal elder Rod Mason: a highly skilled fire man (in addition to being a senior law-man).

The concept of 'patch-burning' in a 'mosaic pattern' has various names: 'cultural burning'; 'mosaic burning'; 'cool patch-burning'; or just a 'cool-season burn'. It means burning in a non-harmful way at the right time of the year to recycle nutrients, trigger germination of seed and control excess vegetation so as to create diversity and variegation in various patch sizes within the landscape.[19] For our region, autumn is the best time to encourage vegetation function (such as seed/propagule germination) but not create an uncontrollable burn as in mid-summer with a dried-out fuel load.

As we have seen previously, Rod Mason already knew our country: its history, its spirituality, and its biological functions and species. Indeed, the first time I met Rod, on a two-day workshop some years ago, I had been puzzled at what I thought was his overemphasis on fire and the need to always be at 'burning country' (that is, to constantly but creatively use the 'fire-stick'). But now I better understood and appreciated his focus on using fire to build – to 'make' and nourish country.

So on that day in April, we burnt two large patches, and I intend to make this practice a regular part of my management.

Recently, Rod told me, 'Making country always starts with fire. And that's because many plants and vegetation communities need a burn. You can walk through healthy country, the soil is cool and the ground is spongy. It is the charcoal that makes it cool. And you can create health – in your rockeries, herberies, ferneries, shrubberies, thickets, grassland, and in making forests of straight and not crooked trees.'

Hard-line scientists would have challenged many things on that cool-burn day, yet somehow the gelling of two knowledge cultures seemed to work – as it always does if the conditions are right and people come with an open mind. And it seems to me we need this blending of the 'two toolkits' if we are to creatively manage and regenerate country.

It is too early to tell the results, but I know we need greater plant diversity; I know we need to further revive our landscape functions; I know there is too much exotic grass seed lying in the soil surface and subsurface (which fire can kill); and I know that by creating patch mosaics we can enhance biodiversity while reducing the risk of impact of an outside encroaching bushfire sweeping dangerously across our land.

Rod has the firm belief that regular, planned and managed ancient fire practices of his ancestors would have left an imprinted pattern on the landscape (one of corridors, mosaics and niches); that this is still there, however faint, in old charcoal remnants; and that over time we can help the landscape become more alive and functioning again if we can discern and part-replicate much of this. I am sure he is right.

After my decades of experience with the 'red steer' or 'red devil' in Australia, there is one more thing I am sure about: I have barely left kindergarten in this area. Clearly, the old adage that 'what is old is truly new again' has weight. But equally (if not more) important is that a partnership using regenerative fire-stick farming with our Indigenous people – where each is involved in regenerating country together – provides a pathway to mutual understanding and endeavour and thus a route to reconciliation also.

That leaves the other great archetypical experience of Australian land management: drought. But here too we 'new arrivals' in this country have much to learn from our Indigenous predecessors who lived in and managed the same landscapes we occupy today. One of the lessons is that you can't 'beat' or 'fight' a drought but merely react to it while minimising long-term damage. Inescapable is the fact that droughts will occur. And they have had a major impact on our landscapes and especially on the Mechanical minds of us recent arrivals.

When I first encountered a major Australian drought in 1979, it proved a gulf too far. This was because I was inhibited by two major mental shackles. The first, as discussed in Chapter 2, was my inculcation into an apparent 'best practice' Mechanical mindset of industrial agriculture in the 1970s. The second was linked to the first: I had no concept of what our landscape had been, and could be, like.

The difference between me and John Robertson back in 1840 was that he had seen the country – a fire-stick-managed Aboriginal parkland landscape – in its natural, healthily functioning, diverse glory. But such a healthy landscape was long gone for me, as I learnt to farm on the vastly modified Monaro. Indeed, I grew up thinking that its degraded state was the norm and a situation we just had to live with. Only decades later would

I come to realise that it was possible to mentally reconstruct what the land-scape was like pre-Europeans arriving, and, in what was an incredibly exciting epiphany, that much (though never all) of the original ecosystem and its functioning could be regenerated. In the interim, however, I was constrained to live and work in an impoverished landscape, and unwit-tingly and mistakenly I busted my guts doing hard work that initially only impoverished it further. That is, it was only later that I came to fully appre-ciate the profound observation by historian Lynn White in a famous paper in 1967, 'Our ideas are part of the ecosystems we inhabit.'[20]

And so it took me nearly two decades of battling before I fully, deep-down, understood that there was a disjunction in Australian farming between applied agricultural practice and our intrinsic physical and cli-matic environment. Significantly, part of my awakening was encountering that non-European phenomena of a prolonged, El Niño-driven drought.

The 1979–83 drought had crept in like a slow-growing cancer. Until then, I had worked off my father's predictable and regular annual plan that had performed well for over two decades: plough at such and such a time for oats or pasture, then cultivate weeds at these other times, followed by sowing at times x and y. But all that was blown to pieces as the variable and then increasingly dry years followed one after the other. Sown-pasture after sown-pasture failed, leaving bare, exposed, 'weed'-infested paddocks. By the shocking year of 1982 (when only a third of our annual average rainfall was received), the entire farm was a dust bowl.

For the first time in our family's recent history, we had a major debt from drought-feeding stock, and having been forced to sell over half our animals, prospects for future income looked bleak. By early 1983, I have no doubt I was mildly depressed, and certainly physically and mentally exhausted. I was also angered by the mocking steel-blue of empty skies, the scouring August winds, and the relentless expenditure of mental and physical energy lumping feed bags, cleaning waters, fixing broken pipes and running endless calcula-tions on grain-feed quantities and a budget ever deeper in the red. At nights, I lay half-asleep, my mind wrestling with seemingly insurmountable prob-lems, my body aching in what felt like every bone and sinew. Trained, and by then confident, in modern agricultural science-based farming management, I had attempted to 'fight and beat' the drought. But like Mother Partington attempting to sweep back the ocean with a mop, in my hubris I had lost. As a consequence, my family would be saddled with debt for decades to come. At times, this made me angry, as I resentfully mulled over the mantra that it was all 'because the bloody rains wouldn't come'.

One day in the darkest depths of the drought, and after weeks on a tractor when I had observed dry storms circling the horizon and willy-willies buffeting dust and grass debris in spirals across my ploughed paddock, I watched a thunderstorm build through the afternoon. At one stage, it swung in over high, wooded hills to our west and I was sure would sweep down across our land, as had occurred in the past. But suddenly it swung to the north, following a high ridge. A few cruel drops of rain on my face and their cratered indents on the dusty surface of the tractor was all we received. I knew by the dark-blue sheets of rain that 'lucky' neighbours were receiving a potential drought-breaker. Whether it was this perceived inequity or the tantalising smell of moisture-laden air, or even the years of unanswered hope and expectation combined with mental and physical exhaustion, I don't know, but I came close to breaking then. At that stage of my life, I was not particularly spiritual, and I suddenly stood up, gesturing with violent, aggressive energy at an invisible and seemingly indifferent God as I used my full repertoire of sheep-yard four-letter words in telling him into which part of his anatomy he could stick his rain.

Today, I realise that the predicament I was in was no one's fault but my own. The reality of Australia is that it has regular droughts. The land, soils, microorganisms and other creatures and vegetation are adapted to this. And when irregular rain comes, it is often heavy and brief. To manage this 'normal' Australian occurrence requires a completely different mindset, and a different management approach and techniques to those of northern-hemisphere agriculture. In other words, the fifth landscape functional component becomes absolutely vital.

Only through intense pain, and after decades of block-headedly trying to 'fight' perceived droughts, did I finally come to terms with this Antipodean reality. What was equally sobering was my discovery that some others were much quicker learners than me.

One such person lives only three hours to our north and is named John Weatherstone. In the same early 1980s drought, he was like me – attempting to fight and 'beat' the big 'D' – and with the same result. However, in his case this drought galvanised him to initiate transformative change.

———

I first met Jan and John Weatherstone of Lyndfield Park near Gunning, New South Wales, in 2010. John Weatherstone was a man of gentle and quiet demeanour, short-cropped hair and weathered face and hands. As we began talking, he gazed out his kitchen window. I glanced out, and on a prominent birdbath in his garden two yellow-faced honeyeaters, a silvereye and a goldfinch were sipping water: the lively honeyeaters clearly bossing the situation.

Bringing his eyes back to the table, John refocused, staring at his hands. He immediately recalled that searing drought of 1979–83.

Lyndfield Park had been in the Weatherstone extended family since 1925, run originally as a merino sheep property. At the start of this tenure, the 880-acre (356-hectare) farm had been heavily timbered. Jan and John had taken over management in 1975, but the farm was barely a viable economic size in the 1980s when John decided to change the direction of his farm management and consequently his life.

On top of that, by 1982 the land was particularly degraded. There was virtually no remnant native vegetation left, and all the remaining dead but wildlife-valuable hollow trees had been packed and burnt. Salinity and acid soils were increasing; red-legged earth-mites were eating new pastures; erosion and weed infestation were rife; pests and diseases were increasing; and there were few remaining native grasses. Moreover, the variety of native birds, plants and animals was decreasing; there was little shelter for livestock; any spare organic matter such as grass was eaten, or, if crop stubble, was burnt; the few remaining trees were being clobbered by Christmas beetles; and remaining biodiversity was dwindling fast. As John recalled, 'To that list we unintendedly added a loss of soil structure, low soil organic matter levels, and increasing problems with new pests and diseases.' These problems, concluded John, 'were brought about by traditional practices such as intensive cultivation, stubble burning, heavy fertiliser application, and intensive grazing' – a practice encouraged and supported by government and private extension and research bodies.[21]

Indeed, John and his father in 1963 had become clients of a well-known consultancy firm which pushed them into high production so as to recoup high outlays in pasture improvement and fertilising. 'We were into cash cropping,' recalled John, 'and one year I had continuous cultivation from the northern to the southern boundary. Problem was, we now know this isn't cropping country. And then we'd get summer storms, and I'd be shovelling dirt and standing fences up again year after year ... but, while we were starting to recognise the problem, we didn't have any clue what the solutions were and so we just kept going on that same path.' In short, John had described a landscape in which all key landscape functions had either collapsed or were in the process of collapsing.

So it is hard to conceive that things could have become worse. But the situation reached rock bottom for John near the end of the horrific 1980s drought on Christmas Eve 1982: a day he calls 'the worst day of my working life'. At that stage of 1982, Lyndfield Park was a barren dust bowl, as was all of eastern Australia. There was no fodder nor agistment left anywhere with which to preserve valuable breeding stock. If the drought didn't break by April lambing, then John was faced with knocking every lamb of his 600 ewe flock on the head so as to save the ewes. Then came the day from hell, as John described: 'Hot, howling north-westerly gale, dust so thick that it was like a heavy fog, visibility down to 200 metres [660 feet]: the day we saw pictures of dust rolling into Melbourne.' Over the fence on John's southern boundary was the Hume Highway. In the poor visibility, limbs from trees were falling on the highway. One brave farmer nearby had at least preserved his humour, in putting up a sign near his property that read 'Beware – paddocks crossing road'. But for the Weatherstones, this Christmas Eve had no festive feel about it.

'Anyway,' John recalled, 'after lunch that day – and I still wonder what made me do it – I took my camera and went out to take a few photos: of this bare place, our topsoil just flying past, blowing away. I ended up near the highway, and I wanted a particular vantage point to take a photo, and I jumped over the fence onto the highway boundary, and of course it hadn't been grazed. It was dry grass this high [pointing to his knee], and when I jumped off the fence it was like landing on a sponge. And a lot of our own organic matter had helped the natural pile.'

The shock hit John. 'So I stood there looking at that and the contrast between the roadside and my own, hard, bared paddocks, and all this precious organic matter under my feet. And at that point in time I'd been working for twenty-four years and I'd put my youthful enthusiasm and energy and blood, sweat and tears and everything else into trying to make this farm something to be proud of and there it was, a bloody disgrace!'

The moment was John's 'Road to Damascus' experience. 'It is a wrenching experience,' he concluded. 'And I stood there perhaps a few minutes, and let that sink in and thinking about it, and I said to myself, "If we survive this, I'll do everything in my power to see that this farm never looks like this again." It was a life-changing moment, and that decision has just totally transformed my life because little did I anticipate the impact that taking those few photos was to have on my life and the future of Lyndfield Park.'

Transformation did not happen overnight for the Weatherstones, of course. There was a tough phase-in period as John began revegetating and regenerating their land and taking other measures once the drought broke in March 1983.

As John said, some steps were obvious and involved taking pressure off the land so that landscape function could begin regenerating. Others just evolved as the Weatherstones' minds moved more and more to the ecological and away from the extractive industrial-productivist.

Key immediate steps were lightening stocking pressure and beginning modified rotational grazing so as to increase soil organic matter; initiating tree planting for soil and livestock protection (in time, this would rapidly evolve into a major biodiversity drive that included planting a wide diversity of trees and shrubs); reducing chemical use and cropping and cultivation treatments; not burning stubbles but seeking to recycle them into the soil; treating erosion areas and preventing further erosion; aiming for more perennial pastures; and seeking greater enterprise diversity. The ultimate hope was that, in addition to reducing damage to the landscape, a regeneration of its functions would actually increase production (which in time did spectacularly occur).

However, as they changed from an exploitative to a regenerative mindset, John encountered two main problems. The first was a lack of available knowledge and experience at this early founding period of regenerative agriculture. 'We turned our back on a lot of what we had been doing,' recalled John, 'but we didn't know where we were going to go, and there was nowhere you could turn for help.' So through trial and error, adapting others' work, and talking to a range of people, John nutted out his pathway.

The second difficulty was the loneliness of going against the grain in what is a conservative industry. This was especially the case with his tree planting. One neighbour had spent a lifetime killing every tree he could get at, and his son likewise. One morning, when chatting over a boundary fence with this neighbour, he aggressively said to John, 'You're one of the blokes that likes planting a few trees, aren't ya?' Local rural merchandise people alluded to regenerative farmers and their reduced chemical use and purchase of organic inputs as dealers in 'muck and misery'.

'The problem was,' said John, 'we were straying from the fold. I learnt later many of our neighbours considered us eccentric.' Moreover, he added, 'our paradigm shift in management also put us at odds with much of the scientific advice … which only added to our feeling of isolation'. What helped to get John through was the support of his wife, Jan, in combination with his own quiet but independent strength, unsullied by overtraining in education institutions. 'I only went to Intermediate Certificate,' he said. 'No formal training. My parents sent me to Yanco Ag. College, but I hated every minute of it, and after two terms persuaded them to let me leave.'

In the course of our conversation (and many subsequent ones like it with other regenerative farmers), I kept asking myself one vital question: what were the key factors that led brave pioneers like John to sail a different course from all those around them in the uniform seas of industrial agriculture? John thought about this. He then suddenly recalled a crucial moment a few years after the end of the Second World War, when he was about six years of age. On this particular day, his father placed him in the passenger seat of an old farm vehicle. The boy had to sit on a cushion to see out of the dirt-encrusted windscreen as his father set out to drive to the far corner of their farm.

John remembered that, by this particular time, the farm had been ruthlessly cleared, and 'you could've jammed all the remaining vegetation into just one half-acre'. Their neighbours were of similar mind: native vegetation was an impediment to 'increased production', and the axe and bulldozer were still familiar tools of post-war heroic farming endeavour.

As the man and boy bounced across denuded paddocks and creek beds, the boy wondered where they were going, because all his father had said was, 'Come along. I'll take you for a drive and show you something.' Eventually they stopped below a long, steep ridge and got out. John knew the spot. He had hunted rabbits here in what was the furthest western corner of the farm, a section of poor, skeletal soil on a sedimentary shale ridge rarely visited by human or domestic animals.

The man and the boy climbed towards the top of the slate ridge. Before the crest, the father stopped, pulled a piece of cloth from his pocket, blind-folded the boy and then led him to the top. Then, father leading son by the hand, they walked over and down for a short distance. John's father removed the blindfold. When his eyes adjusted, the boy found himself in the midst of a carpet of variously coloured orchids and lilies: an entire hillside of some ten to fifteen species, including donkey, caladenia and green-hooded orchids, along with wax lilies – all bowing their heads as they swayed in a gentle breeze. When recalling the moment some sixty years later, John Weatherstone said it was as if he had entered another world: a magical wonderland of gorgeous pastel blooms set against a waving russet carpet of kangaroo grass in seed.

That boy grew up to become an inspiring trailblazer in revegetation of degraded farmlands. While, as a boy, he didn't ponder the stark difference between this patch of native beauty and the rest of a beaten-up farm, John partly attributed his later awakening to that magical day when his father blindfolded him and led him into another world. The empathic feel for nature and love of its beauty that this helped engender in him was part of what led to his farm regeneration.

There is a postscript, however. When I asked him about the fate of the orchids, John said, 'The sad thing is that as soon as we started putting super [superphosphate] out, that totally wiped them out.' We now know that many delicate Australian plants that have long coevolved with Australian landscapes and their phosphorus-poor soils cannot cope with high phosphorus levels when exposed to the super. Among other things, ancient mycorrhizal relationships with the plants (and ultra-efficient cycling of low phosphorus levels) are immediately blitzed by crude agricultural techniques such as massive fertiliser overloads.

As it turned out, the Weatherstones had already made a first move towards regenerating their farm before the 1982 drought. Having realised there wasn't enough income for two families on the farm (Jan and John, and John's parents), John had decided to start a plant nursery in 1977. Undoubtedly, his orchid experience as a six-year-old had contributed, but other factors were his father's interest in nature and his mother being a keen gardener. So, with some tough trial-and-error experiences, John launched into the nursery. This new enterprise received a boost with the UN declaration that 1985 was 'The Year of the Tree'. Suddenly John's knowledge and tree stock were in demand.

This led to him shifting the nursery from some ornamental-type retail plants to 'providing mainly native trees for farmers, mine reclamation, Landcare projects and so on', said John: what constituted a crucial role at the time. 'In some ways,' he recalled, 'we became a sort of father figure of local farm tree nurseries ... So by the mid-1990s, we were growing 120,000 trees a year.' At the same time, from 1984, John began collecting native tree and shrub seeds as a commercial supplier: initially for other seed companies. 'I always had a fascination with native seeds,' he told me.

By the 1990s, their nursery business had become their biggest enterprise. In time, as Greening Australia and others refined direct-seeding techniques, John became increasingly involved in seed supplying for this regeneration work that grew in importance and popularity from the 1990s – with private landholders but especially Greening Australia and Landcare groups. By 1996, John was able to get rid of his sheep enterprise and turn the woolshed over to seed drying and packaging. Then, due to rising costs, he progressively wound down the nursery operation so as to shed increased labour. His seed business now became the main focus. However, the Weatherstones continued running the nursery – largely to provide seedlings for their own on-farm tree plantings.

By this stage also, the fruits of the Weatherstones' courageous shift to a regenerative agriculture were becoming visible. The farm became famous nationally for its magnificent tree planting and transformation. As John said, when they began their tree nursery in 1977, 'We didn't know at the time but growing trees would become the central pillar of our lives at Lyndfield Park.'

Shortly before the 1982 drought, the Weatherstones' own, on-farm tree planting began when John ripped and planted a bare ridge of three acres (1.2 hectares) to provide shelter for lambing ewes. Then the cathartic Christmas Eve experience convinced John of the need for greater ground cover and vegetation, so the Weatherstones embarked on a remarkable regeneration and revegetation journey. Beginning slowly, they moved from planting 600 or so trees a year to many thousands, from narrow, linear shelter belts to blocks, and then, in 1988, adding a central block of forty acres (sixteen hectares) dedicated to biodiversity planting: 'to try and get a significant habitat in the middle of the place' was how John described it.

'This was enough to get birds to both start breeding there and also to start moving out through the linear blocks to the rest of the farm,' he said. Today, over 120 native bird species have been recorded on Lyndfield Park, a number of them regarded as rare or endangered for their region. This is without forgetting beneficial insects and other organisms. Not just the birds provide ecosystem services in the landscape (such as natural pest control and pollination); insects do it too, notes John. These include parasitic wasps, and European and native bees – not to mention microbats, other mammals and various reptiles.

John's recognition that they had lost a lot of their native birds and insects led to a new shift to biodiverse plantings. Here, using wattles and other shrubs, his design was to have at least two species flowering in every month of the year as food sources for native insects, birds and animals. 'We were after a more robust environment and ecology,' John stated. 'This at the start was more an ethos sort of thing … From my bushwalking and stuff, I had a really strong appreciation of natural bush systems, and one of my deep longings was to own a bush block somewhere.'

John also realised he had made mistakes. 'Unfortunately,' he said, 'we "tidied up" old trees, and probably our greatest habitat deficiency today is the lack of trees with hollows.' For he later recognised the value of this habitat in providing nesting, food and shelter for birds, microbats and other creatures such as possums and gliders. The same applied to logs on the ground. 'Tidying up' country is a legacy of the Mechanical mind, for it can take fifty or more years to bring back nesting hollows as trees age.

Notwithstanding this common mistake, as the tree-breaks and vegetation blocks progressively built up and the property (visible from the main highway) was transformed, Lyndfield Park became a beacon of hope: a shining example of how to regenerate land. Inspired were farmers, Landcare and Greening Australia groups, university and school groups, and then other farmers and groups nationwide, as the Weatherstones won major conservation awards. In the meantime, John was active in his own local Landcare group and also in the Yass River Valley Regeneration Project.

Significantly, economic and ecological benefits now rapidly flowed. In his large bush-block plantings and elsewhere, John's decision to not be a purist about local provenance but go wider paid off. He was able to obtain a broader variety of flowering species and gain higher survival rates and vigour, plus huge diversity and health. Within a short time, both bird numbers and species variety escalated enormously.

'In becoming growers of trees, shrubs and seeds,' he recalled, the Weatherstones' shift not only added new sources of income but also 'allowed us to develop skills and experience that have literally transformed our farm and our lives'. Since 1982, the Weatherstones have established more than 100,000 trees and shrubs. Many of John's latest plantings are with a view to seed harvesting: for what he calls 'seed orchards'. More recently, John has experimented with multipurpose food, fodder and timber trees and shrubs, such as exotic oaks, tree lucerne, honey locusts and the Chinese timber tree *Paulownia*.

Other benefits have been astounding. First, said John, was 'a significant lift in the carrying capacity of the land' to nearly double in twenty years. This was notwithstanding upwards of twenty per cent of the place set aside for trees from a start of nearly zero per cent. And then there was increased soil protection, reduced salinity, enhanced habitat, natural weed control, increased livestock forage and more nutrient recycling. In addition, there was vastly improved aesthetics and increased firewood fuel, along with commercial timber products and seed. 'When planning and selecting species,' said John, 'we always try to build multiple functions and benefits into all plantings in order to optimise our return.' But in the end, he said, 'the key principle in a healthy ecosystem is diversity'.

What is interesting is that John's vegetation regeneration activities have been spliced with an ongoing learning journey in regenerative agriculture. For example, he learnt from the Potter Farm Planning Scheme operators and their whole-farm planning ideas (see Chapter 16), and has interacted with leading biological farmers and facilitators. Now, he said, 'I just see healthy soil as a thing of beauty.' This also led to his elimination of nearly all farm chemicals.

A hidden benefit of the Weatherstones' remarkable thirty-year or so regenerative agriculture and biodiversity journey was a large appreciation in the capital value of their farm. Moreover, with sheep long gone and a mob of 200 cows among leafy, shady paddocks, rich ground cover and a diversely vibrant landscape, John had time to reflect on the past. When I asked him what he now called himself, he gazed into the distance. 'Difficult to describe it,' he said, 'but the nearest is cattle farmer and seed farmer' (I could also have added 'landscape manager' or 'visionary recreator'). 'I may have looked like a "greenie" to some people,' he said, 'but at heart I was still a farmer looking to work the land.'

As we finished our conversation around his kitchen table, he concluded, 'The key to it all has been to listen to the land, respond to its needs, be prepared to continually change your approach, and to constantly try new things.'

Those last words admirably sum up the key sentiments, in one way or another, of all the remarkable regenerative agricultural innovators I have been privileged to meet. In the end, the story of their minds opening came down to three key things. The first was they began to understand how the key landscape functions and entire ecological system worked, and how all were indivisibly connected: that none could function in isolation. The second was they got out of the way to let nature repair, self-organise and regenerate these functions. And the third and vital factor was they had the humility to 'listen to their land', to then change but also continue to learn with that same openness.

In this and the previous two chapters of this section on regenerating dynamic ecosystems, we have focused on three remarkable farming families: the Weatherlys of Connewarran in western Victoria; the Taylors from The Hill on the New England Tableland; and the Weatherstones of Lyndfield Park just north of Canberra on the New South Wales Southern Tablelands.

Each family placed major emphasis on regenerating dynamic-ecosystem communities on their farms: actions and considerable investments in time, emotion and dollars that were atypical of their farming communities. Each of them therefore risked ostracism – and some experienced this. Given the unpredictable, self-organising nature of complex systems and the subsequent results when one begins regenerating ecosystems, there is no way any of them could have anticipated the astounding ecological and personal benefits that accrued. In effect, they were stepping almost blindly into the dark.

Yet from getting to know these families, it is clear that each had strong ethical, emotional and philosophical drivers in taking these brave steps. Intrinsically, without knowing all the scientific and ecological mechanisms involved nor where things would end up, they acted because they felt it was the right and prudent thing to do. In the end, they were rewarded across all levels: personally, ecologically and economically.

This was because – and in a refrain I have heard regularly among regenerative farmers – they were prepared to let Mother Nature be in charge instead of seeking to simplify and control her natural systems.

As we have seen in earlier chapters, all four key biogeochemical landscape functions act inseparably and in synergy once one begins to put self-organising natural systems first. Whether it be the examples of regenerating dynamic ecosystems seen in this section; or Tim Wright and fellow holistic grazers regenerating their solar-energy function (via growing more green leaves for longer periods on a wide diversity of plants); or Ron Watkins, Peter and Kate Marshall, and Peter Andrews regenerating and healing their degraded land and salted water cycles (through fixing increased soil organic carbon using more deep-rooted plants); or Geoff Brown regenerating his rich Liverpool Plains soils and soil-mineral cycle (by enhancing a healthy living soil with no bare ground): all have recouped multiple benefits as their other three complementary landscape functions were rejuvenated and regenerated. Finally, the function that drove these transformations – the human–social – has in turn been regenerated by them, as farming families, consumers eating their food produce and many other people in many different places have all benefited.

Given the indivisible synergy of the five landscape functions, we turn now in the next section (Chapters 15 to 17) to that most crucial and fifth element: the human–social function.

Around 150 years of continuous grazing and burning prior to new owners in 2006 led to compaction, bare ground and erosion (and loss of profitability) at the Pearce family's 4000-hectare subtropical 'Bannockburn' cattle station near Rockhampton, Queensland. This 'before' photo was taken during the dry season (August 2006) before the owners began holistic and cell grazing management under the Resource Consulting Service's guidance. *Photo by Catriona Pearce.*

This 'after' photo was taken on the Pearce family's cattle station during the wet season (March 2013) a little over six years later. In that time, much of the landscape function had been restored, full ground cover regenerated and carrying capacity had increased 50 per cent. *Photo by Catriona Pearce.*

This photo shows Norman Kroon's 'Kariegasfontein' farm in the Karoo region of South Africa (*left*) and a neighbour's paddock (*right*). The right-hand paddock was how Kroon's country appeared when he began working with Allan Savory in the late 1970s. Today Kroon's farm is an example of regenerated landscape function due to holistic grazing management. *Photo by Norman Kroon.*

No-kill cropping, developed by Bruce Maynard in Australia (see Chapter 10), involves sowing crops into dormant grasslands using coulter discs and no industrial chemicals or fertilizers, underpinned by the use of holistically grazed livestock. The photo shows summer-active grassland and full ground cover at summer harvest. *Photo by Bruce Maynard.*

Bruce Maynard's son Liam. Bruce writes: 'This picture is of an Australian boy aged sixteen who planted, grew and harvested this crop without harming the ecology, using no fertilizers or chemical sprays of any kind, and with a tenth of the fuel use of conventional methods.' *Photo by Bruce Maynard.*

Pasture cropping innovation developed by Colin Seis and Darryl Cluff in Australia in the early 1990s. It involves sowing crops into dormant C4 grasslands using tined seed-drills. The photo shows the harvesting of pasture-cropped oats in 2005 with emerging C4 grasses underneath (see Chapter 10). *Photo by Colin Seis.*

Colin Seis harvesting high-value native grass seed after the grain has been harvested: one of many "stacked" enterprise options now available. *Photo by Colin Seis.*

An across-the-fence soil comparison of Seis's soil (*left*) and that of his conventional cropping neighbour (*right*). Seis's soil after pasture cropping: Soil carbon (over 204%, with 78% in stable humic form); greater water infiltration and water holding capacity (e.g. more than 200%); improved soil structure and soil nutrient cycling; all trace elements higher by an average of 172%; massive increases in microbial life (e.g. fungi 862%; bacteria 350%; protozoa 640%; nematodes more than 1000%.) This is on top of huge cost reductions and increased production. *Photo by Colin Seis.*

Ian and Dianne Haggerty developed natural intelligence agriculture on largely low rainfall sand country (see Chapter 7). This is Haggerty's crop after harvest. Note the total ground cover, maximised lush green summer growth and the presence of perennials. *Photo by Dianne Haggerty.*

Haggerty's neighbour's conventional crop as seen through the fence after harvest. Note the high percentage of bare ground and minimal plant growth, let alone perennials, on the neighbour's farm. *Photo by Dianne Haggerty.*

Regenerating dynamic ecosystems. This series depicts early results on a 250-hectare degraded wheat farm now part of a 1000-kilometre long reconnection corridor in Western Australia. Revegetation was via both direct-seeding and planted seedlings with many organisations involved. This 'before' photo shows Yarrabee at the time of planting in July 2007. *Photo by David Freudenberger.*

Yarrabee in April 2010 at three years of age. *Photo by David Freudenberger.*

Yarrabee in April 2013 at six years of age. *Photo by David Freudenberger.*

Another example of regenerating dynamic ecosystems. David Marsh's rehabilitation trial of a saline drainage line was started with direct-seeded trees and shrubs, and then holistically grazed. This 'before' photo shows dryland salinity in 1990. *Photo by David Marsh.*

This is the same reference point eighteen years later in 2008. The large dead tree in the centre is the same large dying tree in foreground of the previous image. *Photo by David Marsh.*

Peter and Kate Marshall, Australia, evolved an original water repair and reticulation system to capture, regenerate and heal eroded, degrading landscapes (see Chapter 8). This 'before' photo reveals what the Marshalls began with: typical active erosion and collapsed landscape function due to continuous over-grazing. *Photo by David Marsh.*

This 'after' example of water cycle regeneration shows over fifty hectares of wetlands and rehydrated pastures and forests that now store millions of gallons of water off 4000 acres of degraded catchment above their farm. The photo shows 'Swan Lake', *Phragmites* reeds, other vegetation and some water lilies from Monet's Giverny garden (descendants of material purloined in WWI by an Australian soldier). Reed warblers were singing in the reed beds. *Photo by Fiona Massy.*

When the Marshalls began, millions of gallons of water would be lost through their land in twenty-four hours during a big rain. Today the regenerated landscape allows healed creeks to flow all year round, nourishing those below. *Photo by Fiona Massy.*

Rowan Reid's outstanding agroforestry farm 'Bambra' is a forty-five-hectare outdoor class-room. This 'before' photo shows the waterway when he purchased the land in 1987 with erosion from over-grazing and over-clearing. Note the causeway bridge as a reference point. *Photo courtesy of Bambra Agroforestry Farm.*

In 2010, after twenty-three years of work, Rowan Reid, extracted a eucalypt log selectively harvested from the forest he planted along the same creek. Reid is crossing the same cause-way bridge as in the previous photo. *Photo by Cormac Hanrahan.*

CHAPTER 15

A Mysterious Dialogue

We can never know how wide a circle of disturbance we produce in the harmonies of nature when we throw the smallest pebble in the ocean of organic life.

GEORGE PERKINS MARSH, *Man and Nature*[1]

The day was hot, the ice cold, slippery and grubby, and the glare off the surrounding mountains intense. It was December 1974. My companion, an experienced New Zealand alpine climber, was setting a relentlessly steady pace up the Tasman Glacier. We were en route to a precariously sited alpine hut and hopefully some climbing after midnight: climbing that I, a young and inexperienced uni student, looked forward to with a mixture of excitement and apprehension.

After hours of tortuous meanderings, dirty, sweaty sunscreen, and the shedding of numerous clothing layers, my companion called a break. Having thankfully slipped the pack off my aching shoulders and taken a brief rest, I examined the large patch of coruscated ice we were standing on. For I had noticed it had a reddish tinge. Using my ice axe, I chipped into the sugary layer, thinking the colour came from some type of algae.

To my enquiry, and with an air of aggressive accusation, my companion said, 'No, that's not algae. That's *your* Mallee drought of the early 1930s!'

And he was right. At the same time as the North American Dust Bowl disaster was playing out, with 'black blizzards' sweeping thousands of miles east to cloak Washington DC and New York, Australian farmers and pastoralists had created their own version. An enormous tonnage of precious topsoil, swept by relentless drought winds, had become airborne: and not just from the cropping soils of Victoria's Mallee but also from wider areas of semi-arid grazing and cropping lands in three states. Filling the stratospheric heights of modern air travel and weighing a million jumbo jets' worth, this 'russet blizzard' crossed more than 3500 kilometres of eastern Australia and the Tasman Sea to turn New Zealand's Southern Alps red. Forty years later, I was chipping into a melted layer that revealed one of Australia's darkest, yet virtually unacknowledged, ecological disasters.

Reflecting on this memory, what surprised me was that until only a few years back I had given little thought to that Australian red earth embedded in Kiwi ice. Instead, my main memory had been of a detour my companion had made earlier that day to a steep side glacier falling into the main Tasman ice-river. Here, after a century of millimetric slow progression, the calving ice had revealed the remains of a dead climber, swallowed by a crevasse many decades ago and now exposed to daylight again and just visible through the ice. Silently we had inspected remnants of old hemp rope, khaki woollens, some rusted iron equipment, and the odd recognisable thigh bone and ulna.

Concerning this ghoulish diversion, my thoughts that day and indeed since had been more about my own and others' mortality. But now, over forty years later, it was the russet ice that penetrated my Australian soul. Here was evidence of the fleeting mortality of not just mere humanity but also an entire human civilisation. Oft quoted, but no less true nor powerful, President F. D. Roosevelt said after the heartbreaking years of the American Dust Bowl that 'a nation that destroys its soils destroys itself'. As we have seen in this book, many fine historians, trawling the evidence of over 6000 years of human civilisation, have confirmed this inalienable law. Yet in my own mind, and in the collective mind of the Australian polity, Roosevelt's fundamental observation had remained both uncomprehended and unheeded. Australia's red dust bowl may as well have never been, such is our collective – and, until recently, my own – amnesia-induced blindness.

The point of this story is the power of that fifth landscape functional element: the human mind. It was the imbibed and inherited attitudes of the Mechanical mind that caused Australia's red earth to become airborne.

Since Europeans landed in Australia (or, for that matter, on the American prairies, southern Africa's Karoo and Patagonia's mighty grasslands), such an environmental result was inevitable.

From the moment that the early ploughs broke through Goyder's Line of rainfall in South Australia, and once cattle and especially sheep were set-stocked on fragile landscapes 'to eat the heart out of the country', airborne red dust was also inevitable. This pattern was repeated into the twenty-first century, where Europeans have cleared mallee, pine, eucalypt, box and brigalow to make way for the plough. Both preceding and accompanying them were the ruminants: sheep, cattle, goats and others relentlessly grazing in an uncontrolled manner. That red Australian dust on New Zealand ice has been airlifted to New Zealand on at least five occasions since European settlement. Dust-storms still regularly occur to this very day. Such is the power of the obtuseness of the human mind and its embedded worldviews and mental constructs.[2]

The stories hitherto in this book reveal two things. First, such behaviour and consequences are preventable. And second, leading regenerative farmers have been able to overthrow the Mechanical mind to the extent where, in their oft-repeated words, they see their role as to get out of the way of Mother Nature. As we will see in Part III, specifically Chapter 18, what they are alluding to is allowing the processes of self-organisation and mindset shifting to occur. Indeed, as we will also see in Part III, I believe this entails an entirely new and post-Mechanical mind.

Sometimes, even when not pointed out, particularly sensitive or observant farmers and land managers have the blindfold of entrenched industrial-agriculture paradigms and training ripped off – allowing them to see what is really going on in their landscapes and to recognise its human causes.

The stories in the preceding chapters of Part II are powerful testimony to this capacity. However, in other cases the way our ideas and worldviews – and thus our management – impacts our landscapes is far less obvious. Indeed, these factors can remain invisible for long periods until they become apparent or there is a paradigm shift. One such case is the interaction of landscapes and domestic animal genetics.

One of the fruits of partaking in over forty years of breeding merino sheep and sometimes cattle and goats in the same environment is that I have learnt the folly of trying – in Mechanical-mind mode – to stamp my authority and preconceived ideas on both our domestic animals and the landscape. To the contrary, I have learnt there is a truly magical yet complex, mysterious dialogue going on between us, our animals and the landscapes we all occupy. Again, it was hard lessons that taught me this.

The first lesson was inevitable: on beginning farming I knew nothing of the extraordinary interrelationships within a landscape between the key functional elements that I have highlighted in the previous chapters. Moreover, as my journey continued and further lessons accrued, ever deeper layers of surprising complexity began to emerge. Included in this were seemingly mysterious forces that impacted us and our landscapes. In the end, I came to realise that, just by existing and stepping out my back door, I was in a continuous dialogue[3] with my landscape and all its functions, systems and organisms, with the animals I managed on my land and, perhaps most remarkable of all, with the collective wider environment and universe surrounding us.

Like the brash, young, confident climber that I was so many decades ago, when I began farming I didn't know enough to realise how ignorant and lacking in wisdom I was, which is the crux of this entire story. For, in the previous chapters of Part II, I have explained how four of the five key landscape functions operate, and how our landscapes and the interconnected networks in our environment interact. But indivisible at every step along the way is *the* most influential of all the landscape functional components: us humans. The greatest of all determining factors on the healthy regeneration or else degradation of those very landscapes boils down to the way we think, what we believe, and how we model in our minds the way the world and our landscapes work.

So the rest of this final segment of Part II (Chapters 15 to 17) will focus on elements of regenerative agriculture where the human–social aspect is a key focus or influence. However, as with all the previous examples in this book, each of the five landscape functions is indivisible from the others. All should work in synergy.

It was more than four decades of primarily breeding merino sheep in our tough Monaro environment that fully woke me to the complex and marvellous interactions involved in our own and our animals' mysterious dialogue with nature, and with our landscapes and our surrounding environment. For over thirty years until 2007, I developed and then managed a merino sheep stud on our farm, which annually sold up to 300 or more rams. This required upwards of 1500 stud ewes as a breeding base.

I found that breeding and managing a merino stud – where animals had to perform in a challenging environment – involved an endless journey of ongoing observation and learning. But it is only in recent years, as I have encountered wonderful thinkers and scientists in the regenerative agriculture world, and as new genomic knowledge has emerged, that I have come to appreciate just how vital this key factor of managing genetic packages of animals in an environment is. This area I call 'adaptive landscape genomics'. By this, I mean the genetic and behavioural interactions between either human-managed domestic animals and their environment, or else between we humans and our own environment.[4]

This plastic and fluid genetic–behavioural interaction relates directly to this story in a number of ways. First, subtle but important landscape–genetic interactions impact not just a farmer's animals (and their fitness for purpose) but also their landscape, themselves and their profitability. Second, both domestic and wild animals are constantly learning and adapting their behaviour when grazing and browsing in their environment (for example, in relation to the different chemicals that plants contain).

And third, as we will see in Chapter 21, we too are ancient animals who coevolved in landscapes and became genetically and physiologically hard-wired for dynamic biogeochemical interaction with these landscapes. Thus, the food we buy or eat off landscapes (including from the animals we run that graze these landscapes) and its degree of nutrient density, variety and quality (i.e. its integrity) has a massive influence on our own health and on our own genetic expression, not to mention on our learnt and reflexive behaviour. The fact is, modern industrial agriculture is destroying the inherent natural function of plants growing in healthy soils, and this is grossly denaturing our food. This contravention of millions of years of coevolution is therefore reaping disastrous consequences for human and societal health.

However, to arrive at an understanding of the nature and implications of adaptive landscape genomics, I had to undergo a tortuous journey of trial-and-error learning. This long story I will not relate here, except to say that it began when breeding my own rams in the late 1970s, and then continued as I developed a complex stud-breeding operation that used the latest tools in biology, molecular genetics and breeding technology. By the end of this journey, I had come to the painful realisation that domestic animals grazing in our landscapes are not mechanical boxes we can simply manipulate in the expectation that they will adapt to any and every environment across the country.

In the process of this animal breeding and genetic marketing role and of pioneering some innovative wool marketing to leading global textile customers, I had also come to realise that both our businesses and our landscapes are not immune from another crucial human factor: the intervention of worldviews in the areas of both breeding and statutory wool-marketing politics.[5]

Eventually, I came to my senses and fully committed to regenerative agriculture, but I had lost over a decade of ecological progress. However, the big lesson for me out of this period was just how important the human mind was in our behaviour: that we land managers (and everyone else, for that matter) are driven and captured by our belief systems, which can make it extremely difficult (indeed impossible, in many cases) to see an alternative.

By the time I came to finish examining this issue in my PhD, I saw the key to farming behaviour as being the learnt concepts, understandings, inherited and trained beliefs, mental models and paradigms we carry in our heads. For it is these that determine how we perceive the way the world works and our landscapes function. This in turn determines the management and thus health of these landscapes. Crucially, it is our thinking that therefore determines the health of both domestic animals and the humans dependent upon them. In other words, I realised that the human–social factor ultimately is the most critical of all the landscape functional elements.

I also became convinced of another reason why this fifth component of the human–social should be included in the ecological-literacy model: the mysterious dialogue that we humans have with our livestock and our landscape. Physically, mentally, emotionally and (for some of us) spiritually, humans are an indivisible part of the very environment, landscape and universe in which we exist and function.

This was brought home to me in 2011, when I crossed the Snowy Mountains and drove down to Holbrook: a small rural town located on the eastern, moister edge of the New South Wales Riverina.

Every second year since 1999, a number of regenerative farmers, plus invited scientists and visiting speakers, gather for a two- or three-day conference at different venues around Australia. Their focus is Australia's native grasslands. The convening body is called Stipa (the generic name of a key native grass genus, commonly known as speargrass, and sometimes corkscrew). This organisation – on the initiative of Darryl Cluff – was formed in 1997 by a group of around forty interested farmers, scientists and natural resource managers at a breakfast meeting in Dubbo. All of them felt it was time to promote the enormous importance of Australia's hitherto neglected native grasslands.

I had taken the four-hour drive specifically to hear a visiting American professor from Utah State University called Fred Provenza, as his work and thinking had intrigued me for some time. I was not to be disappointed, for Provenza spoke at length at a number of sessions. What captivated me about this smallish man (whom we will meet again in Chapter 18) was his deeply holistic understanding of grazing animals that had coevolved in landscapes along with the very plants they grazed. Using his lifetime of research in the prairies and rangelands of North America and in the laboratory (and latterly with colleagues in Australia), Provenza was able to draw out the implications of this coevolution for landscape function; for regenerative managers; for a deeper understanding of the way entire ecosystems functioned; and for both animal and human health.

Subsequently, I would get to know and befriend this gentle, humble and deeply insightful man, who had an uncanny ability to think and feel himself into the landscape and the animals grazing therein. But on this particular day, it was one slide in particular in one of his PowerPoint presentations that stuck in my mind. It became for me a motif for this story and chapter. 'The genes of animals and humans,' Provenza said, 'are constantly in dialogue with their environment.' At the time, however, I did not appreciate just how significant this statement was, nor the deeply mysterious yet profound nature of that dialogue.

One of the early books I studied as I learnt my trade was *Animal Breeding* by the famous Dutch geneticist A. L. Hagedoorn. I came to rue the fact that, at the time I first read it, I didn't know enough to appreciate this statement penned in 1939: 'The animals and plants, and the human race breeding them, belong in one symbiosis-group, adapted to the country and climatic conditions. It would not be possible to exchange members of one group for those of another group without disturbing the equilibrium.' However, what I later came to understand is that this connection between ourselves, our landscapes and our animals has a long history of coevolution that goes back millions of years.

It was only later when mulling over Hagedoorn's words and then Fred Provenza's presentations that I realised they were talking about what I had called 'adaptive landscape genomics': that genetic and behavioural interaction between human-managed domestic animals and their environment, and also between humans and their environment. The importance of this factor is this: I have found that only a few leading animal breeders and land managers place a specific emphasis on the plastic adaptation, evolution and selection of animals and plants best suited to their environment. They do this via a range of mechanisms, including allowing for epigenetic functions and by placing an emphasis on other, non-additive gene functions and also on interactive learning between grazing animals and their natural environment.[6] In many respects, Hagedoorn had intuitively anticipated this. Moreover, if allowed for, this interaction between humans and their animals can lead to a landscape manager enhancing the adaptation of a population of animals to a particular landscape (or else to counterproductively triggering the converse).[7]

Such an approach to animal management within a farming landscape has not been fully considered, let alone maximised, even in regenerative agriculture. The fact is, each farm environment or landscape (and even different environments within a farm) varies greatly in terms of biogeochemical composition and many other variables, and these variations can have a huge impact on managed, let alone wild, animal populations. Much of the field of genomics is very recent knowledge, arising from the capacity to describe the entire DNA or gene sequence of different organisms. In addition, we now know there is a close dialoguing between animal genes and their environment and human genes and our environment. To ignore this constant interaction and genetic and physical adaptation is to ignore the opportunity to select animals best suited to a particular environment, and the same applies to learnt behaviour in environmental interactions. Crucially also, to ignore this plastic and ever-changing interaction between ourselves and our own environment – and especially regarding the food we eat and the water we drink – is to ignore a key but subterranean issue as to why Western civilisation is characterised today by massive and rising levels of unanticipated modern diseases (as will be discussed later in this chapter and in Chapter 21).

It is important to understand what I mean by the 'coevolution' of animals and humans in their environment. If we take a thorny bush or wattle in our grasslands, woodlands or rangelands, for example, the bush (such as a gruggly bush on our farm) didn't first appear fully clothed in thorns, nor the wattle replete with high levels of tannins. As the various grazed environments became distinct ecosystems after ice ages or long drought periods, so the adapted vegetation and flora also evolved. Similarly evolving and adapting were grazing and browsing animals, who were consuming the evolving vegetation. So a long and slow process of interactive coevolution occurred whereby when a plant was grazed it responded over time by producing defence measures to deter animals: measures such as thorns, small and hard-to-access leaves, or vegetation with unpalatable chemicals such as tannins and also poisons. Thus, in time, plants in a grazing/browsing landscape didn't just develop physical and also behavioural structures to handle a challenging physical environment; they also developed vast numbers of defensive chemicals to handle the grazing environmental factor. These chemicals, because they are plant-photosynthetically derived, are called phytochemicals or phytonutrients.

But this is where the long process of coevolution kicked in. With each plant evolutionary response, the grazing and browsing animals in turn genetically adapted and evolved to handle, detect and/or digest these different chemicals. Animals learnt to use some of the phytochemical compounds to medicinal and nutritional advantage (for example, tannins to kill intestinal worms, and other phytochemicals for crucial metabolic function). And so, along the way, these animals evolved not just the physiological mechanisms to handle and digest the chemical compounds but also the chemical receptors in different parts of their body to detect hundreds of these compounds that had become integral to their health. This is a classic example of coevolution.

Significantly also, humans – as early and then modern hominids – were likewise coevolving at a crucial moment in the African savannahs. As with the grazing/browsing herbivores, our bodies also respond to plants through becoming hard-wired to detect (via different receptors in various parts of our bodies) a whole range of primary and secondary nutrients and phyto-chemicals essential to our health and nutrients and phytochemicals in both plants and meat (because the latter in turn contain ingested plant-derived nutrients and phytochemicals). As we will see in Chapter 21, when we humans become divorced from a healthy natural environment that should contain foods high in nutrient density and plant phytochemicals (along with chemically balanced water), then the non-activation or disuse of our ancient coevolved nutrient and chemical bodily detection systems can have major, and indeed horrific, implications for human health. And so when we link this background to epigenetics, and then place both animals and humans in maladapted environments, then we should be prepared for both unexpected and often seriously deleterious results.

So why is our new understanding of this field of epigenetics so import-ant? Well, in regard to coevolved animals and humans, we are only now in a position to begin more fully appreciating the implications of much of the new work in modern molecular genetics. Such work has thrown out many of the simplistic assumptions that followed the unravelling of the molecular architecture of DNA by Watson, Crick and ilk from 1953 and the resultant 'central dogma' of molecular biology. This dogma held that information only flowed one way in the genome. Communication was thought to be from gene to its effect, and not the reverse. In turn, linear, mechanistic, post-Enlightenment thinking led to the conclusion that 'one gene coded for one protein'. But much of this has now gone out the win-dow, with huge implications for both livestock managers and ourselves.

We now know that not only is there a layer of gene regulation that directs form and function in complex organisms, but also that this is incredibly fluid and can occur as a two-way interaction with the environ-ment. Ironically, this new knowledge has overtones of the long-debunked thinking of Jean-Baptiste Lamarck: the controversial French contempo-rary of Charles Darwin.

This latter issue of a two-way flow (or dialogue) between genes and the animal's environment is exemplified by epigenetics. By this term is meant the situation where external environmental factors can directly trigger the way genes are expressed or function, but without changing the actual DNA or gene code. That is, what is affected is the activation or silencing of a gene: the on–off switching mechanisms that control gene expression on a chromosome (the string of genes that shape the individual).

In other words, though the underlying DNA sequence of an organism is not changed, non-genetic environmental factors can nevertheless cause the organism's genes to behave differently. This occurs through, for example, chemical control mechanisms such as DNA methylation. Crucially, epigenetic changes to an animal's genome can be far more rapid than standard DNA mutations or changes in evolutionary pathways, and they can be more easily reversible. This only serves to enhance the adaptability or 'evolvability' of a species. Another way to regard epigenetics is to view it as a form of 'gene ecology': whereby there is no separation between genes, the organisms and the environment.[8]

A clear Australian example of epigenetic effects working practically in the field was demonstrated by a group of scientists in Western Australia who placed ewes on saltbush for the full term of their pregnancy. Exposure to saltier forage led to dramatic changes in both form and function of the ewes' lambs post-weaning. Thus, by comparison with lambs *in utero* whose mothers were on different feed, those lambs off the saltbush country showed a capacity to more efficiently excrete salt (the 'function' component). When their kidneys were examined, they showed clear changes in renal structure (the 'form' component), which enabled this higher salt-excretion capacity.[9]

Earlier, when talking about keen horse-breeder and 'water-dreamer' Peter Andrews in Chapter 8, I related his story of a visit to England's leading thoroughbred stud of Beech House, Newmarket (owned by Lord Derby). Andrews recounted how on this visit he discovered 'there was a traditional belief … that once a field was ploughed and reseeded the pasture wouldn't be any good for young horses for five years and was unlikely to produce a Group One winner for ten years'.[10] My bet is that at least part of the reason for this is the impact of epigenetic factors.

But surprisingly (given that the vast array of life is microbial – including in our own bodies and those of the animals we breed and the plants they graze), this epigenetic story assumes even greater significance when we begin talking about the role of the micro-cosmos in human–animal–plant interactions in an environment and attendant adaptability and coevolution.

The fact is that when talking about the broader context of genetic plasticity and adaptability in us and our animals, perhaps the most crucial – but hitherto unappreciated due to its invisibility – element is our gut microflora, or what is called the *enterobiome*. The population of fellow travellers in both humans and animals – the microbiota on us and in us – far outnumber our own cells. Consequently, when an animal senses and interacts with an environment (especially via the ingestion of food), the actual front line of such interaction is this microbiome, given – as my friend molecular biologist Professor Steve Hughes of the University of Exeter explained – that 'the dominant portion of the genetics which interface with the environment is microbial' and that this in turn almost certainly involves vast amounts of epigenetic behaviour.[11]

There are more microbial genomes than human genomes in our bodies, and the same applies to our grazing animals. Logically then, this would seem a key place for the enterobiome to play a hugely important but unappreciated role in this broader context of genetic plasticity and adaptability. Moreover, as Professor Hughes further points out, 'Plants and their endophytes, epiphytes and rhizosphere microbiomes are subject to the same logic. And of course, further complexity,' says Hughes, 'comes from microbe–microbe interactions, for example in surface biofilms. This can help us to explain why selections [both animals and plants] for one environment may not work in another until locally hybridised.'[12]

In short, as animal breeders managing animals in varied landscapes, the truth is we know little about this microbial world – except that it is probably the biggest player of all in human, animal and plant adaptability and function in our environments. The key message? We therefore need to step back, have humility, and not think we can play God in the breeding and management of animals and plants, but instead let natural plastic enablement and adaptation act out: that is, let nature's self-organising capacities unfold unfettered.

In the light of this new thinking on epigenetics and plastic genetic change and expression, I now look back on my early stud merino breeding with dismay. I am astounded at how mechanistic my thinking was: that I simply thought I could be single-mindedly focused on wool weight to construct animals that I assumed could fit into every environment across Australia.

What is even more sobering now is that I have also come to realise that epigenetic effects have huge implications for human diets deprived of nutrient integrity and phytochemicals due to either poor landscape health or agricultural and other cultural food and business practices. This important issue I will illustrate with a story at the end of this chapter, but it will also be addressed in Chapter 21, when I touch on the role of our human enterobiome and its crucial role in our own health.

The functional element of adaptive landscape genomics therefore has three key components relevant to this story. The first concerns the population of domestic animals we managers use as regenerative tools in the landscape. But it can also mean the plants we breed and/or manage in the landscape (as we will see at the end of this chapter). This includes both their genomes but also functional elements such as epigenetics. The second component clearly is the farm or landscape environment the animals run in and where the plants grow (which physically and genetically interacts with the first component); while the third component is the breeding and management goals of the humans overseeing this interaction between animals, plants and landscape – the latter being that fifth but overwhelmingly vital human–social landscape functional component.

But over and above all these factors, and at the heart of regenerative agriculture, is the process of 'self-organisation', which I alluded to earlier. This vital factor will be explored more fully in Chapter 18, but very briefly it involves the larger propensity of not just living systems but also other complex adaptive systems to regenerate and seemingly spontaneously provide solutions to triggered problems or destabilisation. The amazing process of self-organisation has only begun to be articulated and understood in recent decades, but it is unsurprising to me that the insightful and empathic thinker Aldo Leopold early on defined 'land health' simply as 'the capacity of the land for self-renewal'.[13]

Nevertheless, because we are dealing with dynamic ecological and self-organising functions – and functions, in the case of agriculture, part-initiated or triggered by humans – we can never quite know where the landscape will end up. It is almost as if the landscape is also learning and adapting. Indeed, the amazing processes involved in self-organising systems could be interpreted by some as expressing a form of ecological-landscape 'intelligence' or cognisance. What is most certain is that a crucial dialogue is going on between us land managers, our landscapes and all the other organisms residing there, and also with the energy and nutrient and matter flows within the whole ecological caboodle.

In the case of the best regenerative agriculturalists I have met, this mysterious relationship between humans, animals, plants and the landscape constitutes a true interactive dialogue: each appearing to be learning from the other. This of course requires a great degree of openness and humility. Sadly, however, the capacity to be open and humble is not universal. We humans now constitute an overwhelmingly dominant and increasingly dangerous determining factor in the destiny of this planet. This is because we have arrogantly placed ourselves outside the functional parameters of Earth's operating systems, and it is why I have made the human–social factor our fifth – and in fact most vital – landscape functional element. At this crucial juncture of the life of our planet, do we still believe we are in control of nature, or can we be open to humility, wonder and enchantment? In short, can we evolve past the Mechanical mind? As I hope this book is showing by now, I believe the answer is 'Yes'!

It is to illustrate this element of openness and humility in our 'mysterious dialogue' with nature that I conclude this chapter with one family's regenerative agricultural journey. For it is a journey that involves listening to the land and its natural systems, and which in the end yields both a healthy landscape, animals, crops and people but also vastly healthier food for human consumption.

The shift to regenerative agriculture is not always triggered by mind-cracking, 'Road to Damascus' experiences. Instead, it can be more gradual and can be an accumulative manifestation of philosophical, spiritual and other personal factors that have freed farmers from the Mechanical mind. Openness to a post-Mechanical way of thinking underpins the story of Dayle and Terri Lloyd, who come from the cropping and grazing lands of south-west Western Australia.

The Lloyds live in the south of the reliable Great Southern cropping belt near the small town of Dumbleyung. What attracted me to their story is that they are a good example of successfully taking modern farming back to previously more balanced and natural approaches. This in turn is having a major impact on food consumers because of the wonderful healthiness of their produce. They are returning nutrient density, variety and balance to our staples: in this case, cereal flour, sheep-meat and merino wool.

Dayle and Terri Lloyd and son Campbell own farms totalling over 5000 acres (2000 hectares) plus a stone-grinding flour mill: all of which collectively goes under the label of Eden Valley Biodynamic Farm. Before Dayle and Terri married in 1987, each was idealistic and committed to a healthy environment, with an abhorrence of chemicals. Holding strong Christian values, they cared for the Earth, each other and humanity. Dayle first introduced biodynamics onto the family property Eden Valley in the mid-1980s after he heard a talk by Alex Podolinsky.

Podolinsky's focus on healthy soils explained for Dayle why he had chronic and debilitating ryegrass toxicity in his sheep and increasing salinity in the creeks on his farms. On swinging over to biodynamics, Dayle saw rapid and demonstrable benefits in soil, pasture, crops and animal health. He married Terri just as he fully swung over to the new system, and Terri, who increasingly became active in bodies such as Greening Australia, took little convincing.

At that time and into the early 1990s, cropping agriculture was in the throes of changing to direct-till and increasing chemical use, plus increased stubble burning. After the collapse of the wool floor price scheme also, many farmers shifted to continuous cropping. 'We just hated the chemicals,' Terri told me in 2010 when I visited their farm for a yarn, 'and we were very interested in the health of the family [they had three children] and because there is cancer on both sides of our family. So the increasing use of chemicals was close to the bone.'

As with a number of leading regenerative cropping innovators, grazing animals were integral to soil health and resilient crops and landscapes under their system. Dayle's biodynamic approach hinged around his locally adapted merino flock in a three-year pasture phase rotation between crops, and careful stimulatory tillage instead of chemicals. 'Unless you've got livestock and animals as part of your system,' said Dayle, 'you won't have a complete system: you've lost a vital component.' Significantly, and ignoring the new 'high performance' and genetically modified crop varieties that emerged after the 1970s, the Lloyd family deliberately used old barley and wheat crop varieties that they had maintained or sourced going back sixty years: strains that had been bred for more natural conditions.

Owing to this regenerative farming approach, one of the things Dayle noticed was that, under their healthy soil/biodynamic operation, the yields of these old crop varieties were as good as or better than the new, high-production dwarf and synthetic-chemical-addict varieties. Dayle found they were getting up to fifty per cent greater yields in his healthier soils with these old varieties.

Dayle said that the old varieties he chose were bred to utilise natural nitrogen and biological processes. But put them under a modern chemical regime and their yields would crash. Conversely, in a natural system, Dayle and Terri's old wheat and barley strains had great natural disease resistance, and they were finding increased demand for their seed because growers were getting increased production, sweeter hay and edible stubbles when leaving the industrial treadmill.

Furthermore, the Western Australian Department of Agriculture (allegedly there as 'public servants') refused to give Dayle access to older barley varieties that he wished to trial. Subsequently, Dayle discovered that the Department, without consultation, had ceded up to forty-nine per cent of their plant-breeding program to Monsanto. Nevertheless, he was able to privately source some varieties.

Dayle has also been influenced by the unique work of South Australian wheat grower Don Whiting, who conducted four decades of trials with over 200 heritage and modern Australian wheats on his farm. Dayle trialled twenty-two old wheat varieties, spanning the 1890–1996 period, and found some of the old varieties performed exceptionally well. He said 'they seem to have a genetic capability to improve' and so were eminently suited to his 'natural farming system'.

In this system, Dayle selects plant varieties that have the following traits: genetic and physical ability to adapt naturally to his system and changing climatic conditions; early vigour so as to maximise growing conditions and to establish ahead of weeds; a leaf canopy that shades the soil (less evaporation, plus weed suppression); drought tolerance (waxy/hairy leaf surface) and being able to come into head as the season dictates while maximising available moisture; excellent hustling for nutrients in his healthy soils and being able to cope with major fungal diseases; being suited to his individual management; and finally high-quality grain, suited to customer end-use.

In short, this is a list that demonstrates the enormous flexibility and adaptability that arises from working 'with' nature as opposed to the Mechanical mind's attempt to dominate and control it. In the former approach, farmers and the community are empowered and made healthy. But in the latter, it is the dominant power base of agronomists, scientists and multinational product suppliers, traders, food manufacturers and so on who are the key beneficiaries.[14]

Having set up a regenerative farming system, the Lloyds' next step became logically apparent. As the cropping revolution took off through the 1990s, and as they stuck to their one-year-in-four cropping cycle (as opposed to continuous cropping elsewhere), they realised that to remain viable they needed to value-add their products or else they would be forced to join the chemical, constant-cropping industrial treadmill.

The result was twofold. First, they started value-adding their wool, to the requirements of individual customers under the biodynamic Demeter certificate label. But the grain and flour business was their big break-through. This originated in 1994 when a local mill processed their grain. Soon they realised the miller preferred their healthy grain and was getting the benefit of its quality but under the mill's own label, for which the mill had begun purchasing all its grain requirements from the Lloyds.

So, in 2001, the Lloyds bought the mill, and they have since developed a high-quality label and a keen clientele for their flour (marketed to organic-orientated outlets in Western Australia, South Australia, Victoria and New South Wales). The reason for this demand is the high quality, taste and healthy nature of the flours. This fitted in well with their life-style and business. Terri was able to focus on regenerating biodiversity on the farm while partnering Dayle in their emerging flour-mill and stock-feed business.

The Lloyds have not just secured economic survival but also see their supply of a variety of unadulterated, healthy flours as a strong ethical commitment. They are keenly aware of the impact on modern health of both nutrient-bereft foods and chemical contamination. To address this, they utilise a traditional, 200-year-old stone-grinding mill, and also a modern stone-grinding mill. This process enhances their flour quality, because in industrial flour milling the fuel, temperature, high pressure, speed and so on tends to kill off many of the crucial healthy benefits, nutrients, enzymes and other vital elements in grains. In addition, sixty per cent of the mill's power needs are supplied from their own twenty-five-kilowatt solar system, which also covers one hundred per cent of their farm's energy needs.

Their enthusiastic previous mill manager Jeremy Matchem proudly described the chalk-and-cheese difference between the two different worldviews. In addition to the careful, nurturing but varied milling techniques they employed to preserve the intrinsic quality of their grains, Jeremy stated that it is also what doesn't have to go into the flour that's equally important. 'We've no need to use any "flour improvers": no colourings or bleaching agents, no protein enhancers, no vitamin or mineral supplements, no preservatives. There's just no need to add anything,' he said; otherwise, they 'would deny people access to a natural and unadulterated food source'.

It goes without saying that, owing to the Lloyds' healthy farming practices and also their healthy silo storage practices (whereby they have developed a unique system to naturally fumigate their silos for storage), their grains and flours have no synthetic chemicals involved. Crucially, in catering for the home baker, their grains 'aren't over-milled, blended and reformulated with additives to suit the commercial baking process', said Jeremy. There is no compensatory fiddling to cover up for the poor quality of nutrient-empty modern grains.[15]

Therefore, in so many ways, the Lloyds epitomise much of the best of the regenerative agricultural movement. In short, you can't simply separate one action from any other in a complex dynamic system: everything is connected to everything else, and if one wishes to evolve an enhancing, self-organising and regenerative system (as opposed to a self-harming and non-sustaining system), then one needs to tread an ongoing pathway of perpetual evolutionary adjustment and interconnection.

The Lloyds' story demonstrates the vast gulf that separates much of the philosophy of industrial agriculture from a new regenerative agriculture approach. In her seminal work of 1962, *Silent Spring*, Rachel Carson observed that 'man is a part of nature, and his war against nature is inevitably a war against himself'. But the Lloyds, Wrights, Weatherlys, Seises, Maynards and others discussed in this book are saying, 'Enough! Let us declare Peace, Love and Reconciliation with Mother Earth.' And as demonstrated so powerfully by the Lloyds, this is best achieved when all five landscape functions are indivisibly working in synergy – and in this case, particularly that of the human–social.[16]

Moreover, the Lloyds' example clearly illustrates that when social systems are linked to ecological–biological systems, then regeneration of Earth and her systems occurs. This eventuates because, to echo John Muir's words, such farmers are truly 'hitched to the universe'.

CHAPTER 16

Design with Nature

It is the connections, not the oppositions, that form the basis for a creative mind and an adaptive society.

VALERIE BROWN and JOHN HARRIS,
The Human Capacity for Transformational Change[1]

'Ecological design' is the careful meshing of human purposes with the larger patterns and flows of the natural world, and the study of these patterns and flows to inform human action.

DAVID ORR, *The Nature of Design*[2]

This morning, late summer, as I walked to my office trying in the dark not to spill my coffee, I paused as always to gaze up at the great river-stream of the Milky Way.

Suddenly my attention was captured by the syncopated flittering of near-silent wings. Four microbats were working the air channels and alley-ways around some candlebark gums, sometimes sweeping only centimetres past my head. As I continued walking, I pondered why the bats were so active around these gums at this particular time of day: that pre-dawn hour just before starlight fades and the first birds begin to stir and call. Did the bats have other niches and food-zones at different times of the night? There was so much I didn't know about these extraordinary but small aerial mammals. Yet, like the vast, dark-deep night sky, these mysteries were also enchanting.

Shuffling my notes at my desk, I glanced over at a bookshelf as I gathered my thoughts. Serendipitously, my eye was attracted to a book whose spine and cover were placed inwards for some reason. On pulling it out, I recognised the black, grey and white cover whose design always struck me as eerie and mysterious. Stamped across the whitish mist in bold black was the title *Design with Nature*, and below that 'Ian L. McHarg'.

Design with Nature had rapidly become a classic in the field of urban and regional landscape architecture and land-use planning upon its publication in 1969. This was because it pioneered the concept of ecological planning, as ecology was understood at the time. But its fame also spread further afield to influence thinkers in other areas. The book appeared at a time of rising environmental consciousness (just seven years after Rachel Carson's seminal work *Silent Spring*, for example). It was also when scientists such as the Odum brothers were explicating new thinking on ecology, and when Donella and Dennis Meadows and others were challenging the world with new systems-thinking about the sustainability of the planet.

On campuses across the USA, McHarg became a renowned personality who argued very persuasively for the need for a new integration of human and natural environments. In essence, McHarg was attempting to bridge the gulf between environmental 'conservation' and the 'wise use of resources' that had so famously opened up in 1887 between the naturalist John Muir and the forester and politician Gifford Pinchot.

McHarg was forty-nine when he wrote his classic, which, according to philosopher-historian Frederick Steiner, went on to influence such fields as 'environmental impact assessment, new community development, coastal zone management, brown-fields restoration, zoo design, river corridor planning, and ideas about sustainability and regenerative design'. A Scottish landscape architect and Second World War veteran, McHarg went on to found a famous landscape-architecture course at the University of Pennsylvania, to become a TV personality, and to become a partner in a famous architecture firm that played a leading role in the development of the American planning and urbanism movements.[3]

As typified in his famous book, McHarg, fuelled by his passion for respecting the living land, laid down principles whereby any planning (urban, city, highway and so on) should first take into account nature and its features: topography, vegetation, waterways and hydrology, wildlife and any other natural features. Thus, McHarg was synthesising two ends of the design spectrum: garden design and regional planning. This was illustrated in his book by beautiful black-and-white and colour photos, and by graphical cross-sections of landscapes and their ecological communities. All had resulted in a visually exciting mapping approach. This format taught a step-by-step regional, landscape or site planning method using 'layer cakes' of stacked transparencies to build all the different natural features and human needs into a composite 3D design. McHarg's approach became de rigueur for environmental-impact reviews. It also helped inculcate an ecological view.

As a child, McHarg grew up near the ugliness of industrial Glasgow. Despite this, on long walks he was able to experience the beauty of the wild Scottish countryside and pretty farmland. On his return from Harvard after the war, he found all this destroyed. 'Lark and curlew, grouse and thrush had gone, the caged canary and the budgerigar their mere replacements,' he said years later. Nearest to Glasgow there once had been 'the Black Woods, only a few square miles in area but of great richness'. But on his return, McHarg recalled, he found 'the City of Glasgow had annexed this land and made it its own. Each hill had been bulldozed to fill a valley, the burn was buried in culverts, trees had been felled, farm houses and smithy were demolished ...' All replaced by ugly, tall apartments and asphalt.

Though imbued with a new landscape-architecture ethic, McHarg realised, 'I was too late. Memory that had been pleasant was now a goad.' Importantly, and triggered by Glasgow City Council's actions, McHarg realised that only a small contingent of people had 'seen areas redeemed by conscience and art'. Thus began a major revolution in planning and design.[4]

McHarg concluded the introduction to his book with this exhortation:

> Our eyes do not divide us from the world, but unite us with it. Let this be known to be true. Let us then abandon the simplicity of separation and give unity its due. Let us abandon the self-mutilation which has been our way and give expression to the potential harmony of man-nature.[5]

At a conference in 1971, McHarg delivered a paper titled 'Man, Planetary Disease'. Because of the way Western society viewed nature, he said, we were not guaranteed survival. Hence his conclusion that 'man' was a 'planetary disease, who has lived with no regard for nature'.[6] Soon after, in another address, McHarg told the gathering of Fortune 500 executives that the time had come for American industry to be 'toilet trained'.[7]

In the Foreword to *Design with Nature*, Lewis Mumford (another giant thinker, who worked with McHarg) stated:

> In establishing the necessity for conscious intention, for ethical evaluation, for orderly organisation, for deliberate aesthetic expression in handling every part of the environment, McHarg's emphasis is not on either design or nature by itself, but upon the preposition *with*, which implies human cooperation and biological partnership.

Quite rightly, therefore, Mumford places McHarg in a unique canon containing Hippocrates, Henry David Thoreau, George Perkins Marsh and Rachel Carson, among others.[8]

For a long while in agriculture, this element was missing: of designing agricultural activity and the farm to be in harmony with a functional landscape. Meanwhile, McHarg had railed for decades against the arrogant and destructive heritage of an urban-industrial modernity, calling the style 'Dominate and Destroy'. As modern industrial agriculture proliferated from the 1960s, this term also perfectly described its impact on landscapes and the natural world.

Clearly, therefore, an approach that 'designs' with nature and which is simpatico with landscapes and their functions is another key element of that fifth landscape function, the human–social, that can be incorporated in the struggle to counteract the Anthropocene.

Over the centuries, there have of course been rural landscapes that evolved in relative harmony with nature, and especially before the creation of the modern machine and an exponential increase in horsepower. In many cultures, farmers and villagers lived close to the land for centuries, even millennia. They built dwellings of local materials and, though initially degrading many agricultural landscapes, their capacity to cause further rapid ecological harm was limited. Many regions of Italy, Greece and other parts of Europe, Asia, Africa and the Americas typify this.

The beautiful Chinese landscape paintings of the Tang (AD 618–907) and adjacent dynasties, steeped in a Taoist tradition, reveal a philosophical view where humans are but a small part of a big landscape. Consequently, their houses, farm buildings and agricultural plots appear in harmony with a vastly bigger landscape. Both traditional and some modern Asian and Pacific water-based agricultural systems reveal stunning and beautifully designed landscapes. In what is known as a Terraquaculture approach, where water is designed to be stored and absorbed on the land via natural recharge–discharge processes, the result is human-directed springs, streams and aquifers onto farm terraces, paddies and ponds. In these are grown a variety of crops and animal enterprises, such as fish, duck, pigs and so on. However, dictated by slope and contour, the approach (such as the 5000-year-old natural farming approach of the rice terrace-fields of mountainous north-west Vietnam) can lead to stunningly beautiful, human-designed landscapes.

Variations of this ancient Terraquaculture and its close symbiosis between humans, their religions and nature is also seen, for example, in Bali's ancient Subak water-use system, and in many other major Asian and some Pacific landscapes. As student and practitioner of these systems Tané Haikai says, 'Operating as an active and adaptive part of the ecosystem (not the landlord or ruler imposing their own design) terraquaculture farmers aim to work harmoniously with natural seasonal cycles, creating symbiotic relationships with their landscapes.'[9] From such a holistic, nature-empathic philosophy inevitably flows good aesthetic and harmonious design.

In terms of a more recent Western tradition in landscape design, perhaps best known is the English landscape garden tradition. This emerged in the early eighteenth century in England among wealthy landowners who were reacting to the formal symmetrical gardens of the baroque French garden tradition known as the *jardin à la française* of the seventeenth century. The picturesque English tradition, with its focus on the visual and sensual (and in keeping with the wider Romantic movement), was a renunciation of the Mechanistic Enlightenment view of the 'subjugation of nature'. This English tradition rapidly spread across Europe and was led in

England by such famous designers as William Kent, Charles Bridgeman, Lancelot 'Capability' Brown and Humphry Repton (who called himself a 'landscape gardener'). They spread English gardens across whole, sculptured, beautiful landscapes. However, for Ian McHarg and other students of the style, what was missing was a broader ecological sensibility to interweave human needs with the landscape's functional needs.

That is, in the context of the emergence of the disciplines of basic landscape ecology, particularly after the 1950s (let alone a more comprehensive understanding of systems ecology, which came later still), what becomes clear is that good functional landscape design relates back to one's philosophical view – which in turn impacts on one's design philosophy. That is, yet again our fifth landscape functional element – the human–social – is critical.

So it is no accident that, beginning in the mid-twentieth century, a bifurcation in farm landscape design emerged at the very moment of explosion of industrial agriculture. The 'thesis' of the Mechanical-industrial mind almost immediately gave rise to its own antithesis, in the form of a number of nature-focused landscape designs and implementation. A famous example of this was P. A. Yeomans' Keyline Planning system. Indeed, much of Yeomans' work can be seen as a counter-culture reaction to industrial agriculture and, at the end of his career, even to urban design, as he encapsulated in 1971 when he stated simply that 'Nature is the Master Designer.'[10]

Beyond the practical extension of this approach to regenerating farms, (as we saw in Chapter 7) in 1971 Yeomans published a little-known book that nevertheless stands as a landmark work in Australian farm and city landscape design. Called *The City Forest*, this was the first book in Australia to promote a more ecological approach to whole-of-landscape design. It also included the incorporation of urban areas. Here, Yeomans had a vision way ahead of his time for the enhancement of aesthetics, energy use, and the recycling of water, sewage and waste so as to also grow food and timber crops.

The ideas in *The City Forest* represented a grand utopian vision, and an ecological one. 'Instead of farming land being the greater polluter it is now,' said Yeomans, 'it would become the perpetual guardian of the health and balance of the environment.' That is, Yeomans' full vision of the redesign of the whole farming (and indeed urban) landscape based on his ecological planning approach was that this would lead to a 're-union of mankind with the natural landscape and the ways of Nature'. The new city forests would also be multi-use: not just for building soil, biodiversity and hydrology management but also for food, agroforestry and aesthetics as they became part of playgrounds, parks and gardens.[11]

At the end of his book, Yeomans concluded that it was only regenerative farming that could turn things around. 'The important people,' he said, 'are the rebel farmers who have not been bastardised and who have stuck to the healthy and meaningful way of working with their soil and of farming with Nature.'[12]

In turn, it was Yeomans' ideas that helped spawn the even more famous permaculture movement.

However, before I deal with what became a major international landscape design movement and set of practices, there was one other landscape design approach that arose slightly later than permaculture, which came out of concern over ongoing land degradation as industrial agriculture began ramping up. This approach – the Potter Farmland Plan – was also combined with concern over the reductionist utilitarian approach of land-class planning. The latter came to prominence in 1950s Australia at the very moment P. A. Yeomans was popularising his Keyline Plan. At this time, various Australian state departments of agriculture introduced the American system of land-class capability classification under the rubric of 'whole-farm planning'. Such planning had its birth in the American Dust Bowl trauma of the 1930s and the resultant formation in 1933, under Franklin D. Roosevelt's administration, of the Soil Conservation Service. The system that was carbon copied and adopted in Australia was based on an eight-class land capability classification primarily aimed at soil erosion control and purely built around the assessed capability of utilitarian land use. For example, Class 8 land was deemed totally useless, in contrast to the hopeful and positive attitude of regenerative farmers towards such land.[13]

Behind Yeomans' thinking in *The City Forest* was the fact he found land-class planning and 'soil conservation' limiting. He irreverently observed that, 'the great cry of the soil erosion campaign was "Save the soil that is left"'. Consequently, he described this as a 'false and utterly pessimistic teaching' that largely suited the politicians and 'the chemical fertilizer manufacturers and the financial institutions.'

That is, Yeomans and countless generations of organic farmers had shown healthy, living soil could be built in a few years and not millennia if regenerative farming practices were adopted. 'Soil conservation was absurd,' said Yeomans, because 'CONSERVATION IS NEVER ENOUGH'. He believed traditional land-class planning was part of this paradigm.[14]

Another person sharing concern over the limitations of the reductionist utilitarian approach to landscape planning was Melbourne philanthropist Sir Ian Potter. Through his Ian Potter Foundation in 1984 he presciently initiated a three-year program that, through a refined landscape-planning approach, was designed to demonstrate that sustainable farms could be developed which combined both ecological and economic considerations.

Fifteen farms were chosen in Victoria's Western Districts where considerable land degradation had occurred. The idea was that the Ian Potter Foundation would throw in $250,000 per year for three years, and the farmers would pay one-third of the costs of the works on their farms: including tree planting, pasture establishment and fencing. Various federal and state government departments, farmer groups, and conservation and other bodies were initially involved, and a key component of the entire plan was that farmer attitudes were the crux of the issue.

The person chosen to manage what became known as the Potter Farmland Plan was a young forestry graduate called Andrew Campbell. Mark Wootton – who, from 1996, with his wife, Eve, and others bought two Potter farms in the process of putting together a large amalgamation – remembers working with Campbell. 'He was brash and brazen and everything that would piss people off,' said Wootton, 'but he was also everything you needed to actually have a breakthrough.'[15]

So, led by the energetic Andrew Campbell, the first step was farm planning using McHarg-type overlay transparencies to redesign each farm. But this time the exercise was not just using land classes to dictate land use, but examining riparian areas, topography, erosion, salting, the lack of trees and so on in order to completely redesign, even re-fence their farms. In the end, 200,000 trees and 200 kilometres of new fences were installed in the protection of important sensitive and degraded areas, and for high-quality pasture-improvement areas.

Results came quickly, and soon the farmers themselves became strong advocates of the new approach, and especially as production, aesthetic, biodiversity and other benefits accrued. In addition, many good natural scientists (like Rod Bird of the Hamilton Institute and pragmatic foresters like Bill Middleton) were associated with the project. In time the plan was extended to farms in Queensland, South Australia and Western Australia. The results – due to demonstration on practical working farms – and the new, 'greener' approach to land planning then began to attract a lot of interest and was strongly promoted.

Besides advancing the concept of whole farm planning and influencing public policy, the Potter Farmland Plan project directly led to the inspiration of the Landcare movement – with Andrew Campbell as its first national Landcare facilitator in 1989.[16]

Prior to the Potter Plan, however, a movement that would soon burst onto the scene to become world famous was quietly gestating. Unsurprisingly, it had its genesis in Australia's first major original landscape planning and regeneration movement: that of P. A. Yeomans' Keyline. This daughter of Keyline was to become known as permaculture.

Permaculture was a child of its time – sprung from the loins of the nascent environmental and ecological movements and also the post-1960s counterculture, with its rebellion against the Vietnam War, consumerism, greed and industrial agriculture. Flying in the face of the latter, permaculture emphasised preserving landscape function and using the patterns of natural ecosystems to grow healthy food and fibre in sustainable human communities and on urban areas or farms – both large and small – based around fundamental elements of design.

In this respect, permaculture differs from other regenerative agriculture approaches, as it comprises a broad church of ideas and principles built around design, so as to enhance all landscape functions but also to be flexible enough to be applied in any and every environment. The idea, as Mulligan and Hill summarised, was to design and build 'complex, integrated, even multi-storeyed systems within which all organisms – primarily perennial rather than annual crops – perform not single and competitive, but multiple and mutualistic functions'.[17]

I was an ANU undergraduate in the early 1970s and one of the first to enrol in human ecology. Our enthusiastic lecturers Stephen Boyden and Jeremy Evans, plus tutor Val Brown, almost daily introduced us to an abundance of new books and ideas emerging on the scene. Unsurprisingly, therefore, as soon as the groundbreaking book *Permaculture One* appeared in 1978, it found its way onto my bookshelf. On the cover of this A4-sized, intriguing book were two names: Bill Mollison and David Holmgren.

David Holmgren left school and entered university around the same time as I did, but on the other side of the continent in Western Australia. He is a distinctly individualistic man, originally of Swedish descent, and from a household one could describe as counter-cultural.

I first met him in 2015 at a Landcare Conference we were both speaking at on the Murray River at Echuca. The next day, Fiona and I dropped in to Melliodora, the permaculture home and garden of David, his life partner, Su Dennett, and son, Oliver. Located just on the edge of the spa centre of Hepburn Springs in central Victoria, their two-and-a-half-acre (one-hectare) property has become a famous demonstration site from when the Holmgrens purchased it in 1985. Back then, it was an over-cleared, blackberry-covered wasteland, but now it is replete with diverse and abundant food gardens, orchards, dams, livestock and revegetated creek. All this is overlooked by beautiful yellow box trees, *Eucalyptus melliodora*, by which the farm is known.

Fiona and I walked around to the back door. We passed laden fruit and nut trees and glimpses of tiered vegetable beds and berry bushes past a compost toilet. In an alcove at the back door sat baskets of freshly harvested food. Inside the house, we chatted around furniture and beneath beams all milled from gorgeous native Australian and exotic timbers, from such woods as Monterey cypress (*Cupressus macrocarpa*) and blackwood hickory (*Acacia melanoxylon*). And midway through the conversation, David's mother, Venie, who then lived on the farm (but is sadly now deceased), dropped in for a chat: an alert, serene lady in her early nineties, then still active in publishing poetry on environmental and peace activist issues.

The influence of David's parents on his thinking was significant. 'I've always been a conceptualiser, and building houses and gardens and forests kept me grounded,' he said, 'because I grew up in a family with a different version of geopolitical understanding. For example, my father left school at fourteen and he was probably the best political historian I've come across.'

David's mother was of Jewish background, and his father of Swedish descent. One could see the strong, broad-faced, wide-mouthed and sensitive features of his handsome paternal race in him, with his longish hair tied back in a bun. It turned out that each of his parents (first-generation Australians both) were rebels who broke from their respective Jewish and Protestant upbringings to become utopian socialists or communists: a path many intellectuals trod in the 1930s. Though they left the party in the 1950s, David said he grew up in a household committed to social justice and political activism concerning issues such as the Vietnam War, and his parents had friends who were pioneers in the early environmentalist movement. His parents also encouraged his creativity and interest in design.

Born in 1955, David also grew up in the natural world, which was possible in a Perth suburb in the 1950s and '60s. His parents had a vegie garden, and as a kid he could scoot off and explore his suburb's surrounding native bush and catch tadpoles in the Attadale Swamps. 'But when the swamps were filled in for housing development in my childhood, and when they filled up the foreshore with a rubbish tip,' said David, 'that was certainly part of my early environmental awareness.' Another influence was when his parents obtained a holiday shack on an abandoned orchard at Denmark in the southwest of Western Australia. 'I realised later this was also a major influence on me,' he concluded. 'We could go out discovering all this food, and nature, and the forest, not being sort of human managed. It wasn't neat and tidy.'

Unsurprisingly, David was emerging not just as a distinct individual of enquiring mind but also as an outsider in traditional society – and especially at school. 'My Swedish name and more particularly my Jewish heritage gave me reason to understand differences and victimisation,' he said. 'I took for granted that we did not belong and was more or less comfortable and proud, even arrogantly proud, of the questioning of orthodoxy of every kind.'[18]

His secondary schooling occurred during the emerging counter-culture period of the 1960s: a time that further highlighted his dissident thinking, and so David left school feeling an 'outsider'. But he also wasn't ready for university (despite an interest in architecture), so he spent a year hitchhiking around Australia. During this time, he discovered Tasmania, which greatly attracted him. While there, he heard about a unique design course in Hobart.

And so, abandoning his plans to do architecture in the west, David returned to Tasmania in 1974 to enrol in what he described as 'the most radical experiment in tertiary education in Australia's history': the environmental design course at Hobart's College of Advanced Education. Set up by Hobart architect Barry McNeill in 1970, the course – a conversion from architecture to environmental design – was radical both in what it taught and in its pedagogical methods. David found himself amid an incubating environment of new ideas. But he still didn't feel he belonged there. 'I was highly critical of the design professions. I thought they were fiddling while Rome burnt,' he recalled.

In his first year in Hobart in 1974, David moved into a spare room in the house of Bill and Philamena Mollison, and so began a unique and very brief collaboration. Bill Mollison was a generation older than David – aged forty-six against David's nineteen. Yet the two shared similar interests and a friendship developed, built around spirited daily discussions on how to build a better, more sustainable and equitable world. 'Mollison and I had sort of surprisingly similar views on a lot of the big controversial issues,' said David, 'even though we were very different in our approaches.'

And different they were. David was a quiet, contemplative, deep-thinking person, whereas Mollison was a bit of a larrikin: a knockabout raconteur, jack of all trades and extrovert polymath. Like many similar innovators, Mollison's sizeable ego and enormous self-belief meant he could be prickly and difficult. At different times of his career, he had been a researcher and scientist in forestry with the CSIRO, author, teacher, naturalist, politician and more. He was born in Stanley, in Tasmania's unique north-west. His roles with the Wildlife Survey Section of the CSIRO and later the Inland Fisheries Commission of Tasmania led to much time spent monitoring the life of these ecosystems – which had a formative impact on his philosophy. As one author put it, 'Mollison's life has been as unruly and abundant as his garden.'[19]

By his mid-forties, when Mollison graduated as a psychologist, this pipe-smoking radical in a polo-necked jumper with his bushman's beard became politically active. He then ran for parliament alongside Bob Brown for an early 'green' political party. And, in the meantime, Mollison had become a founding member of the Organic Farming and Gardening Society in Tasmania.[20]

David Holmgren met Mollison when the latter had become a lecturer in environmental psychology at the University of Tasmania. By this stage, as David recalled, his interests had gravitated towards how landscape design, aesthetics and ecology 'could be applied to agriculture'. Then he heard Mollison expressing ideas at an environmental design seminar, and though he was unaware of Mollison's 'infamous reputation in Hobart academic circles', the two engaged. 'This,' said David, 'was the beginning of an intense and productive working relationship.'[21]

However, authors Mulligan and Hill said, 'Mollison was the expansive, divergent thinker and explorer ... In contrast, Holmgren was the convergent, critical practitioner who did most of the actual work of design and maintenance. Mollison was the mentor with challenging ideas, and Holmgren the questioning, reflective and practical student.'[22]

David recalled, 'In our discussions and then refining my ideas, there was that element of me in rejecting contemporary society, and in both of us of "bugger struggling against the world we don't want, we're just going to go out and create the world we do want".' The result was the birth of permaculture.

To complete his course, David was required to write a thesis. Inevitably, he used ideas originating from his interactions with Mollison and their joint visits to different sites in Tasmania. 'I submitted a design for a property – a sort of case study on a self-sufficiency design,' said David. 'And I had the audacity to submit my own unpublished manuscript as my primary reference.' This manuscript became the famous book *Permaculture One*. But the manuscript was very radical for the times, examining energy from an ecological perspective, non-monetary economics as an essential part of the future, and the need to reconnect the consumption and production cycles. Surprisingly, David passed the course.

Having written the draft of a landmark book, David stopped there and soon severed his relationship with Mollison. 'I put my passion into more practical work,' he said, 'building, hunting and gardening, and I handed the *Permaculture* manuscript to Bill for edits and additions that eventually became *Permaculture One*, published in 1978 to substantial fanfare and even acclaim. But my days of working with Bill Mollison were limited to those three years that my self-effacing youth allowed me to ignore his difficult personality. I had already concluded that an evolution into a truly collaborative relationship was not possible, so I quietly disengaged.'[23]

As far as David was concerned, his manuscript was a collection of ideas that then needed ground-proofing and refinement. But not for Mollison, who had already begun spreading the gospel of permaculture (a term they had coined in 1975). Therefore, when *Permaculture One* appeared, as Mulligan and Hill described, for Mollison 'it was the answer to the numerous demands he was receiving for a book to back up his growing media commitment, lecture circuit and workshop presentations, at which he was rapidly becoming famous for his outrageous claims about the wonders of Permaculture'.[24]

So Mollison's worldwide career was launched. He travelled widely in Australia and overseas (Mollison claims to 160 different countries) for more than two decades, spruiking the new planning and living approach to food growing on the land. In 1979, he wrote *Permaculture 2: Practical Design for Town and Country in Permanent Agriculture*, which became the basis of his many workshops (updated in 1988 with his *Permaculture: A Designer's Manual*). And today, permaculture principles are practised by tens of thousands of adherents in dozens of countries.

However, the essence of permaculture is that it is more a set of design principles than it is a series of fixed practices. Even Mollison recognised this. What he never fully acknowledged was David Holmgren's crucial role in the concept's evolution – giving him 'inadequate credit', according to Mulligan and Hill. Luckily, and as proof of the significance of permaculture's design foundations, after some years David returned to his permaculture origins and has since contributed greatly to its further development.

David told me that the word *permaculture* 'was coined by Bill Mollison and myself ... to describe an "integrated, evolving system of perennial or self-perpetuating plant and animal species useful to man"'. However, he said that a more current definition of permaculture that was widely used was of 'consciously designed landscapes which mimic the patterns and relationships found in nature, while yielding an abundance of food, fibre and energy for provision of local needs'. Nevertheless, David said, 'this is too broad and that more precisely I see permaculture as the use of systems-thinking and design principles that provide the organising framework for implementing the above vision'.[25]

David added, 'people, their buildings and the ways they organise themselves are central to permaculture. Thus, the permaculture vision of permanent [sustainable] agriculture has evolved into one of permanent [sustainable] culture.' It is therefore not hard to see, as David acknowledged, that P. A. Yeomans' Keyline ideas and design principles were a major influence on their thinking.

In essence, however, the permaculture concept was all about a novel, ecological and community-based design system that integrated

agriculture, landscape, architecture and people. Above all, it required practitioners to learn with and work with nature. The Holmgrens' own
two-and-a-half-acre (one-hectare) property on the edge of town was a living example of a regenerated landscape, and a lifestyle lived according to
permaculture principles. Incorporating passive-solar building design and
use of local material, composting toilet, planned humus and compost production, clever siting of a variety of perennial and annual food plants,
shrubs and trees according to slope and aspect, clever use and conservation
of water, the integration of biodiversity plants in aesthetically pleasing and
functional ways, and their own integration with a vibrant local community
of like-minded people seeking to change the world, it all 'walks the talk' of
permaculture's guiding principles.

While the permaculture name might have been inspired by J. Russell
Smith's classic earlier book of 1929 *Tree Crops: A Permanent Agriculture*, for
David there were many other influences at this peak period of the count-
er-culture movement. This included many classics in the regenerative,
organic and biodynamic agriculture field. For example, David also found
Rudolf Geiger's *The Climate Near the Ground* (1950) and Howard Odum's
Environment, Power, and Society (1970) particularly influential.

David told me he was not surprised that Tasmania was permaculture's
site of origin. After all, he said, 'it is a place where modernity and nature
collide, both destructively and creatively'. In addition to Tasmania's temperate climate and individualist nature, and an ancient but tragic Aboriginal
history, there were such key factors as its diverse European food culture, its
largely rural nature and physical beauty, and especially a rising green consciousness. At the same time, Tasmania was the touchstone for widespread
environmental indignation at the obscene destruction of one of the world's
most beautiful and pristine places: Lake Pedder.[26]

As with many things in nature, diversity is a healthy marker of a resilient
system, and the fact that David Holmgren spun out of Bill Mollison's orbit
to forge his own way was crucial to the enrichment and ongoing development of the permaculture movement. This is because after 1978, when
they parted company, David (whom I regard as one of the deepest thinkers
in regenerative agriculture) went on a path of experiential and philosophical development. Ultimately, this helped take the permaculture movement

considerably further, making it more applicable to both small-scale, self-sufficient operations and also broader-scale agriculture.

David left Mollison and permaculture for nearly a decade as he sought to test his theories and gain experience – and also to separate himself from his former colleague. 'I was really suspicious of Bill's approach that "through the media you can work magic over people", remembered David, 'because I thought that could so easily degenerate into ideology and false thinking.' He also felt some of Mollison's assumptions were flawed.

Nevertheless, David acknowledges Mollison as the first of three crucial influences in his evolving thinking. The second was the brilliant work of pioneering ecologist Howard T. Odum (and specifically his work in the area of energy dynamics), and the third the New Zealand permaculture practitioner and eclectic thinker Tané Haikai.[27]

David was focused on going out and doing physical things – such as building a passive-solar house and establishing a functioning permaculture system for his mother on the south coast of New South Wales. In the early 1980s, he also got involved with the revolution then occurring in putting different tree systems back onto farms, particularly in central Victoria. This meant he was around at the genesis of the Potter Farm Planning project and subsequently the birth of Landcare. He therefore interacted with leading thinkers in this field and with agroforesters. Nevertheless, it was clear his heart had never left the ideas and thrust behind permaculture.

Unsurprisingly, therefore, unfinished permaculture business kept calling. And so, in 1983, David established his Holmgren Design Services, which soon became based at the home and land he and Su purchased at Hepburn Springs. Here, David and Su could at last put into practice all their thinking and experience as they developed their famous and healthy functioning permaculture gardens of Melliodora. There followed increasing design consultancies, including for well-known projects such as the Fryers Forest Eco-village. Beginning in 1993, David then began teaching permaculture design courses at his home.

This re-engagement with permaculture and his practical experience thus allowed David to continue evolving his wider ecological thinking. The result in December 2002 was the publication of a major new work: *Permaculture: Principles and Pathways Beyond Sustainability*.

David laid out twelve key permaculture design principles: a structure, he said, where 'each principle can be thought of as a door into the labyrinth of systems-thinking.' This 2002 work thus cemented an increasing recognition that David had become again a crucial intellectual force in the now global permaculture movement.[28]

In describing permaculture as 'a design system for sustainable living and land use that's concerned both with the consumption and production side', David said that definition was why he had also seen the limitation of McHarg's work. This system was conducted, he said, 'in highly modified urban landscapes and a consumer economy on steroids'. The latter was typified, he told me, by 'an American landscape that was completely emasculated of healthy food production in a way far greater than what happened in the Soviet Union'.[29]

That is also why David went public in correcting the record and formally acknowledging the influence of P. A. Yeomans. 'Yeomans' work and things like *The City Forest*, it absolutely rang a bell for me,' said David as we sipped home-grown and -made fruit juice with his partner, Su, and mother beneath beautiful timber beams in his long, north-facing living room. 'Mollison and I were looking for examples of ecologically functional landscapes and design. Yeomans' stuff on soil development and water harvesting et cetera and ideas in *The City Forest*, it was the only work we could find in the world that was of consciously designed, modern functional landscapes rather than traditionally evolved ones.' David thus saw the permaculture design principles as being integral to regenerative farming approaches, such as organics, biodynamics, agroforestry and their general landscape design.

There is a whole generation of new landscape designers and consultants now coming through, who have been trained by Mollison and/or David. Leaders such as 'Keyliner' Darren Doherty and permaculturalist Geoff Lawton are involved in teaching and also work overseas, and all are constantly evolving as they learn to read each different landscape and trial and test new approaches.

The title for this chapter was taken from Ian McHarg's famous book. But while a pioneering work at the time, when compared with leading permaculture thinkers today, the more urban-orientated nature of McHarg and company seems relatively superficial. It was based more on obvious geophysical and hydrological parameters than on a deep understanding of how landscapes function, and especially in the context of wider functional and systems elements such as coevolution and self-organisation. Included now in such an approach is increasing interest in the fact that Mother Nature herself is the best designer of all. For example, Janine Benyus's *Bio-mimicry* (1997) has popularised the idea of eco-design, whereby living organisms and ecological communities over aeons of 'tinkering in elegant, efficient and ecologically sustainable ways' have invented and laid out beautiful and workable design solutions.[30] This is because, as the physicist Fritjof Capra and natural scientist Pier Luisi encapsulate, 'ecodesign principles reflect the principles of organization that nature has evolved to sustain the web of life'.[31]

There has been some permaculture-designed work on broad-acre farms in Australia and elsewhere. Following Bill Mollison's initial influence, David Holmgren has laid out an excellent plan that has been well executed over decades by David Watson and Judith Turley at Bungendore in New South Wales (including the strategic use of exotic oak varieties and other tree species to suit a tough environment). Tim Woods's farm near Wellington, New South Wales, is another applied example of broader-acre design work. However, there is clearly still a large opportunity for the wider application of permaculture design principles on broad-acre grazing and cropping farms. In the true spirit of empathic ecological design and regenerative agriculture, and in the words of David Holmgren, if this occurred, a larger-scale permaculture vision could be achieved. This would embody 'the conscious design and maintenance of agriculturally productive ecosystems which have the diversity, stability and resilience of natural ecosystems'.[32] And this is because, as David told me as we finished our yarn at Melliodora, 'so much of those aspects of the regenerative agricultural revolution are still waiting to happen'.

How true, I thought, as I reflected on my own bumbling, uncoordinated and halting efforts at farm and landscape design. I remembered how David Watson and Judith Turley had visited our farm in the early 1980s just as they were engaging with Bill Mollison and beginning to collect thousands of acorns from different hardy oak varieties on Canberra's streets. Through holding a vision and applying a clear plan, today they are reaping ecological and financial rewards, having worked with David Holmgren on farm planning in 1990.

There are of course other paths beside permaculture to landscape design for ecological regeneration. Peter and Kate Marshall near Braidwood-Araluen on the New South Wales southern tablelands (as discussed earlier in Chapter 8) are path-finding a unique planning, landscape function regeneration and farm forestry approach. Peter calls their approach 'Successional Land Repair', and it is not built on a rigid plan upfront nor severe earth construction contours or bunds, but on an evolving scientific, technical and ecological approach to retaining water in the landscape; restoring all landscape functions; and then using diverse tree and shrub species (including 'pest' exotics like Scotch broom and blackberries) in a successional sequence to constructively play a role in regenerating an entire farm landscape.

The upshot, as we saw briefly in Chapter 8, is a water-rich and rehydrated landscape of engineered billabongs, 'chinampas' and Australian 'ciénegas' or water-meadows, chockfull of *Phragmites* reeds and other aquatic vegetation. Water is now running clear and healthily, biodiversity at all levels has hugely increased, and superimposed on all this is not a livestock operation but a profitable farm forestry operation built on largely exotic species for long-rotation timber and high value truffle-growing. In short, the Marshalls have evolved a unique, flexible landscape design approach with huge implications for many degraded catchments in Australia.

Though feeling I have been a laggard when compared to people like the Watsons, Marshalls and other innovators, I console myself that we are now catching up, even if it is the next generations who will see the benefits. To that end, I came back to the present and reached for the telephone. For my immediate task was to organise a Yeomans plough to undertake deep-ripping for an impending extensive biodiversity and food tree and shelter planting. The obvious components for this would be shrubs and trees – those anchor plants for both ecosystems and permaculturalists.

CHAPTER 17

Agri-Culture:
Source of a Healthy Culture,
Society and Mother Earth

A new vision and a new energy are coming into being ... Yet as so often in the past, the catastrophic moments are also creative moments.

THOMAS BERRY, *The Great Work*[1]

I t is late December, a time of intense, baking heat. I go for a long walk across our landscape, where a once green, diverse grassland sward smells like freshly cut hay as the plant sugars dry. Listening to the morning bird chorus, which is overridden by the warbling of magpies, I contemplate the connections and interconnections of our world on the farm: of flowering trees and shrubs, insect hatchings, temperature, atmospheric pressure, wind currents, aerial odour streams, and thousands of other factors that trigger and enable this ancient movement of birds, insects and other creatures across our landscape.

By mid-morning, on the older parts of the landscape, large patches of vanilla lilies unfurl their pastel flowers beneath candlebark and snow gums. Around them, bees buzz, bathed in rich scent, while orbs of rich yellow button daisies present their curled silver-side leaves to the thirsty sun. Beyond them, open grasslands are stippled with white flecks of flowering *Danthonia*. This is butterfly, insect and arthropod time.

A common brown butterfly among the candlebarks falls like a dead gum leaf, twisting to earth – only to suddenly spread wings and glide away, while painted ladies delicately flit from flower to flower. Other insects fill the air: the subdued hum of flies; the loud buzz of demonic, single-minded Christmas beetles; the quavering thrum of grass cicadas; the clustering of lost bogong moths on a fence post; and the slow waking of black prince cicadas, red-eyed and torpid in the morning, but then, as they warm, their full foot-song beginning in scratchy fashion and then rising *en masse* into a deep, humming buzz that resonates in the eucalypt canopies.

On the pads of sheep tracks, black wasps choose this time to bore nesting holes in hard ground, while under the shaded cover of huge granite slabs, ant lions – the personification of patience – dig their inverted cones of slippery death in the dust, as they lie invisibly in wait for careless prey.

Returning this morning, I watched a blue-tongue lizard swim across the edge of a dam, while high above a wedge-tail circled slowly in the washed-blue sky. And close by, as I dissected the grasslands, rosellas hung sideways off thistles, knapping their seed, and further on a wagtail rode a sheep's back, plucking gently for nest fibres.

Finally, after breasting the hill and seeing the homestead and beyond it a broad, healthy, rolling landscape of bronzed patches mixed with khaki, yellows, whites and dun browns, I was prompted to ask myself, 'What is it that makes a landscape? And how, in the face of Australian summers that now, year by year, seem to be fiercer and more desiccating, do I continue to manage and regenerate this extraordinary world around me?'

The answers, I know now, reside in our heads – for the latest global and national records confirm both my gut-feel and also some plain indicators. Here on our landscape, for example, for the last five years or so our native corkscrew grass now regularly ripens three weeks earlier than previously. Because, yes, the summers are hotter, the heat fiercer, and the drying more intense and rapid. And this is just one of a number of key indicators of a world changed by humans.

It seems incredible that the 1.3-kilogram mass of brain matter, tissue and neurons within our heads is now in the driving seat of a vehicle that can take us for a frightening plunge over the cliff's edge or else carefully steer us away from the abyss. Yet the overwhelming majority of the human species is not aware of the imminent danger, and the reasons for this are twofold. On the one hand, billions of people face a daily struggle for survival in the developing world. On the other hand, the vast majority of us who live in the developed world are too busy pursuing our consumptive lifestyles to care or become informed about the greatest crisis to ever confront humanity.

Even more worrying is the fact that for those few of us who have been appraised of the life-threatening consequences of our ongoing human behaviour and are in a position to do something about it, many are at best sceptical (despite the overwhelming evidence) and at worst downright dismissive of the unpalatable and inconvenient truth. At the same time, those of us in the best position to be well informed – our political and business leaders – carry on as if they live in a parallel but removed universe, so deeply are they embedded in the core ruling precepts of the Mechanical mind.

By arrogantly having placed ourselves outside the functional operating parameters of Earth's systems, we now see nature and the Earth as separate from us and so just available for use – while all the time the panacea for our ills and insecurities is seen to be endless consumption and growth.

Such is the power of both the individual and collective human mind, which is why I have focused these last chapters of Part II on the key fifth landscape function of the human–social. However, to emphasise the optimistic side of this factor, there is a faint but strengthening glimmer of hope: a gathering insurgency against the above stupidity. This rebellion is coming from those closest to Mother Earth: the farmers (as this book reveals). But it is also coming from their urban cousins who seek better health, tastier and more 'wholesome' food, a closer connection to Earth, and a life lived more simply and meaningfully in a sharing community.

That is why the stories told in this book are cause for optimism, for they are just a minute selection of many good works being done across the globe to regenerate Earth. For me, when I think about the big-picture stuff and the overwhelmingly important role of self-organisation in our complex systems, these acts of agricultural and urban-food and lifestyle defiance against the ruling Mechanical mind are more than acts of insurgency. They are the emergent properties that enable all self-organising systems to direct themselves towards ongoing survival and life.

———

Robert Rex and his wife, Caroline, run a sheep and cropping operation near Darkan in Western Australia's Great Southern wheat–sheep belt. Two decades ago, their operation was typical of farming in Western Australia, being based on the past clearing of ancient, coevolved vegetation layers off some of Australia's oldest landscapes and often on fragile leached or sandy soils underlaid by salt.

The Rexes were part of a leading industrial farming advisory group. 'When the chemical revolution and no-till started here in 1986–1987,' Rob told me in 2010, 'we got full-on into it – into direct-drilling ... It was fantastic; we could do all these great things so quickly ... just got full-on into N, P and K. We pushed things hard. But, after that, the wheels began falling off. The costs were going up, and the weeds were increasing; we were putting more fungicides out, and I just thought, "No, this isn't working." What's more, the stock feed we were growing just had nothing in it for ruminants; it was like eating cardboard. You see, the soil was getting worse; everything was getting worse; the salt was getting worse; everything was more compacted and it was just crap and we just weren't feeling good about it. And so then we started looking.'

Their 'looking' led to them attending a Resource Consulting Services (RCS) 'Grazing for Profit' course, and in short time they shifted to regenerative agriculture and an open-ended, exciting learning journey. 'That led to more questions,' said Rob, and soon that journey became a classic learning-change story.

When I visited Rob in 2010 (Caroline was working off-farm that day), he told me over a cup of tea that 'the big thing that drives me is water, rivers'. He then recounted that their family had a small holiday house down on the Walpole River estuary in the beautiful south-west of Western Australia. The Walpole River is one of a number that drains the cropping region of the Great Southern. 'We go there every year,' he said, 'and we used to catch prawns in the river and catch oysters in the estuary at Walpole; go down with a super bag and we'd just fill it up within half an hour, just mud oysters, and we'd go home and cook them.'

Then he told me that, about fifteen years ago, just after chemical agriculture had fully ramped up, 'we went down there for our holidays again. There were no oysters ... The whole lot were dead, every one of them gone, and they've never come back. And that was after a really wet winter when there was just bare ground everywhere and everyone got into growing canola and insecticides, weedicides and all that sort of shit. And the prawns are gone too.'

In short, Rob had described the collapse of an estuarine ecosystem – due to upstream poor landscape management – in less than ten to fifteen years. It was a sobering lesson in timescales and fragile ecosystems.

It was this experience, concluded Rob, that was the final trigger to change. 'When you sit on that river now and you think of all this crap coming down the river because of what we're doing up here, that's what really gets me. The other farmers don't understand it. They think once it's run down the creek it's gone, it's finished. But that sort of thing – I realised then that if we get it right here, this is where it starts.'

Getting 'it right' can mean many different things to regenerative farmers. Sometimes this can mean ideas and practices so simple and logical that upon reflection you wonder why you didn't think of them. But at other times ideas and their application can be so left-field and, on the surface, so counter-intuitive that one wonders how the ideas and their application were able to emerge in the first place. This latter situation certainly applied to another Western Australian couple I met in 2010.

On first meeting Di and Ian Haggerty you would not guess that here are two of the most radical and open-minded thinkers in world agriculture. For they seem an incongruous couple. Ian is tall, lean, dark-haired and laconic, with a lazy Western Australian drawl. Di, much smaller, is his complementary opposite, and her brown hair frames a round face. Neither have time for fancy clothes or an obsession with appearance, as life is far too busy and exciting. But when one begins to probe the depth of their thinking, all appearances disappear, for I soon realised as I unravelled their remarkable story that they made an extraordinary team in the fullest sense.

Di and Ian own Prospect Pastoral Company, whose HQ is based near Wyalkatchem, some 190 kilometres north-east of Perth in the central Western Australian wheat belt. I first met Di in 2010 when I visited their aggregation of farms, and then subsequently met them both on regenerative agriculture courses, and again when they came and stayed on our property in early 2016. They began farming in 1994 on a mere 1630 acres (660 hectares), having only been able to place a meagre down payment on a block of land, and against the best farm advisory counsel. Today, not only has this farm been transformed into one of regenerative-ecological health and profitable yields, but they farm a conglomeration of properties in three districts over a hundred kilometres apart and totalling over 30,000 acres (12,000 hectares). And as with the first advice they were given, they have challenged every basic platform of modern cropping, grazing practice and commodity marketing so as to develop a holistically integrated operation that merges cropping and grazing into a world-breakthrough, new farming system suited to semi-arid country. This they describe as a form of 'biological farming', but it deserves its own appellation. At this stage, they are calling it 'natural intelligence agriculture'.

Though both Di and Ian came off farms, there was no room for them on their family operations, and so, prior to buying their own farm, each worked outside farming. Di trained as an occupational therapist, working with people with disabilities such as autism. 'It's a broad, innovative profession,' she told me. 'It's holistic by nature because it's about preventative health care and you've got to think around corners all the time – have to invent and create things on your own. Taught me to be a flexible thinker. But both of us were always holistic, flexible and inventive thinkers.'

For his part, Ian, after leaving agricultural high school, put in five years driving road trains ('triples') on the 3500-kilometre-plus haul from Perth to Darwin. Having long hours to observe and think was a side-benefit. 'I saw the cycle of changes in the Kimberley and Territory over that time,' said Ian. 'I saw grass higher than the bonnet of the truck and skinny cattle, and I saw bare dirt and stock dying on the side of the road, and I thought about how to do it better.'

On getting married, the Haggertys bought and ran the Derby Road House in the Kimberley. It was a formative time. 'First, we had to dig really deep in ourselves to survive,' said Ian. 'When you have to do that, it strips you back to honesty and genuines you out, and you learn to value the small things. You learn to judge character too. And second, it gave us different business skills.' In a statement that seems to define many of the leading regenerative farmers in regard to a high moral and ethical drive, he said, 'but especially the time at Derby allowed us to follow our values. That meant asking the question: "How would we want to be treated?"' Consequently, by forming strong and trusting relationships with Aboriginal leaders and their communities, and with white pastoralists, they developed a profitable business – providing enough, when sold, to put down the deposit on their first farm in 1994.

Being rookies, they initially followed conventional farming, which at that stage had shifted from cultivation to direct-drilling based on ever-increasing chemical use. Their cropping country lies in the marginal, semi-arid agricultural belt, in only eight- to twelve-inch (200- to 300-millimetre) rainfall areas. Some of it bounds salt lakes and has inherited dryland salinity issues; other farms comprise a variety of soil types. But it is the strong observation skills of both Di and Ian as they regularly travel through different lands under different farmers' management that have enhanced their learning and ongoing cycle of innovation. In the process, each has become highly ecologically literate: able to quickly and accurately read the state of health and function of a landscape.

At the start, though, it was their mistakes that were the big teachers. One important early lesson was traditionally cropping a paddock that had not been cropped for nearly two decades. It possessed some natural vegetation and was noticeable for a profusion of mushrooms after an autumn rain. But upon successive cropping attempts, the yields plummeted and the mushrooms disappeared. After attending an alternative 'healthy soil' workshop, the Haggertys soon realised they had progressively plundered and then destroyed the natural capital of a functioning soil biological system and its built-up nutrient bank. It was an epiphanic moment. 'We blew the system,' said Di, 'and it was probably one of the best learning curves we ever had.'

This experience was followed soon after by the Haggertys attending further soil-health workshops, which confirmed their desire to go down a biological and not chemical path.

As their thinking and systems began to shift and evolve, Di experienced another seminal moment. She booked into a two-week course run by Elaine Ingham in northern New South Wales near Lismore. Many will claim there is no such thing as 'coincidence', but on the first plane leg to Lismore, Di found herself seated next to Rhonda Daly, who, with husband Bill, later formed the leading on-farm compost-making and educational business Ylad ('Daly' spelt backwards). On the second plane leg, Di was then seated next to Rachelle Maddock (later Armstrong), who was part of the family business NutriSoil, based on worm juice from earthworm vermicastings. 'We bonded together,' said Di, 'travelled around together and called ourselves "the three blind mice".' Later, both Rhonda and Rachelle would supply the Haggertys with the two biological ingredients that are a key part of the platform to their evolving new farming system.

Through trial and error, making plenty of mistakes, and farming and running sheep on different blocks of country totalling nearly 100,000 acres (owned and leased) through their career, the Haggertys' new revolutionary cropping approach took shape. The key to this approach was their intense study of varied landscapes and the varied responses of these landscapes to different cropping regimes (both their own and others').

This approach is two-pronged. The first involves using microbial organisms and the rest of the soil food web to provide a nourishing, healthy and self-organising soil base in which to grow a crop. 'If you get the microbial community doing all the work, well, bang, there's a heap of costs gone and there was no new machinery required. They are the workforce,' said Di. And so, as Di further explained, the second prong 'was utilising the livestock as the machinery on top. In effect, they are our inoculators, our first line of defence. They can really heal country.' This livestock component was vital and incorporated a high degree of adaptive landscape genomics (see Chapter 15). As ever, however, necessity was the mother of invention. 'When we started,' said Ian, 'we had no money, no machinery, and we said, "We've got to do things differently, but how?" And Di said, "Well, we've just got to do it with animals."'

The way the couple work is that Ian does the cropping and Di manages the livestock, but they closely plan and integrate these activities because each component is a synergistic partner. What woke them to this was the intriguing consultant Jane Slattery (neé Hinge), who originally ran an animal-nutrition-based business out of Victor Harbor in South Australia. Hinge became a key advisor to Di, as she understood not only animals but also subtle energies. 'Jane had a lot of understanding about the animals' self-selection and their role in re-fertilising the soil specifically, and about moving nutrients from areas of high fertility to low fertility,' said Di.

Aiding her learning curve was that, when they were at Derby, they met the unusual American manager of Birdwood Downs station, Robyn Tredwell. She showed them how to think holistically about animals in a landscape, and so how to use livestock as 'leaders, weeders, seeders and feeders' in order to regenerate grasslands and landscapes. This lesson (plus Jane Slattery's advice) alerted the Haggertys to the holistic nature of a functioning landscape, and that especially monoculture cropping would be severely handicapped without animals in an integrated system.

In time, the Haggertys have evolved a unique strain of robust, highly fertile, easy-care and productive sheep that need neither drenching nor mulesing. They let the sheep adapt to the landscape, and critical is the fact that in harsh times, when the soil microbial world shuts down, the sheep carry similar but invaluable microorganisms in their gut. This helps in reseeding the soil microflora and also enables the sheep to readily break down and utilise the nutrient-packed stubbles off the Haggertys' healthy soils. When new country is leased or bought (invariably previously sprayed and cultivated-out, with shallow soils on a hard-pan and little microbial and biological life), the sheep become a key regenerative tool in helping restore that land.

The Haggertys are aware of the complexity of interrelationships in a coevolved landscape. 'As the plant emerges in our system, it's coated with such a diversity of microbes as it comes out of the soil,' said Di, showing me a photo, 'and you can see there some sheep manure and its got fungal strands wrapped round and round like a ball of wool – that's a happy association.'

'Yes,' Ian confirmed, 'it all works in unison, and when the green starts in the autumn it's either the sheep saliva or their microbes and similar signals that stimulate what's there on the plant to crank it right up again.'

Di summed up this entire integrated approach in a way that had now become very familiar to me: 'We're always trying to enable the natural system to bring forth the produce: that is, we have to get out of the way of the natural system as much as possible, just remove the hindrances to nature. Mother Nature then just transitions the country to health.' To a holistic landscape ecologist and systems-thinker, this is a classic description of allowing emergent behaviour to occur and self-organisation to operate. Aiding this, and rounding out the fully holistic and integrated nature of their farm, the Haggertys' regenerative farming approach is bringing back previously unseen native grasses and understorey. In addition, they are catering for stock needs, land salination prevention and biodiversity by undertaking extensive plantings of both forage shrubs and other species aimed at enhancing biodiversity.

The above elements, of course, are typical of many regenerative farmers. It was how they upended and reinvented traditional industrial cropping that marks the Haggertys' extraordinary, global breakthrough. Ian explained how his approach uses almost the same machinery as traditional cropping but is driven by a biological system. 'We're creating a furrow with the tine, but into that we're liquid injecting at low pressure our biological component. I like to do 110 hectares [270 acres] without a stop, and to do that I would have had to buy massive air-seeder bins or run two air-seeder bins because you've got it half full of fertiliser and half full of seed. But now my air-seeder bin is totally full of seed, and I just drag a 7000-litre liquid cart behind the air-seeder and that's got our biological component. Vastly more efficient.'

The real secret is the biological liquid. This took years of attending workshops, training and reading, but they eventually cracked the secret. 'We really looked at it,' said Ian, 'and we nutted it out by thinking, "Well, what's going to mimic nature to stimulate degraded soils in cropping phase in arid countries?" Then all of a sudden it became really simple. Jane's advice in passing once was that, "If you are going to put anything on your paddocks, it needs to have been through the gut of a worm first" – advice supported by Elaine Ingham. So, after, meeting Rachelle at Lismore was like finding a lost piece in a jigsaw puzzle. Rachelle's father, Graham, had developed NutriSoil – but we haven't got enough worms in this dry country and we can't cart them out here. So we got the product, vermi-juice, from the Maddocks. It was full of all the goodies and liquid which made it possible then to integrate into our cropping system.'

'Yes,' agreed Di, 'If you want to mimic nature, let nature do it.'

In short, in properly extracted vermi-juice where worms are fed on truly healthy compost, there are all the microbial goodies (in dormant form) along with most of the valuable derivatives such as enzymes, proteins, hormones and such, and especially humates. The Haggertys discovered that the secret was to place the biological liquid straight into what becomes the rhizosphere (that crucial micro-zone around plant roots).

The Haggertys still source their vermi-juice from Di's original serendipitous contact of Rachelle Maddock at NutriSoil. Moreover, said Di, 'the Maddocks feed their worms like thoroughbreds'. In addition, the Haggertys came to understand that the dormant microbes and other biology needed food in order to wake up and function in a year-round resilient system with minimal dormancy. So somehow they needed to prime the worm juice at sowing with a healthy compost that served as both worm- and soil-biology food. The problem was, as they quickly nutted out, you couldn't afford to put on the vast amount of raw compost required, let alone have the time and resources to make it. But they found you could extract virtually all the goodies out of compost. So now they source around seven tonnes of Rhonda Daly's healthy compost (Di's other serendipitously found colleague) and, using an imported American compost converter, make their own liquid compost extract.

At sowing, the two liquid ingredients are combined (essentially dormant bugs and goodies with the bugs' tucker and other required inputs). Ian's machinery then carefully places the liquid 'ecology fertiliser' around the seed (using only twenty-eight litres to an acre). That is, the seeds are microbiologically coated, and the seed and plants hit the soil running as the rhizosphere develops. Their aim here was to utilize as many processes that nature engages to build soil fertility and cycle nutrients. That is, they already had the livestock component; the NutriSoil was helping to improve the soil environment to be more 'microbe friendly,' and to provide the earthworm impact, they wanted to incorporate a 'compost' component – as the latter was also relevant in high functioning natural ecologies.

The combination immediately delivered outstanding and consistent results but also, as happens when we work with nature, unanticipated positive side-benefits and synergies. These extra synergies come, first, from the sheep component via their dung, urine, microbial and other animal impacts (such as their hooves breaking crusted ground or disturbing hardpans to create a seed-bed), plus redistribution of nutrients. But then also, as Di explained, when the combined liquid fertiliser is placed around seed in the soil, 'what they [the sheep] are doing is facilitating the environment for the indigenous species [microbial and other biota] to get going'. In short, the Haggertys were facilitating various forms of coevolved function, along with self-organisation in a natural system.

As a result of this innovative breakthrough, the Haggertys now reap multiple benefits from their healthier soils. They get viable yields in annual drought rainfalls as low as four to six inches (100 to 150 millimetres) because their retained soil moisture is vastly higher than traditionally, and in country with horrendous summer evaporative loss. 'In a traditional system,' said Ian, 'when the tractors are out seeding, you can see the dust cloud from twenty kilometres away. But we don't have that, and you can smell the moisture in the air when we're sowing.' Moreover, due to their efficiencies and low inputs (and thus less costs), the Haggertys have a commercially viable crop even if it yields 280 to 320 kilograms per acre (700 to 800 kilograms per hectare). An industrial cropper needs around double this yield to be viable.

Another huge benefit of a healthier biological soil is disease protection. In what scientists are now calling 'the fungal superhighway', we know that through mycorrhizal fungi's hyphae networks there is a whole underground communication system. This can be between microbe/fungi in the soil aggregate and the plant; between the cereal plant and competitive weeds; and between the plants themselves. Moreover, when, for example, an insect or pathogenic fungal or bacterial pest strikes the edge of a crop in healthy soil, 'news' of this immediately spreads via the fungal superhighway. This enables plants to pump up defence chemicals in response. The Haggertys now enjoy extraordinary levels of disease protection in their naturally grown crops.

Other new inventions and variations on traditional practices are part of their innovation package. For example, so consistent are their crops now that in health monitoring they continue to use a refractometer for tissue testing to monitor crops. This includes assessing the level of plant sugars and nutrients (such as healthy oils, amino acids, proteins and minerals) via the brix approach. Their tissue testing has shown a gradual improvement and consistency in nutritional balance over varying soil types. The Haggertys also mix strain varieties when sowing crops, having found that seed diversity yields greater resilience in multiple ways. Moreover, Di in particular has become highly skilled in detecting and utilising subtle energies for their farming and daily decision-making, and a dowsing pendulum is her constant companion.

A further benefit of not having chemically induced high nitrate levels is that, compared with chemical farming neighbours, the Haggertys generally suffer less frost damage in their crops, and they frequently attain higher grain quality. In a wet harvest season in particular, traditional industrial farming crops suck in moisture due to the high nitrate levels, which leads to spoliation. Such grain is then frequently downgraded in price (often by at least half) to only livestock feed-wheat status. At such times, the Haggertys are often the only farmers attaining high-quality hard grain levels. Moreover, another vitally important quality not recognised in industrial farming but present in the Haggertys' grain is high nutrient integrity – indicated by their high brix readings. Derived from their natural biological farming, this indicates healthier but also tastier grains and food.

Also, because their healthy soil biology digests their stubble, the Haggertys are able to use less fossil-fuel energy by harvesting only the crop heads. Cost savings are huge on chemicals and synthetic fertilisers anyway, and when they do use glyphosate for weed control, their non-nitrogen-gorged weeds are more susceptible to lower rates and then the soil life appears to rapidly break down the chemical. As we have already seen, the Haggertys therefore have no chemically resistant weeds. As also seen earlier, by having vigorous, healthy, non-drug-addict crops, the Haggertys find that their plant-root vigour quickly begins to penetrate and break down the hard-pans of any land they purchase or lease.

The best way to fully comprehend the Haggertys' revolutionary new cropping approach is to first examine the approach of standard modern industrial chemical farming in broad-acre regions such as the Western Australian wheat belt and then compare this with natural intelligence agriculture.

First, in industrial agriculture, giant tractors and other machines with all the latest gadgets (including electronics and computers) are used to create efficiencies over broad acres. The Haggertys use the same, just modified to their biological system.

Second, the industrial system is dependent on large applications of modern synthetic fertilisers, such as superphosphate and synthetic nitrogen-based products. The Haggertys instead use biological components, including micro-amounts of humates, minimal nitrogen as an ignition fuel to kick-start the biology at sowing, plus micro levels of foliar sprays of NutriSoil and compost extract.

Third, the industrial system is highly dependent on liberal herbicide, fungicide and insecticide use – such as glyphosate and sulphur-urea-based chemicals, which persist for a long time. Believing over-tillage is a massive destructor of soil microbial communities, the Haggertys still use some herbicide (confident their biological activity breaks it and toxins down) but no fungicides or pesticides. Independent laboratory testing has demonstrated nil residues of any chemicals, heavy metals or other toxins: a result of their robust biological system detoxifying (as nature intended).

Fourth, due to the chemical nature of industrial agriculture and associated sterile soils and thus little breakdown of stubble (which means soil organic matter and thus carbon are not being put into the soil), on industrial farms stubbles have become an impediment to effective tillage. They also harbour fungal diseases in a system now bereft of any biological-control capability. Moreover, because the unused stubbles are in the form of labile carbon (not long-lasting), on the next tillage event this carbon is lost to the atmosphere. An associated problem is that when the season does break and crops are sown with their fertiliser and nitrogen drugs, any remnant soil microbiology (invariably only bacterial now) pulls much of the nitrogen from the young crop in an attempt to break down the undigested stubble. This in turn requires more harmful nitrogenous fertiliser and greater cost. The Haggertys don't have these problems because their healthy soil biology digests the stubble within six to ten months (or even quicker if there is moisture available).

Fifth, associated with high amounts of artificial fertilisers in industrial farming has been a proliferation of weeds that thrive on these inputs, plus the shift from a fungal/bacterial-balanced soil to one that is bacterially dominated. This has fed ever-increasing chemical use and thus universal and ever-developing weed resistance (the latest tally of such resistant weeds in Australia now exceeds forty species), escalating further the required volumes and strengths of chemical inputs. 'Roundup' rates, for example, have now increased sevenfold, and this without new varieties of stronger-strength Roundups. The Haggertys are reversing the treadmill.

Sixth, the above treadmill situation in industrial farming has led to the next problem: that to try to solve the stubble issue of residual weed seeds described above, as well as the carryover of fungal disease, many farmers have resorted to the old technique of stubble burning in windrows or 'black burns' in whole paddocks. So having destroyed the original natural capacity to digest stubble, and thereby encouraged weeds and disease, they use the blunt and further damaging tool of fire. All this has achieved is the loss of any remaining carbon while in the process destroying any protective ground cover and killing any remaining surface soil microbial life. In turn, this requires more inputs, and on and on. Again, due to their natural biological approach, the Haggertys have avoided these problems and costs.

Seventh, the constant chemical tillage in industrial farming means the drug-addict crops that are so dependent on humans and not natural systems therefore have weak root systems. The result is a common hard-pan of non-penetrable soil only a short distance below the surface – requiring high intensive mechanical and chemical interference methods to part-alleviate the issues. Again, on buying new country, the Haggertys quickly heal this via healthy soil and vigorous, deeper root action.

Finally, the above combination of industrial factors constantly leads to ever more intervention to solve ever-increasing problems. The result is continually escalating inputs and costs associated with these increased interventions, and this is accompanied by decreasing yields and quality, plus plummeting natural resilience as the climate becomes increasingly unreliable.

Instead, through enhancing biological systems, the Haggertys have improved yields and product quality. They have robust disease resistance and extraordinary climate-variability resilience, and have largely eliminated costly industrial inputs and so enhanced profitability, while virtually eliminating synthetic chemicals. Following a decade and more of highly variable climatic conditions (including frost) that have put much of the Western Australian wheat belt deep in debt to the banks, the Haggertys by contrast have expanded their farming enterprise over twenty-fold.

The overwhelming importance of their system and its results (such as successfully and profitably growing crops in drought years on four- to six-inch (100- to 150-millimetre rainfall) is that they have evolved a low-cost, viable cropping operation for semi-arid, marginal country that has worldwide implications for vast areas of agricultural lands where cropping and pastoralism is practised. This doesn't just mean for the developed world, but for the developing world too.

In many respects the development by Di and Ian Haggerty of natural intelligence agriculture constitutes a world breakthrough. Though the Haggertys will tell you that there are other leading biological farmers around the country following similar principles, theirs is revolutionary for a number of reasons. They are not just leaders in scale but have also developed a farming system built around multiple innovations that hinge on natural function and that is suited to all farm sizes, to high rainfall and semi-arid country alike, and to rich soils and poor, sandy, low-nutrient and even salty country.

Moreover, their system of making and using biological inputs has wide application, including for peasant farmers in the developing world or pastoralist-croppers in the semi-arid African Sahel and other lands down to eight inches (200 millimetres) of rainfall and even lower. Besides the implications of their system for global cropping and integrated farming, plus for societal, environmental and climate change, natural intelligence agriculture is a recipe for weaning farmers off expensive, ultimately poisonous industrial chemicals and synthetic fertilisers and high fossil-fuel use. At the same time, such a new cropping system plays a major role in addressing not just climate change but also other Anthropocene boundary crossings (such as the nitrogen, phosphorus, water and land-use cycles and systems) by putting long-lived carbon in the soil and consequently impacting on all the key landscape functions. This builds resilience into farming systems.

In short, the Haggerty family example (they now farm with their son James) is a classic case of the best utilisation of the fifth landscape function: of free and flexible human minds using nature's principles to create an innovative system that regenerates landscapes to health and consequently regenerates rural communities. Another huge benefit in regard to this fifth landscape function and the urgency for change in the Anthropocene is that, whereas the key elements of pasture cropping and no-kill cropping can prove 'a bridge too far' for many traditional croppers, the Haggerty approach is clearly a more easily adopted bridging approach for croppers and their existing technology.

It seems inevitable – given their ethical/moral foundations and the open-minded, innovative nature of their thinking – that many regenerative farmers sooner or later go to the next level of helping to change others. This is because their journey has allowed them to see the bigger picture and the need and urgency for change. The Haggertys are no exception here. They feel obliged to spread their ideas: not in a top-down manner, but rather through helping others and acting as non-threatening exemplars. This is because the full implications of their approach go way beyond landscape regeneration and enhanced farming resilience. Their use of adapted cereal varieties under a biological cropping regime that mimics natural diversity, in combination with the virtual elimination of chemicals, opens up enormous opportunities to seek differentiated higher-value markets and thus spread the benefits of their farming system further. This extends to improving human physical and mental health and rejuvenating rural and city societies via enabling access to healthy, nutrient-dense food and a reconnection to Mother Earth.

In addition, the Haggertys' leadership involves them working with fellow farmers and young trainees to train them in their systems. In the future they plan to engender support from financial institutions with a global climate conscience to enable energetic, passionate farmers without an equity base the opportunity to begin building their own business and land base working on regenerative 'natural intelligence' principles to grow food and fibre and restore health and diversity to the landscape. In turn, a community of knowledge and people is growing – spreading to leading croppers even in Sweden. Their latest move to focus on segmented direct marketing of healthy food to urban people, connected to associated education, has led to partnering with a young protégé to form a company called Bio-Integrity Growers Australia. This company is developing tools and protocols to provide meaningful grain and meat nutrient-status measurements for both farmers and consumers.

What is exciting is that the Haggertys are not just healing but also transforming Earth, with a consequent enormous capacity to also transform associated human systems and society. And it is all because they are working with, and not against, Mother Nature. That means empowering her coevolved functions such as capacities for self-organisation. In this respect, the name 'natural intelligence agriculture' truly seems appropriate.[2]

Never, in the entire history of agriculture over the last eleven or more millennia, has the 'culture' component ever assumed such critical importance.

I recently read Dan Barber's fine book *The Third Plate: Field Notes on the Future of Food*. This outstanding chef and writer reconnects the importance of diverse and regionally based healthy farming ecosystems with food nutrient integrity and thus distinctive regional or 'provenance' taste through skilled chefs creating great cuisines. But on reflecting further, I wondered how this might apply to Australia. First, concerning food demand, we have less than one-tenth the sized population of the USA on a similar-sized continent. This means our broad-acre farms are primarily export-focused. Second, while having a rich diversity of immigrant backgrounds, we don't have as long a post-indigenous settlement history nor a derived, old peasant culture, which, in the USA and places like Europe, have allowed time for more distinctive regional cuisines to evolve. Third, while Australian Indigenous nations evolved forms of proto-agriculture over millennia, they had not developed as rich and diverse a range of domesticated foods as existed in the USA or European settlements (and they were led by corn).

However, the final point is key: aside from a higher socio-economic strata of society who can indulge in a chef-led cultural resurgence via distinctive regional or local cuisines, the rest of us are prisoners of the industrial food-growing, manufacture, distribution and marketing food systems – underpinned by chemically based industrial agriculture.

That is, as Dan Barber points out, 'Modern agriculture has thoroughly separated the agri from the *culture*.' This is especially manifested in cropping, where, says Barber, 'We're producing grain strictly as a commodity, without its cultural heritage. Uniformity has replaced excellence.'[3]

Despite all this, however, the leading regenerative Australian farmers in this book have opened a pathway to change this. Yes, we don't have the population to domestically gobble up most of our produce, but that doesn't prevent the creation of new niche markets (both local and export) for grains and other crops and food with great nutrient integrity: the latter due to ecological health and the basis of superior taste and the underpinning of healthy nutrition and tasty cuisines – markets that Barber calls the 'middle in agriculture'.

Moreover, Australia's own demand for 'organic' bulk grains is fifty per cent unmet. Therefore, the Haggertys and their natural intelligence agriculture, the Lloyds and their biodynamic, stone-ground old provenance grains, the Seises and their pasture-cropped combination of grains and stacked multi-purpose crops, and the Maynards and their 'Grassland Grains' have presented us with the opportunity to once more provide healthy, tasty foods that allow a diverse range of customers to reconnect the 'agri' and the 'culture' in an integrated food and social system.

However, in terms of our collective futures, this reconnection has led to one further and overwhelmingly vital factor that the Haggertys epitomise in terms of the human–social landscape function: they clearly exhibit a new way of thinking. Such thinking, as we shall see in the next and concluding Part III, constitutes, I believe, a new evolutionary step in the human mind.

It is this thinking and a capacity for transformative change that gives me confidence that through regenerative agriculture and its connection to like-minded urban people, we humans do have the capacity to address and turn around the Anthropocene era and thus pull back Earth's operating systems to safer and longer-lasting levels.

PART III

TRANSFORMING OURSELVES - TRANSFORMING EARTH

CHAPTER 18

The Big Picture: Co-Creating with Landscapes

The universe is revealed to us as irreversible emergent process ... We now live not so much in a cosmos as in cosmogenesis; that is, a universe ever coming into being through an irreversible sequence of transformations moving, in the larger arc of its development, from a lesser to a great order of complexity and from a lesser to great consciousness.

THOMAS BERRY, *The Great Work: Our Way into the Future*[1]

The rolling, semi-arid, shrub-dotted country of the Great Karoo, west of Graaff Reinet in South Africa, somehow reminds me of the Western Australian Kimberley in the dry. They are both varied, desiccated lands with high mesas, dark 'kloofs'[2] and shadowed flanks, and the same stark southern light. As we pulled into Norman and Jennie Kroon's Watervlei farm, their landscape breathed health and vibrancy – unlike what we had been driving through for the previous few hours.

In Chapter 6, I first described this visit to one of Allan Savory's earliest clients. What was striking as we first drove in was that there was nearly total ground cover, while various Karoo bushes of all colours, shapes and sizes dotted the paddocks. Flitting between them were strange birds with nervously ticking tails.

That evening, Fiona and I sat around a dining-room table with Kroon family members. The scene could have been in any Australian farmhouse (Jennie herself having come from Australia). Chewing on a thin steak of flavoursome and juicy dark meat, our host Norman asked me to guess where it had come from. I plumped for beef, but I was wrong. It was a kudu (antelope) bull that Norman had shot some weeks back and then allowed to season.

In a matter-of-fact way, without boasting, Norman told us about the steep kloof tucked under the head wall of a mesa where he had found the kudu. It turned out these large and beautiful creatures frequently visited this distinctive area on his farm to browse on some shrubs there.

The context is that, in many parts of southern Africa, farmers have a big problem with bush cattle-ticks, necessitating frequent and costly mustering and plunge-dipping. However, Norman noticed that the kudu he occasionally harvested up in this corner of his farm – and unlike elsewhere – were free of ticks. So next time he was up that way, he quietly poked in and observed what was going on. He found that the kudu were attracted to one particular bush (*Aloe ferox* – Cape or Bitter Aloe). Repeated observations led him to send away some leaves of the bush for analysis. The report came back showing that the bush was full of a host of chemicals with names that read like a jumbled alphabet. But Norman acted on a hunch. He started harvesting the sap from this bush, boiling this down, and drenching his cattle and sheep with the potion. As a result, his animals were now free of both ticks and intestinal worms.

The next afternoon, as we left the Kroons and headed towards the Eastern Cape, I got to pondering the kudu and ticks.

What was going on here in this ancient landscape? How did the kudu know to poke up into that kloof and browse on that particular shrub? Why was that shrub full of so many chemicals, and indeed why was it different in composition from other shrubs? And how did the kudu detect the difference? Moreover, why was that shrub growing in that particular environment anyway, and what made that microclimate at that particular altitude and aspect different from areas only hundreds of metres away? Did it come down to differences in soil, minerals, water flow or underground microbial function? And what was the role of the dynamic complex ecosystem communities up there in the kloof? These and many other questions whirled about in my mind as we headed east.

By the time I had completed the journey of researching this book, I felt I could partly answer just a few of these questions, but with scores more queries unanswered. This was because I now appreciated that the landscapes we farmers manage are so extraordinarily complex in function that it was a wonder I ever dared set foot outside the back door to begin intervening in them.

Truth is, our farms comprise a whole series of complex creative systems (a term, as we shall see, used by landscape ecologist Fred Provenza), which range across endless time and from the invisible micro level to whole catchments and regions. Each nested, interactive system is constantly evolving, reacting and adapting in a quest to either readjust to any disturbance or maintain states of stability. This has been achieved through millions of years of coevolution, change and adaptation, underpinned by giant, patterned drivers that allow these interconnecting systems to continually self-organise towards greater complexity, diversity and resilience.

Throughout this book, I have attempted to describe a basic ecological-literacy toolkit that allows us farmers and others to get a handle on how our landscapes work. We have encountered the solar, water, soil-mineral and dynamic-ecosystem landscape functions, and how these are impacted by the fifth function: that of the human–social. We have also seen how none of these acts in isolation but that all are indivisibly connected, and that if we destabilise or simplify one, then all the others are destabilised and in time degraded.

But now I discovered there was an even bigger picture: that these five landscape functions and cycles sat within a far larger matrix. This matrix comprised the operating influences, constraints and functions of nature in general.

The first and most obvious of these is the constant operation of the mechanisms of natural selection and its related processes of evolution, coevolution and factors like epigenetics. No doubt, there will be other components of this that modern science is yet to discover too.

A second is the process of self-organisation. Previously, when discussing what leading regenerative farmers such as Tim Wright, Ron Watkins, Colin Seis, Richard Weatherly, Bruce Maynard and the Haggertys are doing, I mentioned that time and again I heard the refrain, 'My job is to get out of the way of Mother Nature.' When they did this, all sorts of amazing regenerative effects occurred. This is because nature is able to take care of, and to organise, itself.

As I described in Chapter 15, self-organisation is the inherent capacity of systems (and especially natural systems), following disturbance or in the normal course of events, to move towards greater complexity and thus greater stability and resilience. It is an amazing capacity of complex creative systems that has only begun to be articulated and understood in recent decades. Healthy, functioning living systems behave in a way that is not merely imposed by the external environment but that is established by the system itself.

New understandings about complex creative systems have come out of a range of contemporary thinking: from 'systems-thinking', 'chaos theory', 'complexity theory', and new ecological, biological, physics, chemistry and other research. Great philosophers and scientists of the past intuitively felt there were some deeper, inner drivers of life and the Earth's (even the universe's) systems, from Darwin, Wallace and Einstein to Popper, Ilya Prigogine and many more.

In his groundbreaking book *Cosmos: A General Survey of the Physical Phenomena of the Universe*, begun in 1845, Alexander von Humboldt (the empathic visionary plus anticipator and influencer of Darwin) mused that the Earth was 'a natural whole animated and moved by inward forces'.[3] The intellectual giant Charles Darwin stated in 1859 in his *On the Origin of Species*, 'From so simple a beginning endless forms most beautiful and most wonderful have been, and are being, evolved.'[4] Nearly seventy years later, in 1926, Jan Smuts, in his prescient book *Holism and Evolution*, also pondered these big issues. At the time, he saw holism as being 'an operative factor in the universe'. He used the word 'holism' to describe 'the ultimate synthetic, organising, regulative activity in the universe which accounts for all the structural groupings and syntheses in it, from the atoms and the physico-chemical structures, through the cell and organisms, through Mind in animals, to Personality in Man'. 'Holism,' he said, was 'the fundamental activity underlying and co-ordinating all others', thus revealing 'the universe as a Holistic Universe'.[5]

Not bad, given he wrote much of this by 1908, and considering the century of extraordinary research and thinking that followed it. Owing to remarkable discoveries in successive decades in all disciplines, we now know more of the detail of this 'holistic' story – and of how, as Smuts expressed it, we live 'in a universe of whole-making'.

In 1977, along with Australian scientist John Eccles, philosopher Karl Popper was honest enough to admit, 'The greatest riddle of cosmology is that the universe is, in a sense, creative.'[6] In the course of writing this book, it has become clear to me that the processes of emergence and self-organisation (each embedded in 'systems-thinking') are the 'big engine' that underpins and structures the functioning of our world. As such, we landscape managers need to be cognisant of, and simpatico with, this over-arching functioning principle. As revealed in this book, the leading regenerative farming innovators and practitioners most definitely are: even if they don't fully understand it.

Writer and philosopher on this issue J. J. Clarke, in his book *The Self-Creating Universe: The Making of a Worldview*, stated there is a central theme that is 'the idea of creativity', and that this theme reveals 'how at all levels, from the cosmological to the human, the world demonstrates a universal inclination towards the emergence of new and unpredictable forms of order'. Moreover, the new knowledge and thinking surrounding emergence and self-organisation is positive, as it dispels the nihilism of a purely reductionist, mechanistic world and replaces it with an open-ended view of creative evolution. As Clarke puts it:

> Where once there was a sense of alienation from a world which seemed at bottom no more than the mechanical motion of dead matter, causally determined and in principle completely predictable, *the new thinking … locates us in a world which is open and constantly transforming itself, creating new and astonishingly complex and beautiful forms.* [my emphasis]

One of those 'astonishingly complex and beautiful forms' – perhaps even the most astonishingly complex and beautiful – is the human mind. Nowhere more relevant to this story and humanity itself, therefore, is the idea that self-organisation is a powerful property involved in the never-ending development of the human mind and of consciousness itself.

Furthermore, in a statement that powerfully accords with the findings in this book, Clarke states that out of the 'creative exuberance' of the processes of self-organisation 'there arises … a worldview which makes much sense in an epoch which is often dominated by scientific thinking yet which yearns for meaning and values and for a spiritual dimension to life'.[7]

This book is about the very processes of transformation: from the deeply embedded Mechanical to the post-Mechanical mind – or to what I propose calling the Emergent mind (see discussion in Chapter 19). So, in every respect, but beginning with the most fundamental – that is, with agriculture and its impact on the planet, human society and human health – this story becomes one of transformative change in our landscapes and selves via this giant story of emergence, creativity and self-organisation.

It is imperative we farmers have a basic working knowledge of the nature of complex creative systems, because that is exactly what we are managing. What once were assumed to be inanimate ecological systems are instead extraordinarily complex, biogeophysical natural creative systems of our landscapes. On top of this complex challenge is the fact we are also managing a social–ecological system. That is, upon the biogeochemical system is superimposed our belief systems and inherited land-use practices, plus their connection (and that of the farm) with the global economy, politics, family, and so on.

So the only way to get our heads right for managing complex creative systems is to think holistically, flexibly and openly. This is radically different from the traditional reductionist, linear way that 400 years of science has inculcated, and which is drenched in the mechanistic way we practise industrial agriculture. Instead, this new way is called 'systems-thinking'. Crucially in this systems-thinking approach, 'the properties of the parts can be understood only from the organization of the whole'.[8] This, as we shall see, means there is a fundamental unity in the structure and function of all life.

Therefore, to get a grip on all this 'complexity', I just want to break down the term 'complex creative system'. I want to clarify what I mean by it and to explain why I favour it over the more common scientific term 'complex adaptive system' (used in fields such as ecology, brain research, physics, chemistry, engineering and computing).

The word 'complexity' has as its Latin base the root word *plexus* – braided or entwined – which leads to *complexus*, or 'braided together', and so to *complexity*. Thus, *complexity* results from the interrelationship, interaction, intertwining and interconnectivity of elements distributed over different hierarchical levels: first, within a system; and second, between a system and its environment. 'Complexity thinking' is therefore based on relationships, patterns and iterations. It is these, when in unstable states, that give rise to discontinuous and unexpected changes – changes that are not explicable by a single causal factor and that feed back into the system to inform its ongoing interactions.[9]

Moreover, a complex system is constantly adapting to the environment: hence 'adaptive' in the 'complex adaptive system' term. And finally in a complex creative system, features *emerge* that cannot be predicted from the current description of the structure. Hence, this whole field of complex adaptive or creative systems is called by some 'emergentism'.

Such complex systems – including the ecosystems we manage as farmers – are never static but always changing: due primarily to factors such as the weather and our interventions. This is where the 'creative' bit comes in because such complex creative systems are doing more than just reacting to 'disturbances' (such as rain, drought, fire, tilling, grazing, spraying a paddock or a vehicle driving over the paddock). Instead, they are involved in actively responding to these factors. A complex creative system is constantly adjusting so as to re-attain or maintain a particular state of existence it needs to function properly. That is, such systems are ever dynamically changing and throwing up responses called 'emergent properties'. Crucially, this creative response is in constant operation as these systems strive to find the best fit with their environment.

Today, scientists, ecologists and others studying a range of complex creative systems have identified about twelve key attributes, which I won't fully list here. But we are already familiar with some of them, including connectivity, diversity, networks or life-webs, and recognisable patterns. The truth is, when we examine life, what we constantly find are networks within networks. It is this connectivity that leads to distinct, recognisable patterns in the system as the feedback is disseminated. Famous biologist Lewis Thomas believed that the 'biosphere is all of a piece, an immense, integrated living system, an organism'.[10] Corroborating this, as systems scientists Fritjof Capra and Pier Luisi describe, is the fact 'that there is a fundamental unity to life, that different living systems exhibit similar patterns of organisation'.[11]

Out of the dozen or so defining attributes of a complex creative system, there are three fundamental 'biggies' that I wish to highlight because I find them crucial to my thinking not just about managing a farm but also about how all ecosystems and indeed Earth and life function. These attributes are coevolution, emergent properties and self-organisation.

If we take coevolution, for example, it was a long process of interactive coevolution that led to the kudu on Norman Kroon's farm being able to detect a particular combination of chemicals in a shrub. This arose because, first, the shrub had evolved the chemicals as a defence due to aeons of grazing pressure. However, this in turn (and also in a coevolutionary process) led to the kudus learning to use the chemicals for self-medication. And then came Norman Kroon – that human–social landscape-function factor – to connect the dots.

Of huge relevance to human health is the fact we likewise coevolved in the African savannahs and similar environments to a hunter-gatherer diet. Consequently, we also are equipped to digest, sense and detect a range of chemicals (phytonutrients) in our food. But when we don't have a sufficiently varied diet or when it is corrupted, that is when severe health issues develop: as seen today in modern Western society and our epidemic of diet-related diseases (to be discussed in Chapter 21). The fact is that any species that has evolved has also coevolved.

But this function of coevolution is connected to the idea of emergent properties. These appear because there is nothing planned or controlled in complex creative systems. Things, organisms, nutrient cycles – the agents in the system – they all interact in what appear to be random ways. Yet these interactions lead to novel coherent structures, patterns and properties that just *emerge*. It is these emergent, novel properties that inform the behaviour of the agents, creatures, cycles, et cetera in the system and thus the system itself. Back in the early twentieth century, philosopher Alfred North Whitehead called this constant genesis of novelty in our world 'the creative advance of nature', meaning 'the perpetual transition of nature into novelty'.[12] Biologist Brian Goodwin recently emphasised that 'the whole spectacle of evolution is this "creative advance into novelty"'.[13]

Another way of expressing this idea is that this undirected, apparently random, constantly adjusting behaviour of a complex creative system throws up a whole series of experiments, or new traits and ways of doing things – the emergent properties. It is the most useful of these that provide the solutions and tools of natural selection and adaptation. This spontaneous emergence of order is not just due to non-linear feedback processes but occurs at critical moments of instability in a system or organism, and is one of the hallmarks of life. Indeed, life itself is an emergent property, as it originally arose 3.8 billion years ago out of inanimate matter on a relatively new Earth.

James Lovelock views Earth as *Gaia*: a super-organismic system of all life on Earth, where Earth is seen as a self-maintaining, self-creating, self-organising system engineered by life itself (and largely microbial life).[14] According to Gaian thinking, therefore, life itself creates the conditions for its own existence. In other words, higher levels of complexity arise out of the putting together of components of lower complexity, and these emergent properties then appear as a result of specific relationships and interactions between these lower-complexity parts.

Given that emergence is one of the hallmarks of life, in Capra and Luisi's words 'creativity – the generation of new forms – is a key property of all living systems'. Concerning regenerative agriculture and our capacity as a species to turn around the Anthropocene, this fact is clearly of great significance.[15]

A simple example of emergence occurring in a regenerating, self-organising system is one regularly experienced by regenerative farmers applying holistic grazing management on different continents. Once they get ground cover up to near one hundred per cent and all their landscape functions working in synergy, many experience the sudden emergence of, for example, perennial grass species, forbs and shrubs that have not been seen for many decades (even a hundred or more years). This is because increased soil biological activity (such as by burrowing creatures in deeper functioning soils) leads to seeds from long-wiped-out plants that have been buried deep long ago now being brought up. In our case at home, I have been thrilled to discover at least ten invaluable native grass species I thought I would never see on our property.

This feature of emergence leads to what I believe is the most important attribute of any complex creative system: self-organisation. In other words, new patterns and behaviours arise from within and not without. As Capra and Luisi describe, the behaviour of an organism or ecosystem is 'determined by its own structure, a structure formed by a succession of autonomous structural changes'.[16]

These processes of self-organisation are why this new systems-thinking is such a challenge to the old reductionist Mechanical mind. We have here the creative, spontaneous cooperation of many parts in an organism or ecosystem but without the need for any external regulating principle. In a nutshell, therefore, self-organisation (sometimes called 'self-assembly') is a bottom-up process where complex order spontaneously emerges at multiple levels from the interaction of lower-level entities. That is why it has become clear to me that the best description for the new, post-Mechanical mind exhibited by regenerative farmers is the 'Emergent mind'. By this, I mean a mind that combines elements of the previous Organic mind with the best of the Mechanical mind and modern science, but in addition has a capacity to respect and encourage the processes of self-organisation, open-ended creativity and thus emergence.

The final point I wish to make here is that a self-organising system is forever throwing up and then utilising (or rejecting) various emergent properties. This means that constant change is the norm. It takes an Emergent mind to be comfortable with change, with creativity and reorganisation being the general state.

Unfortunately, however, we tend to view our farms and paddocks as fixed entities because we are far more comfortable with the idea of things being static and predictable. This was also an old idea in ecology, devolving from former days as scientists sought to understand how ecosystems worked. Early ecologists and foresters such as Henry Cowles and Frederic Clements realised that ecosystems go through phases of definable change: from a starting base (such as after a fire) with pioneer species, thence moving through stages to a mature or 'climax' stage. However, we know today that a climax stage is never fixed. Either humans come along and disturb it, or else a drought, fire, insect attack or fungal disease, flood and so on occurs. And so, the whole complex creative system has to reorganise, adjust and adapt to survive and function. And of course it does this by providing alternative options and solutions (emergent properties), which then allows it to self-organise back to a more stable, functioning, resilient and surviving state.

But this is where we humans step in. We assume complex creative systems are rigid and fixed. We set-stock and overgraze. We clear trees and long-coevolved vegetation in complex systems so as to grow monocultures of wheat and other crops but in country maladapted for this, with resultant dryland salinity in five states. We treat arid and semi-arid rangelands on the assumption they behave like ecosystems with predictable rainfall and seasons, as if on the 'inside' country, but not appreciating their cycles of boom–bust rainfall may be over two, three or five years (thus requiring extremely delicate management). Yet we then wonder why topsoil blows across the Tasman Sea to New Zealand. Or we modify, extract, over-fertilise and overgraze country, as in the New England, Monaro and other tableland areas, and wonder why eucalypt dieback suddenly appears. That is, treated as rigid resource boxes that we can manipulate with levers so as to extract product, such agricultural systems eventually reach tipping points, beyond which they cannot develop enough emergent properties, and beyond which they can't healthily self-organise anymore.

Comprehending that our farms are complex creative systems means they don't need dictatorial control and rigid, extractive rules and behaviour. So, when impacted by us as farmers, or by droughts, fire and so on, we as managers need to be sensitive to their need to self-organise in an unfettered way. This adaptive process is now seen by ecologists and others as a recognisable four-stage 'adaptive renewal cycle', or what is called a 'state and transition' process. But the key point is that nothing is ever static: change and self-organisation are the constants.[17]

What is sobering, is that our self-organising Gaian Earth and its dynamic self-regulating systems are likewise governed by complex creative systems. Humans have now destabilised and distorted these, precipitating us into the Anthropocene. As occurred with John Robertson's grasslands in the 1840s and with Western Australia's salted wheat lands and New England's eucalypt dieback in the twentieth century, the key question now becomes when could Earth's macro-systems likewise reach tipping points?

It is October on our farm: a blustery day as I walk alone, trying to pull together in my head the complex threads of this book. This is hard, as the distractions at this time of year are many, given that October is both a crucial seasonal turning point and also a month of beauty and frenetic biological activity.

Various wattle species are in stages of budding, flowering or seed-setting, while other plants too are in various stages of vegetative reproduction. In the bush, white-flowered tendrils of draping bush clematis have begun shrivelling, but there are splashes of purple with Austral indigos.

Nevertheless, it's the birds that are the real harbingers of change and spring, ubiquitous in activity, sound and display. Moreover, bird watching was my entry point into a slow-growing environmental sensibility. To badly distort an avian metaphor, they also are the 'canary in the cage' of the health of terrestrial ecosystems. Birds are super-sensitive to environmental dislocations (such as habitat loss, delayed or advanced insect hatchings or vegetation flowerings due to climate change, and so on) and that is why, both worldwide and in Australia, bird numbers are crashing. A major culprit is industrial agriculture. Pesticides and other poisons, industrial fertilisers, ongoing land clearing and habitat destruction, the accelerating loss of biodiversity via escalating monocultures, altered seed varieties – the list goes on and on, but all are contributing to loss of food, nesting sites, and secure habitats. As New York nature writer Jonathan Rosen said in his landmark book *The Life of the Skies: Birding at the End of Nature*, 'If we don't shore up the earth, the skies will be empty.'[18]

If I had a favourite bird out of the more than 140 species sighted on our farm, it would be the yellow-faced honeyeaters. Their evocative calls of 'Quick! Quick!' when on the move have great poignancy. These birds – no larger than a delicate wine glass – have been around for at least one to two million years. Many pairs nest in our gardens and many more up in our bush, tree-breaks and mosaic patches where their territorial calls comprise an explosion of liquid, galloping notes. This year the honeyeaters once again have begun arriving in late August and September from their over-wintering habitats on the coast, woodlands and forests of northern New South Wales and the forests and brigalow lands of south and central Queensland.

The poignancy of their arrival is because this north–south movement to the south-east forests of New South Wales and Victoria only a few decades ago involved a million or more birds, often in large flocks of up to 100,000. Today, such flocks are less than 1000 – and perhaps their total population has shrunk to below 100,000. The destroyer is ongoing massive and uncontrolled clearing of the honeyeaters' woodland habitats and food sources, due to the giant machines of modern cropping agriculture. Automation, GPS tracking and the broad wingspans of giant seeders and spray rigs have determined that trees are an impediment.

So we cherish these birds and their happy presence in our garden at home, because 'Quick, Quick' may be their path to extinction if such rampant tree and shrub destruction is not halted.

As I near home, I pause to watch a scene that makes any farmer's heart swell with joy: the sight of lambs playing in spring sunshine. Camped under gum trees nearby are a few hundred ewes in various states of rest. Some are head down and sleeping, others lying or standing as, eyes closed, they contentedly chew their cud. But nearby there is another group that seem slightly perplexed: maiden ewes concernedly watching as lambs gambol on a creek bank.

In mobs of twenty to thirty, the lambs gather atop the creek. Then suddenly they race down and up the other side, long tails trailing, then back again, until a small knot of them climb a granite rock, where bunting and jostling reveals it has become a game of 'king of the castle', until off again they career in a mob – along, down and then up the creek bank once more. It is play, and not unlike in a school yard; and it is *joie de vivre*.

Smiling, I head for home, but nevertheless with questions rattling in my head from the walk. It is the big questions that really confound me, such as how do the extraordinary patterns of micro-life, soils, plants and animals – above and below ground – fit together? How and why did they evolve like this? And how best can I enhance, enrich and sustain them while at the same time growing food and fibre in a healthy fashion – healthy enough that my animals feel sufficiently happy to play games on creek banks and rocks, and yellow-faced honeyeaters continue to bless us with their presence?

The answers, I know, lie within the self-organisation processes of our complex creative systems, but I am happy to leave the reasons in the 'mystery' basket and just content myself with marvelling at the explanation that J. J. Clarke gave, quoted earlier in this chapter: that this new view of a self-creating universe does indeed locate us 'in a world which is open and constantly transforming itself, creating new and astonishingly complex and beautiful forms'.

Of all the work I have encountered, perhaps that which best encapsulates this entire story of functional, complex creative ecosystems underpinned by the big drivers of coevolution, emergent properties and self-organisation is the lifelong research and teaching of Professor Fred Provenza: a man truly of an Emergent mind.

Now retired from Utah State University, where he was a professor in the College of Natural Resources, Fred Provenza still occasionally emerges from his home retreat in the Rocky Mountains to run week-long workshops or speak at conferences. But generally these days he prefers to go on daily walks with his wife, Sue, in the Rockies: just observing and contemplating nature, walking in the forest, watching eagles, buzzards, coyotes and other creatures, and enjoying the changing of the seasons.

In earlier chapters, I mentioned my first brief encounters with Fred. Since then, in the process of befriending him, I have come to appreciate the depth, interconnectedness and profundity of his work in relation to this big-picture story I am telling: work embedded in animal behaviour, biochemistry, nutrition, landscape–animal coevolution and the holistic story of creative self-organisation.

Fred is a smallish, vital, energetic man who overflows with humour. With his bald pate, glasses, Buffalo Bill–type apron goatee beard and moustache, he seems more like a genial grandfather than an academic who has published hundreds of papers and run a large research program.

To sit through a day-long or week-long Provenza course (his serious 'introductory' course used to be ten days!) is to be taken on a gentle, whimsical journey into what he prefers to call 'complex creative systems'. This involves landscapes, their coevolved plants and grazing/browsing animals, and their linkages to human management and human health. It is truly a big-picture journey, built on detail from the ground up. I cannot hope to fully capture that here, but I will briefly attempt to sketch the main themes.

Embedded in sometimes detailed, painstaking research, Fred's slowly evolving but now profound holistic view hinges on 'creativity'. In a recent joint paper, Fred and peers stated that 'organisms don't merely respond (adapt) to the environments where they live, they actively participate in constructing (creating) them. These dynamics,' they continued, 'suggest that we should take our views of evolution beyond an account of how organisms developed from earlier forms during the natural history of the species to include changes occurring within the lifetime of the individual.' Then, in a statement that should be embossed on the front of all traditional science textbooks, they challengingly conclude, 'Life is ever creating its way into the future by engaging in new cooperative relationships.'[19]

Much of Fred's work relates to such an Emergent view of life and to these cooperative relationships: and particularly those involving coevolved plants, herbivores and humans, and thus to attendant creativity.

I have emphasised throughout this book that the leading regenerative farmers work with and not against nature: that they seek to enhance her uniquely complex functions, nuances and directions – even when they don't fully understand them. That is, being Emergent thinkers, they try to enable the big drivers such as coevolved properties, emergence and self-organisation.

The reality is that modern agriculture has simplified natural systems and plant and animal breeding to the nth degree. Therefore, today we simply don't realise how far removed we and our agricultural systems have become from how natural landscapes work. For example, whether in Africa, the Americas, Australia or any other coevolved, herbivore-impacted landscape, plants – in the process of coevolution – have developed to use, concentrate and even produce tens of thousands of primary and secondary compounds.

By primary compounds is meant the macronutrients (water, carbohydrates, fats, proteins, and so on), key minerals (calcium, iron, magnesium, et cetera), vitamins, fatty acids, amino acids, plus various kinds of phytosterols. Many of these primary compounds are essential for the functioning of living organisms such as the plants themselves and, largely through ingesting them, animals (including humans).

The secondary compounds in plants are called 'phytochemicals' or 'phytonutrients' because they are derived via photosynthetic, energy-driven plant function. For example, there are more than 8000 phenolics, over 25,000 terpenes, more than 12,000 alkaloids, plus a huge range of animal medicinal compounds, and so on: perhaps more than 100,000 different secondary compounds. Crucially, following millions of years of coevolution with herbivores in landscapes, there is (or was) a reason why all of them are here.

When we cut this cake further, things get more interesting for us landscape, plant and animal managers. This is because in resource-rich environments (the higher-rainfall, richer agricultural lands in places such as north-west Europe), plants promote their survival by concentrating high levels of primary compounds and low levels of secondary compounds, as this ratio attracts herbivores. However, in resource-poor environments (like much of Australia), the converse occurs. Here in the evolutionary process, plant species are favoured that tend to concentrate high levels of secondary compounds, so as to deter herbivores as opposed to favouring growth. Moreover, this allocation of resources by plants for both or either growth and defence occurs on a daily basis.

Much of Fred's work has been about unravelling the existence and mysteries of these primary and secondary compounds in plants. He has then revealed how, through human management, we farmers can put knowledge of this to use (together with knowledge of coevolved animal behaviour) so as to get the best out of human-controlled grazing–browsing systems. This means to the benefit of landscapes, animal production and consequently human food (and, I would add, of beautiful natural fibres like wool).[20]

A key factor Fred emphasises is that nothing is ever static in a grazing environment. In response to grazing, plants will react by, for example, changing the types and concentration of defence and other chemicals to protect or enhance their survival. Moreover, plant phytochemical concentration can vary through the day and, if animals are grazing a diverse landscape, will vary in the same plants across that landscape. This potentially has huge implications for regenerative agricultural management.

Fred Provenza and colleagues repeatedly emphasise that animal palates (and, as we will see later, human palates also) link animals with landscapes through flavour-feedback associations. As they describe, such relationships:

> involve primary and secondary compounds interacting with cells and organ systems in a dynamic network of communication that guides food selection. They are mediated by aroma and taste receptors linked by nerves, neurotransmitters, peptides, and hormones with organ systems throughout the body, including the tongue, gastrointestinal tract, pancreas, respiratory tract, heart and brain.

In addition, says Fred and colleagues, the lining of the gut of mammals has a range of receptors, including those for odorants, nutrients, secondary compounds and toxins, and because of this, sensory information 'is transferred from the gut to the brain through four effector systems'. Furthermore, and as we will see in Chapter 21 concerning human health, these 'associations are further mediated by the microbiome': that is, our crucial gut flora. This hitherto largely ignored but hugely important gut microbiome plays a major role in many of our modern diseases, from obesity to most of the other modern killers (including the alarming new information on the impacts of glyphosate disrupting our gut flora – and thus our immune systems).[21]

Diets rich in secondary compounds (as occurs in a healthy and diverse landscape) lead to an increase in gut microbial populations that enable herbivores to eat plants they otherwise couldn't eat. In turn, if we humans get caught up, for example, in eating nutrient-empty junk food or diets that exclude meat sourced from healthy landscapes, then we are denying ourselves access to a huge range of essential nutrients that we are coevolved to detect, utilise and thrive on – and which keep us healthy.

What is crucial here (and because it further relates to us humans and fellow mammals, as we will see directly) is that if herbivores are not divorced from their diverse, natural world, their palate is still in touch with their bodies. This enables such herbivores to selectively graze so as to meet their needs for energy, protein, minerals and vitamins, but also to self-medicate.

Furthermore, herbivores (and humans) begin learning even *in utero* to associate the flavours of foods in the mother's diet with their post-ingestive consequences. Then, after birth, these infants begin learning which foods to eat or ingest from foraging with their mothers, and learn to associate unfamiliar (novel) foods with their consequences when eating different foods in a meal. These lessons can last for years.

This extraordinary interrelated process is what Fred calls 'the wisdom body'. By this, he means 'the ability of creatures to grow, reproduce and survive on foods they've become accustomed to eating in the social and biophysical environments where they've become adapted to living. Of necessity, the wisdom body is preserved in all life', from bacteria through to humans.[22]

Clearly, for true animal health, *variety* is the spice of life, via a healthy, diverse environment. This in turn relates to palatability, which is more than a matter of taste. Instead, it is a process that involves dynamic and ongoing interrelationships among cells and organ systems that feed back to the palate to change likings for the various mixes of forage available. The interaction of phytochemicals and flavour-feedback interactions with cells and organs plays a huge role here, as it comprises, in Fred Provenza's and colleagues' words, 'a dynamic network of consciousness and communication that unites digestive, excretory, cardiovascular, respiratory, skin, muscular, skeletal, immune, nervous, endocrine, and reproductive systems in the body'.

There is yet another important issue here: *satiation* or *satiety*. This is where animals get sick and tired of eating the same old foods in the same old places. For reasons explained earlier, they need a variety of foods to maintain health and performance, and from this (due to the internal feedback mechanisms we discussed) comes contentment and a lack of cravings, and thus reduction of over-ingesting certain foods. That is, truly healthy animals need both a variety of foods and also to forage in a variety of places. Implications for obesity in humans are frighteningly obvious here (as I discuss in Chapter 21), as a lack of satiety from good nutrients means the mind is being told the body is still hungry, and so excess eating occurs.

Use of this extraordinary animal wisdom and of an understanding of satiety has enabled astute livestock managers to get excellent performance and health from their animals and even to train them to graze in places such as forests, vineyards and citrus groves so as to: eat foods they would normally avoid; learn to avoid poisonous plants; specifically eat invasive weeds; and rejuvenate landscapes for the benefit of both wild and domestic animals. This is exemplified by the best of modern French shepherding, where herders came to understand the processes involved in food and habitat selection.

In various regions across France, herders use a four-step process to first teach their animals to use the full range of forages they provide them with in a planned process. Second, they train the herd to respect the boundaries of the grazing area. Third, they modulate what they call the 'temporary palatability scoring' of the various forages. And finally, they establish daily grazing circuits to stimulate appetite and intake through meal sequencing. That is, such herders use their understanding of space, time and biochemical diversity to stimulate forage intake in their livestock and to more fully use a wide range of available plants. They do this by manipulating the entire flock to eat six ecosystem 'courses' in a daily meal: say, three courses before lunch, then a rumination during lunch, and then three courses after. These courses comprise appetite stimulators and moderators, first and second courses, boosters and then desserts.[23]

Clearly, this astutely developed French shepherding system maximises plant diversity in different biogeochemical aspects of the landscape so as to enable animals to select nutritious diets while also self-medicating. But, looking further afield, this separate but convergent work of French shepherds along with Fred Provenza and colleagues has huge implications for livestock and human health when we consider that both we and our grazing animals coevolved in diverse landscapes and that we also still exist in complex creative systems. Thus, grazing animals managed in a holistic grazing regime, where they are regularly moved through diverse pastures, shrubs and landscapes, are functioning in close approximation to how they coevolved in landscapes.[24]

This is in stark contrast to animals in an industrial system, where monocultures or near-monocultures of restricted diversity exist; where such feed is pumped full of industrial nitrogen and an imbalanced range of other minerals; and where 'improved' industrial forage species bred by multinational agribusiness and scientific organisations are targeted to have low levels of secondary compounds (e.g. less saponins in lucerne; less alkaloids in grasses such as fescues; less tannins in trefoils; less cyanogenic glycosides in clovers). That is, what these controlling powers have done is to eliminate secondary chemicals that, in low doses, play vital roles in animal (and subsequently human) health.

If we now examine industrial agriculture at its next level – that of feedlot beef and lamb, for example – we have an even more warped and unhealthy situation. Here, feed is mixed, thus eliminating animal choice. Pure grain (which is totally unnatural to ruminants) is fed, with potentially major negative health impacts for gut health (such as lactic acidosis). Because of this health-damaging artificial environment, doses of antibiotics (in addition to ingredients such as growth hormones) need to be added. Little wonder that such meat is bereft of a diversity of phytochemicals and has major negative health implications for humans. In short, humans in industrial agriculture are contravening every major principle whereby healthy landscapes, functioning as complex creative systems, mutually sustain the health of herbivores that are coevolved to graze them in synergistic fashion.

Artificial, monocultural, highly fertilised pastures and feedlots thus constitute an obscene travesty in terms of healthy natural function. The next frightening part is that these artificial industrial systems are largely managed to feed humans. The consequence is food bereft of essential primary and secondary compounds for which we were coevolved on the same landscapes as these herbivores. As a double whammy, our food is also laden with by-product toxins or damaging additives such as antibiotics and hormones. If we then add to this the alarming evidence that major herbicides such as glyphosate are getting into our food, it is little wonder that there is an alarming explosion of modern health epidemics across the spectrum.

Quite simply, modern industrial agriculture has crudely and violently destroyed healthy soils, healthy functioning landscapes and ecosystems, healthy animals, healthy behaviour of animals and humans in landscapes, and healthy people. This is the harmful whirlwind that the Mechanical mind, taken to extremes through industrial agriculture, has visited on Mother Earth.

When Fred Provenza talks about 'health from the ground up' and how 'the health of people is entwined with the health of landscapes', he reiterates, 'landscapes with diverse arrays of plants are nutrition centers and pharmacies with vast arrays of primary and secondary compounds.'

In *The Third Plate*, leading New York chef, agroecological thinker and vibrant communicator Dan Barber recounts a unique meeting he attended in Hampshire, England, in 2005 at the famous Laverstoke organic farm. This farm was founded by the flamboyant and controversial world champion racing driver Jody Scheckter, who invited twelve leading agroecological farmers from around the world to a private think tank.

Barber recalled that his friend Klaas Martens – a raw-boned, Gomer Pyle lookalike, organic grain farmer in Massachusetts – rose to address the august company as last speaker. He opened with the simple question: 'When do you start raising a child?' Both his question and answer, Martens said, came from a Mennonite bishop, who, when Martens couldn't answer it, told him that 'child rearing begins not at birth, or even conception, but one hundred years before a child is born', and 'because that's when you start building the environment they're going to live in'.[25]

But what the Mennonite bishop would not have known is that perhaps most insidious of all the factors involved in industrial agriculture and its attendant modern foods and its manufacturing, distribution and economic systems is that unborn generations are now also being impacted through these radical changes to natural landscape function.

A key factor here is epigenetics. As discussed in Chapter 15, the emerging knowledge in this area has completely upended thinking about the mechanisms of evolution and about the form, function and behaviour of living organisms. Crucially, why epigenetics is such a game-changer is that it reveals we need to expand our view of evolution beyond the story about how organisms develop from earlier forms during the natural history of a species. That is, our thinking now needs to include changes occurring within the lifetime of an individual.

In short, developmental plasticity is crucial in biological evolution. Evolutionary biologist Patrick Bateson explains, 'The combination of developmental and behavioural biology, ecology, and evolutionary biology has shown how important the active roles of the organism are in the evolution of its descendants.'[26] As Fred Provenza put it to me once, 'Individuals are *involved* in the world, which allows them to *evolve* in the world.' In other words, we and our animals have transgenerational linkages to the landscapes we inhabit.

As Muley intuitively articulates in *The Grapes of Wrath*, 'Place where folks live is them folks.'[27]

There is a positive side to this new knowledge field. Regenerative farmers are now beginning to explore and apply the new thinking and research that Fred Provenza and colleagues have revealed. This is because Fred's work has now combined with research in Australia that examined over one hundred different native shrubs for features such as salt tolerance and levels of primary and secondary nutrients for application in the field in areas including nutritional boosts and self-medication.

To this end, Fred and colleagues at Utah State University have linked to scientists, farmers and land managers in Australia and three other continents besides the USA to form an international consortium called BEHAVE (Behavioral Education for Human, Animal, Vegetation, and Ecosystem Management). Behind the BEHAVE approach is the basic goal of understanding how animals learn so as to enable us land managers to train our animals to better fit our landscapes rather than our current approach of manipulating landscapes to fit our animals.

All this is why Fred Provenza sees landscape management as an artistic, multi-dimensional creative process. 'In practice,' he says, 'creativity is manifest when a land manager "paints on the canvas" we call a landscape. Managers paint using understanding of relationships among different "colors" – soils, plants, and animals. The challenge of adequately embracing multiple causes and outcomes over time and space translates into transforming with the landscapes we inhabit ...'[28]

One exciting spin-off of this program is the initiation of a sophisticated series of practices called 'Self-Herding' and 'Self-Shepherding'. These were developed by regenerative farming innovator Bruce Maynard in partnership with leading Australian scientist in this field, ex–CSIRO principal scientist Dr Dean Revell. Maynard and Revell have adapted research and experience in animal behaviour, nutrition, physiology and ecology to define a series of operating principles. These in turn create opportunities for pastoralists (particularly in the rangelands of the Kimberley and Northern Territory, where it is being applied) to positively influence the decisions animals make on where they go and how they interact with each other, with the pastoralist, and with landscapes. These principles, however, can be applied in all country and on different-sized farms. When linked with holistic grazing management, the results are profound for animals, landscapes and humans.[29]

The last time I saw Fred Provenza was at a week's conference in Dubbo in March 2016, where I was a minor speaker on his course. His holistic material being by now familiar to me, I had time to study the man and to note how his final sessions now focused on the ethical and spiritual.

One reason for this, I discovered (besides basic integrity plus a whimsical sense of humour and a never-ending search for truth and personal growth), was a tough series of life challenges, including major bouts of cancer and depression.

Once, on a three-day car trip with Fred, when discussing macro landscape systems, he suddenly turned to me. 'Charlie,' he said, grabbing my arm, 'these systems are made up of connections, not just parts. It's the connections we need to study.'

And I knew he was right. Our training in the modern scientific tradition tends to preclude us from seeing the big picture: from seeing the forest instead of a bunch of trees. Life is about choices, and Fred, faced with a divergence of paths, took the one less travelled.

In short, despite his scientific training, Fred was able to go beyond the Mechanical mind to the next stage: to what is clearly an Emergent mind.

That is, this alternative path allowed him to see the world afresh, to realise that, if allowed, nature throws up emergent properties and is constantly, creatively self-organising. And as Fred had found after years of challenge but also achievement, if the big picture – the whole and not the parts – can be appreciated, then our inner yearning for interconnection with our landscapes, community, loved ones and life will provide true meaning, values and spiritual satisfaction. After all, we also are coevolved for this intimate connection with our living, nurturing world.

In the recent words of Fred Provenza, Dean Revell and colleagues, 'The challenge we face is to avoid illusions of stability, control, and permanence and to embrace inexorable cycles of transformation'. And they are optimistic about meeting this challenge, for they concluded: 'The strange and wonderful irony is that working together with open hearts and minds to transcend the boundaries we create is addressing the "really big issue".[30]

To my mind, the really big issue is saving the planet and its systems from human destruction. As we have seen, this can only occur via personal transformation. So the stories in this book reveal that we have a pathway full of solutions – even if it is currently the path less travelled. All that remains is to swallow our hubris and walk humbly before the wonder of nature's extraordinary coevolved systems.

CHAPTER 19

Transforming Ourselves

A system that cannot deliver the well-being of people and nature is in deep trouble. It invites ideas and actions that are transformative.

JAMES GUSTAV SPETH,
The Bridge at the Edge of the World[1]

This morning pre-dawn, as I walked to my writing office, I paused to listen to the mysterious and penetrating call of a boobook owl. 'Morepoke … Morepoke,' he echoingly cried from a grove of gum trees. Whether he was proclaiming territory or in search of a mate I knew not, but it was a reminder that the season was turning this mid-August. As the days lengthened and light took on an intense spring shimmer, silver wattles were in flower, stippling gold across hillsides, mountain duck and wood duck had taken to tree nests, and crows and magpies called and squabbled as they busily raised broods. Up in our sheltered paddocks, lambing was in full swing, though at this time of pre-dawn the ewes had to share fresh pasture with a mix of kangaroo, wallaroo, wallaby and the odd stolid wombat.

In my office, I turned to the key issue confronting human society: how do we transform ourselves and prevent humanity hurtling deeper into an increasingly unpredictable Anthropocene, and so avoid unimaginable disaster? For I knew that all the research and technology in the world would not rescue us when the solution lay inside our heads: in the world-views that governed our and society's behaviour.

It was this dilemma that led me, in Chapter 15, to pose what I believe is the most crucial question of all: 'Can we evolve past the Mechanical mind?' This was because I argued that it was not just a lack of ecological literacy that lay behind our degradation of landscapes. Instead, and at the most fundamental level, it was also a lack of understanding of ourselves: that as a modern civilisation we weren't even aware of how deeply we were locked in the Mechanical mind.

However, as many examples reveal in this book, despite their early training and inculcation in the Mechanical world view, brave individuals involved in agriculture, science, the urban food movement and other areas have been able to overthrow their Mechanical mind in a process of mental transformation. The result has been a shift to what I call the Emergent mind. This combines the best of the old Organic – namely its reverence, empathy and respect for Mother Nature – with the best of modern, ecologically simpatico science and Earth-empathic thought. But the crucial difference in this momentous shift (as we saw in the last chapter) is that these farmers and others have developed a creative mind that is open to the processes of emergence and self-organisation.

I believe it is in these examples that there resides the exciting promise and template of how we as a society can extract ourselves from our collective Anthropocene dilemma.

There is thus a supreme irony here, for it was the beginning of domesticated agriculture that led to the Mechanical mind. But I believe it will also be agriculture that will help lead us (and already is leading us) to the next stage: that of the Emergent mind – a mind open to the new and positive reorganisation of thought and subsequent action. This is because, as per the quotation from Wes Jackson in the frontispiece of this book, '[s]ince our break with nature came with agriculture, it seems fitting that the healing of culture begin with agriculture, fitting that agriculture take the lead.'[2]

Pondering the entrenchment of the Mechanical mind, I recalled a childhood memory. On driving to our local town of Cooma when I was seven or eight, we would pass a farm about halfway, situated on the edge of timber and open basalt grasslands. It belonged to an old bachelor. Strung up either side of his front gate, stretched wing-tip to wing-tip and thus covering nearly a hundred metres, were twenty-five wedge-tail eagles this man had proudly poisoned or shot. Today, legislation and an increased awareness of the importance of apex predators generally preclude such frontier-like, Mechanically minded excesses.

In one way, such behaviour is at least understandable when we consider the rise of the Mechanical mind and how it arrived in Australia after 1788. However, it was another story that penetrated more deeply and that led to me pondering the intransigence of human behaviour as we move deeper into the Anthropocene era.

Michael McKernan in his book *Drought: The Red Marauder* recounts an interview between a journalist and a First World War–returned soldier who had been placed on a soldier settlement block near Kulwin on the Victorian Mallee. This interview occurred just before Anzac Day 1931, in the midst of both drought and the Great Depression. The man had been attempting for eleven years to farm his pitifully small 320 acres (130 hectares): just over a square kilometre of poor, sandy country in a low-rainfall region that had, when he took it up, been thick with mallee stems and roots. Mercifully unmarried, the man still lived in a hessian humpy; for blankets, he had waggas (wheat-sacks sewn together). He had never been able to buy clothes in that eleven years; threadbare khaki dominated his 'wardrobe'. And while working seven days a week, daylight to dark, and with a mountainous debt, he lived on less than twenty-five per cent of his army pay.[3]

The soldier settler had virtually gone straight into farming on returning home after surviving the horrors of the Western Front. He told the journalist, 'I would sooner do ten years at the War than one at the Mallee.' Yet the man refused to leave his land, even as his very soil capital blew away across the horizon. What was it that made him stay in a place he rated ten times worse than the Western Front? Was it an immigrant's, a political radical's, or a poor man's hunger for land? Was it a fierce desire for independence after the incompetent leadership and mindlessness of the war? Or did he just want to live the agrarian dream of being a self-sustaining yeoman farmer who could feed a family off his own land?

Who knows. But this story reveals something central and of vital importance to this book. Whatever his reasons, attitudes, values, it came down to the fact that what drove him on – drove him to stick out a living hell of suffering and utmost poverty – was his personal belief system: his mental model of how the world worked. That is, it came down to his mental or psychological constructs: to his world view.

This raises key questions concerning far more than Australian land use, for it goes to the heart of our Anthropocene dilemma. The rise and entrenchment of the modern Mechanical mind, I believe, explains where our mental constructs have come from and how they evolved. But that leaves the biggie: why are our personal belief systems so powerful and hard to change?

I pursued this issue – as one farmer expressed it to me, 'of what goes on in that one square foot of real estate between our ears' – for four years in my PhD thesis. The quest continues still and, while a clearer picture is emerging, it is only perceived 'half-darkly'. But I now realise this question doesn't just go to the heart of Australian and American land use and misuse; it is also directly relevant to the fact that the Mechanical mind is what now underpins a powerful and suicidally destructive belief system. This is apparent in the extreme development of modern economic-rationalist society and its relentless plundering of the entire Earth, its resources and its people for the sake of greed, blind consumption and the perceived necessity of endless economic growth and profit: of returning sharehold value.

Crucially, this mindset is manifested in modern industrial agriculture and its giant machines, corrupted technologies and chemicals that poison our food, ourselves and the Earth.

So the vital challenge is shifting a human mind from a deeply embedded paradigm. And such a shift is extremely difficult. Our belief systems are mired deep in our self-identity, psychology and psyche: both conscious and especially subconscious.

The complexity of this issue was regularly brought home to me in the course of my previous consultancy work (sheep classing, ram-client servicing and wool marketing), when I annually travelled up to 70,000 kilometres across Australia's landscapes and across every state. Evidence of widespread and worsening land degradation on these trips was and is self-evident.

It is well known and researched that attempts to apply new remedies in the context of land degradation have been fraught with difficulty and disappointment. Part of the reason for this is that solutions have been imposed in a linear, top-down manner, which is how Western, reductionist, scientific society attempts most solutions. The fact is, however, that those seeking to apply remedies are actually dealing with a group of complex and frustrating problems (because they contain many paradoxes), so these people have frequently either failed in their attempts or made things worse.

What really jolted me into delving into this powerful role of personal psychological constructs in farmers' thinking was reflecting on my own behaviour. Despite my education, love of nature and exposure to regenerative agriculture (plus some mind-cracking experiences such as the early 1980s drought), I had initially missed the boat of regenerative agriculture. Then two more puzzling experiences finally convinced me that the issue of land-use change needed to be probed through an entirely different set of questions. Pat Francis (long-term editor of the *Australian Farm Journal*) told me how he had noticed that some farmers appeared to carry two mental constructs in their head at the same time. At the peak of the Landcare roll-out in the 1990s, such farmers would proudly show Pat Francis healthy, revegetated riparian corridors. But to view these, they had to travel through sprayed-out and bare or overgrazed cropping and pasture paddocks. I too have witnessed this behaviour.

The second experience arose from my extensive visits to clients across the Australian landscape. On such trips, I noticed what I call the 'vegie-garden paradox'. On visiting a farming family's home, I frequently observed a well-managed, healthy, chemical-free, organic home vegie garden. Then we would leave the perimeter of the home garden en route to the fields beyond, and outside the garden fence were piles of chemical drums behind machinery sheds, while the paddocks were heavily modified, sprayed and chemically fertilised. Clearly, such farmers held two different and opposing views of how agricultural practices and natural functions worked.

It became evident that two key factors existed concerning change in agricultural practices. The first was social: farmers' attitudes, beliefs and practices. The second was cognitive: psychological constructs and subconscious processes highly resistant to change. Clearly, what was involved were issues to do with early education and then an immersion in rural society, and later adult learning, history, power connections (visible and hidden), habit and much more.

Yet, as this book testifies, farmers and others have been able to change their entire world view and shift to a completely different, open and creative Emergent mind: as it were, to radically change and 'turn over' that square foot of real estate between the ears. After eighty-odd interviews for my PhD thesis and many dozens more since, a clear pattern emerged as to why this change occurred: notwithstanding the fact it is a pattern that by no means explains all the factors involved.

First, in the majority (around sixty per cent) of those who had enacted a shift to an Emergent mind, the key reason for change was that they had experienced some sort of major life shock that cracked open the carapace of their mind so as to be able to entertain new thinking. For many, including Tim Wright and John Weatherstone, it was the harrowing 1979 to '83 drought.

Sometimes, it took a primary shock to crack a mind and then a second to serve as the final trigger. For leading holistic grazer David Marsh, the early 1980s drought was, in his words, 'the initial "head-cracker" for sure', followed by discomfort with high use of chemicals in his cropping. The final 'cracker' was when he undertook a Myers–Briggs personality test during a university Masters course on sustainable agriculture. He suddenly realised that while he was intrinsically a nature-loving (biophilic) person, his farming was the opposite. He immediately acted to correct this dissonance.

The cracking of minds had many other causes, including: possible bankruptcy; marriage breakdown; the loss of large numbers of stock off-shears; a water crisis; loss of valuable land to insidious soil salinity; the collapse of the wool reserve price scheme and ensuing financial crisis; superphosphate killing native grasses, or various other signs of ecological deterioration threatening their viability. For Colin Seis, it was losing his assets to a bushfire (plus ending up in hospital with severe burns) and having to start again with nothing.

For biological agriculture pioneers Rhonda and Bill Daly, the catalyst was chemical poisoning. While the Dalys were increasingly becoming dissatisfied with declining yields and increasing costs and problems with industrial agriculture, their impetus to make a transformational shift went back to when Rhonda nearly died after being poisoned by arsenic when dipping sheep as a child, as well as being sprayed by deadly chemicals while marking for the aeroplane to crop-spray. In turn, these poisonings subsequently collapsed her immune system, and at the age of forty-seven she lay undiagnosed in hospital for seven weeks surviving on fluids. She came very close to death before she was finally diagnosed with meningitis and encephalitis. As a result of her illness, Rhonda was left with chronic fatigue and suffered from loss of hearing and other afflictions.

It was while lying in the hospital bed, desperately ill, that Rhonda said she received spiritual help from a 'guide', which prompted her 'to heal the soil and help others'. It appeared that Rhonda and Bill were to go beyond regenerating their own farm and establish Ylad Living Soils in 2002 so as to help the industry. Ylad Living Soils is a company that supplies biological, organic and humus compost fertility products as well as composting equipment and tea extraction units. The Dalys also play a leading role in agronomic advice for natural farming systems and humus compost production along with a change-agency role via education.

But there were and are many different triggers that prompt farmers to make a transformative change to regenerative agriculture.

For the remaining forty per cent or so who changed, it was either a series of small, destabilising steps and challenges to their thinking that eventually culminated in a shift, or else serendipitous and individualistic circumstances. Sometimes, this came quickly, as the acquisition of ecological literacy led to a cascade of changes; while for others, people of influence such as family members, friends or mentors persuaded them to attend life-changing courses or field days run by the various diffusion organisations involved in regenerative agriculture.

And sometimes the shift was slow – taking many years. In many cases, these people were already biophilic or sympathetic to an alternative approach but didn't know how to change until encountering new knowledge. These can be characterised by the phrase: 'When the pupil is ready, the teacher appears.'

Leading holistic management educator and practitioner Dick Richardson expressed the insight, 'There are points in a person's life when you're ready for new things and it's not when you've just come out of university, bullet-proof, and everything's simple and you can just reach for this technological answer.' In John Weatherstone's case, in addition to the initial drought 'head-cracker', he got fed up with industrial agricultural advisors pushing greater production and greater inputs and chemicals. 'We reached the limit of our comfort zone,' he told me.

But why is it so hard to shift from a Mechanical to an Emergent mind: from one that is Earth-destructive to one that is Earth-nurturing? The answer, while complex, is also simple: it involves overcoming paradigm entrenchment deep within our heads along with those external powers that have a vested interest in keeping us 'on track'.

It is my experience that a majority of farmers don't understand that they are captured by, or exist and operate within, a particular view of how the world works. This is because the dominant industrial-mechanical world view is subtly inculcated through a prolonged period of education. As a consequence, it is hard for them to comprehend that there can be other, different worldviews from theirs. This is what I mean by paradigm entrenchment. This entrenchment can be very deep because their entire personality and sense of self-worth can become indivisibly entwined in their cognitive-emotive paradigmatic state: in their thinking and feeling. As a result, their entrenched world view often runs so deep that disrupting it or even acknowledging it is perceived as a threat.

Moreover, massive forces are arraigned against farmers wishing to overthrow the dominant paradigm. What makes this situation so challenging for those wishing to change is that any shift leads to that interface where innovators and early adopters butt up against enormous power: given that this issue of power is intimately connected to knowledge and thus language. Crucially, it is usually submerged and unrecognised. That is, this knowledge–power nexus shapes a person's beliefs, habits, understandings and attitudes – in other words, their rationality and self-identity – all of which makes fundamental change so difficult.

In a detailed letter to me in February 2012, Allan Savory reflected on his 'half-century of struggle trying to help people and finding it hard and so often resisted'. It wasn't just changing farmers' minds that was the problem, he said, it was also 'getting extension workers and advisors to farmers to change': the latter being the most difficult group, in his opinion. In this letter, Savory identified two key vested interests. The first is obvious: monetary vested interest. But, he concluded, 'there is a second far greater vested interest at play and this is the elephant in the room no one talks about: people's egos and pride, both farmers and scientist advisors'. I believe he is right – that the 'professional pride' block is crucial for farmers – because their approach to farming and concepts (or lack of concepts) about how a landscape functions is, as seen earlier, indivisibly bound up with their world view and self-identity.

As holistic manager David Marsh told me, despite the clear difference in his farm's health, resilience and productivity to those of his traditional adjoining farmers, 'none of my neighbours have ever asked me what we're doing'. David's conclusion? 'I think humans have this infinite capacity to justify to themselves whatever they're doing is okay.' But there are also other subtleties involved. For leading innovative thinker and instigator of the Dawson Healing Program Peter Downey, those most open to change are those who 'get the body back into a balanced state so the left and right brain are working together again, where you get the magnification of both'.

So, as surprising as it may seem, and apparently without knowing it, we are all unwittingly captured in our minds by our surrounding society, family, peers and colleagues. It is this capture that makes any attempt at transformational change so hard, because, in effect, the advocate inevitably comes across as an insurgent revolutionary, taking on all of society.

The crucial point, however, is that *what* we farmers carry in our heads determines the health of a landscape, whether that land comprises a 125-acre (fifty-hectare) peri-suburban 'block', a one million-acre (405,000-hectare) pastoral lease in the Kimberley or an entire continent.

Not for nothing did leading regenerative agriculture proponent Vandana Shiva write so insightfully about the 'Monocultures of the Mind': where the dominant Western knowledge systems (based on a particular culture, class and gender that have been globalised) 'first inhabit the mind and are then transferred to the ground'. The result, concludes Shiva, is that '[m]onocultures of the mind generate models of production which destroy diversity and legitimise that destruction as progress, growth and improvement.'[4]

In the course of my travels and research, therefore, it became increasingly clear that the Mechanical mind and associated power structures have captured the majority of Western farmers' outlooks: a capture that is psychologically deep and powerful.

Yet, as evidenced by the examples in this book, transformative change is occurring: change to a new, Emergent mind. This is not the place to go into the mysterious depths of psychology, suffice to say that the starting point, as seen earlier, is often a major shock or a process of destabilisation that opens the carapace of the mind to change. Then, when this occurs, the energy derived from it (as in cracking an atom) fuels active searching for information about alternatives to current practices. This is supported by much work in the psychology of learning. That is, for virtually all regenerative farmers who have undergone transformation, the catalyst that sent them searching for new knowledge also led them, however serendipitously, to encounter one of the leading change organisations that have been founded by leading innovators and supporters. These organisations ultimately comprise 'communities of practice'. These are places, connections or organisations where learning occurs through social participation in the practice of a knowledge community.[5]

In my extensive research, I found it was these different communities of practice that played a double and vital role: first in providing new information counter to the unsettled person's previous world view, and then, second, socially reinforcing that transformative learning via participation. In a non-threatening environment, those undergoing change received help from others in recognising perspectives and trusting their own values and beliefs. Over time, they were able to amalgamate the other perspectives and cultures of the group into a holistic, integrated, transcendent approach to knowledge.[6]

Another important element of this process of change is the involvement or elicitation of emotion. This invariably is linked to the galvanisation of a strong ethical element in transformative farmers. In other words, emotion is the critical motivator to change. That is, the cognitive confusion and emotional distress at such revolutionary periods leads people to search for answers or to seek help. That is why transformative learning has been called 'emancipatory learning'. It is also why transformative farmers evince enormous enthusiasm, passion and energy after they have made their big change in thinking, feeling and practice. With this goes an extraordinary hunger to continue on an upward learning path. This too is usually accompanied by increased optimism and positivity – all of which is infectious. Frequent feedback from family, friends and peers after they have made the 'big change' or gone to a life-changing course is: 'What drug are you on?' The process seems to be a re-birthing – a 'born again' experience. Some academics call this 'symmetry breaking'. All of these things characterise an Emergent mind.[7]

My research also confirmed that, through transformative learning, the individual often develops a greater interdependent relationship with, and thus a compassion for, society (and this includes both nature and society). That is, a high percentage of transformative farmers – once they had made the change to the Emergent mind – saw their role as extending well beyond the individual farm level to their own community, region and wider society.[8]

However, as testified by most of the leading innovators in regenerative agriculture, it is initially a lonely and uncomfortable task to take on a dominant paradigm such as industrial agriculture. That is where a supporting network of like-minded people – a 'community of practice' – is so vital in assisting a world-view shift.

Part of the value of social learning is that, in emerging areas of regenerative agriculture, knowledge that is relevant to a specific area and context is not always readily available but has to be constructed by giving meaning to existing knowledge. Once this happens, learning and change generally occur via stories, narration and dialogue, which in turn form connections, provide insights, and impart exemplars.[9] These stories and conversations are also crucial in engaging the emotions and other subconscious cognitive functions so integral to the process.

Therefore, the critical aspect of this form of social learning is that individuals learn by observing others and by their social interaction in a group (which includes having role models). Implicit in this is an iterative feedback between the learner and the learner's environment, whereby the learner is changing the environment. Such changes in turn affect the learner.[10]

If one looks across the field of regenerative agriculture, clusters of communities of practice with attendant social learning are obvious. In holistic grazing regimes, organisations such as Holistic Management and Resource Consulting Services (RCS) and its spin-off Principle Focus now have sophisticated courses and processes to educate, inspire, nurture, support and mentor those making changes. They are therefore places where initial transformative change can occur. Other areas of regenerative agriculture also have similar but various forms of supporting communities of practice: such as the biodynamic group, different strands in biological agriculture, and most other fields. The Otway Agroforestry Network (mentioned in Chapter 13) is an outstanding example of a community of practice exhibiting social learning at its best.

Importantly, the best of these communities are not only vehicles for the evolution of practices and the inclusion of newcomers, but are also vehicles for the development and transformation of identities, whereby people can produce meanings of their own. In other words, communities of practice (or what can be termed 'knowledge communities') offer their members ways of organising and grasping the world. These communities therefore help define what is knowledge and truth. Its members are the bearers of knowledge. By implication then, such knowledge cannot be value-free but comprises communal, value-laden projects.[11]

Another aspect is that through inclusion in a community of practice (often via continuing iteration with a mentor), decades of learning and growth within that particular community can occur. What I have also found is that, as leading regenerative agricultural writer Jules Pretty noted, 'agricultural systems with high levels of social and human assets are more able to innovate in the face of uncertainty'.[12]

In the course of my research, I gathered dozens of testimonies that support the above pathway of transformative change via communities of practice. One person told me, 'All of a sudden there's a whole new group of people to meet, a whole new network, new things to learn. It opens your world up a bit wider.' Holistic grazer and native grassland manager in the New South Wales central-west Chad Taylor reflected, 'For me, RCS was the breakthrough of information and change. More recently, it's been peers who nearly all went through that process.' This was reiterated by Julian von Bibra in the long-settled, once-conservative Midlands of Tasmania. 'Much of my motivation is influenced by the people I mix with; I only choose positive people,' he said.

So, what is the core belief system that differentiates regenerative farmers of an Emergent mind from those in the dominant paradigm of industrial agriculture?

Besides a deep emotional connection to their landscape (characterised by distinctively biophilic and ecological thinking) and a general hunger for learning about other areas, many also exhibited great excitement and enthusiasm for their change in farming practices.

But one of the most distinctive features of those who experience a transformative shift is that such farmers seem to have embraced the old 'organic' components and amalgamated them with new knowledge: confirming a conversion to an Emergent mind. Another feature common to regenerative farmers is a strong belief in local community and healthy food for healthy people. This is seen in involvement in farmers' markets, organic marketing, and the development of their own food or fibre brands.

Due to our more recent national history, Australia's shift to a new regenerative agriculture has not had a long agrarian philosophical and literary tradition to piggyback off. Nevertheless, this gap until now has been filled by a deep and rich American tradition. This includes writers, thinkers and doers such as Aldo Leopold, Wendell Berry, Wes Jackson, Fred

Kirschenmann, poet Gary Snyder and landscape writer Barry Lopez, all standing on the shoulders of earlier pioneers such as Henry David Thoreau, Ralph Waldo Emerson and John Muir. Many don't realise this independent agrarian tradition was begun under Presidents Washington and Jefferson. Even back in 1859, Abraham Lincoln encouraged a group of citizens in Milwaukee to cultivate the art of agriculture because, 'Such community will be alike independent of crowned-kings, money-kings, and land-kings.'[13]

The new Emergent thinking is exemplified by one of Australia's leading and deepest thinkers in regenerative agriculture, David Marsh, who, with his wife, Mary, farms near Boorowa in New South Wales (and whom I referred to earlier in this chapter). David's thinking has evolved through his own experience and has also been influenced by a number of his many heroes, such as Aldo Leopold. It is no surprise, therefore, that for Marsh the way we farm and live has wider global philosophical implications. 'We are just fellow travellers with the other members of the biotic community,' he says, and thus 'natural capital is the only true form of wealth'.

To emphasise this, he quotes another favourite author, Paul Hawken, who states that 'the economy is a wholly owned subsidiary of the environment'. That is, says David, 'when a farmer makes a capital purchase, he is laying a debt on the landscape to deliver a return against that outlay. It is the interaction of water, sun, animals, plants and the biota that must deliver this return.' Furthermore, he says, 'it is possible to be profitable by treading more lightly in the landscape, but if we can harvest the interest and not delve into the capital in the landscape, and even reinvest in the future just by virtue of managing differently, surely that is exciting.' But, concludes David, 'if we abdicate the thought process to someone else, then we can find ourselves playing the game others want us to play and getting on an economic treadmill that is hard to step off.'[14]

In a few sentences, David Marsh thus encapsulates not just the power of enabling all five landscape functions to indivisibly work in synergy but also the operation of an Emergent mind: one open to ongoing disturbance, adjustment and equilibration. He calls his approach 'Simplifying to Complexity'. Putting this in a wider context, in a chapter David wrote in a book on biodiversity in 2008, he said, 'natural ecosystems are self-organising and self-repairing entities expending about thirty per cent of the energy they get from the sun just maintaining their structure and diversity'.

Describing this further to journalist Matt Cawood, David explained, 'The natural capital on our farm is growing, because managing holistically (people, business, landscape) has provided time for the living community to invest in its own future at almost no cost.'[15]

Just here, David captures the essence of regenerative agriculture, while also throwing down an incontrovertible radical challenge to the self-annihilating agenda of the dominant, growth-based, consumptive paradigm of modern industrialised global society.

In the end, this radical difference in thinking, feeling and acting towards nature and the landscape exhibited by leading regenerative farmers reveals a full transformation of the mind and heart. It is this massive reorganisation of a world view and personal mental constructs that leads to an almost born-again view of the world and to totally new and fresh perceptions. The Romantic writer Ralph Waldo Emerson captured elements of this when he stated, 'To the illuminated mind, the whole world burns and sparkles with light.'[16]

A burning question then emerges as we enter deeper into the Anthropocene: 'How can we engender more rapid change in those of us managing landscapes so as to prevent ongoing corrosion of function?'

For it is abundantly clear, as environmental advocate Gus Speth said, that '[t]oday's problems cannot be solved with today's mind'; or, as one Kimberley pastoralist said to me the other day, 'the thinking that got us in the shit won't get us out of it'. In other words, we need to change our mindscapes before we can change our landscapes.[17]

However, I don't believe we can deliberately trigger mind-changing shocks. Instead, as we slip deeper into the Anthropocene, we can only prepare for the likely increased occurrence of such shocks. And so we need to have solutions ground-proofed and ready when people do begin searching.

Fortunately, in Australia and the USA today we do have some excellent examples of leaders triggering change and a move to regenerative agriculture among farmers, and with the portfolios and experience to do so. From my wide research and travels across Australia, two individuals stand out in this field: Rowan Reid in the specific area of agroforestry (whom I will return to briefly), and Terry McCosker, whose work – as we shall soon see – has had, is having and will have a huge impact in transforming farming families both personally and in their regenerative practices.

As discussed earlier, in Chapter 13, Rowan Reid has emerged as one of the leading innovators of a new path in agroforestry. Moreover, he has dedicated much of his life to addressing the key area of increasing the knowledge, resources and confidence for farmers engaging in agroforestry. Along the way, he has helped redefine the very concept of farm forestry.

With helpful collaborators, Rowan has pioneered an enlightened and radical form of farmer-led social learning and extension in Australia, which has led to personal transformation in farmers. The result was and is a revolution in agroforestry in the Otway Ranges near Melbourne, which then spread nationally and later internationally. The flagships of this revolution became the Otway Agroforestry Network, and then, in turn, the Australian Master TreeGrower Program. Each was crucial to regenerative agriculture for two reasons. First, this new form of agroforestry, in engendering concepts around multipurpose forestry as opposed to industrial monoculture forestry, is based not just on restoring landscape function but also on building economic, ecological and social resilience into rural communities and landscapes. And second, led by Rowan, a sophisticated form of social learning that brings about transformative change in regenerative agriculture was evolved and refined.

What aided Rowan in this evolution was perceiving how the top-down forestry extension arrogance towards farmers wasn't working, as seen in low adoption rates. 'Our approach,' Rowan told me when I first met him in a poky little cafe down a side alley off Carlton's famous Lygon Street in Melbourne, 'was to plant a diverse range of tree and shrub species in places where they were needed to protect the farm, their stock and themselves. We were unsure where it would take us.' But by taking such a leap of faith, Rowan ended up confirming, and then laying out, what a new agroforestry might look like. 'I was interested in how forestry can actually be attractive to the farming community and not be seen as a threat,' he reflected, adding, 'It was clear that every farm would develop in a different way, reflecting the different aspirations, interests, concerns and resources of the farmers involved rather than demonstrating "world's best practice". I wanted to challenge farmers to engage with science using trees; to express their own farming aspirations on the landscape just as they would with their fences, pastures and stock.'

In 1989, Rowan was asked to design and run Australia's first agroforestry course at Melbourne University. It was a unique chance to realise his vision, and the popular course began in 1991. While Rowan was developing this course, he also continued evolving his own Bambra farm in the Otway Ranges west of Melbourne, eventually turning it into a stunning demonstration and education site. At the same time, he was interacting with innovative local farmers who would become co-innovators and collaborators (such as near-neighbour and outstanding leader Andrew Stewart). In the process, all their lives changed profoundly.

In 1993, along with Andrew Stewart and a number of other early like-minded adopters, Rowan helped establish one of Australia's best peer-managed rural extension and support networks: the Otway Agroforestry Network (its formation assisted by the National Farm Forestry Program). This network was built around farmers and their farms becoming the sites of learning – aided by subsequent and varied publications.

The result was that, in the Bambra–Deans Marsh area, said Rowan, 'agroforestry is so popular that many in the community are keen to make it a national hub. Out of all that experience, I developed the Australian Master TreeGrower Program as a university outreach program for farmers interested in multipurpose tree-growing.' The secret of this huge shift in a community, said Rowan, is that '[r]ather than trying to change the farmers, we have been working to change forestry to suit farmers: fitting forestry into a farming landscape.'

Today, the Otway Agroforestry Network has over 200 paid-up members and more than forty paid mentors, who have taken ownership of and run the network. This organisation is undoubtedly the most innovative, constantly evolving and forward-looking farmer-driven agroforestry network in the nation: and one that has developed out of the community rather than being imposed on it. In the process, the network has taken on greater significance with time, and this is now transgenerational. This has arisen, said Rowan, because it has 'become the owner of a lot of information and past knowledge', which can be passed on to new owners when land is sold along with its trees. Rowan said this is because 'the trees are actually there on the landscape longer than people are in many cases ... it's building up cultural history in the landscape ... people have got reports and so on. So it's a unique model.'

It is probable that Rowan Reid has had a greater impact than almost any other individual on envisioning, catalysing and laying down a practical path for regenerative agroforestry in Australia: through his teaching at Melbourne University, his catalysing of the Otway Agroforestry Network, and the combined integration of the Australian Master TreeGrower Program's courses (initially sponsored by the Myer Foundation). In the latter, for example, more than 110 courses have been conducted to date around Australia, while the program has been expanded into Africa, Indonesia and East Timor.

The reason Rowan is a fine leader in this important field of changing lives and landscapes for the better is that he has empowered a farmer-driven, bottom-up revolution via social learning that puts farmers first and in charge of their own destiny. Instead of employing the traditional top-down view of extension, Rowan asked the opposite questions: 'How can innovation be changed to fit in with the adopters?' and 'What would farmers do if it was driven by their own interests alone?' The answers revolved around shared learning, inspiration and ongoing innovation.

But then the next phase occurs, said Rowan: 'You get this cyclic thing, where feedback occurs. And the model works, because you've got all these farmers doing things, and the net impact of lots of people doing something will be of benefit to the wider community. It's that diversity thing again.' Rowan summed up with the observation, 'It's the ownership of that decision which is really important – that's what makes it such a good learning experience! It's because they had become comfortable in this new learning environment. That enabled them to do what they wanted to do.'

At the end of our extended yarn, Rowan gazed unfocused out the window of that poky cafe in Carlton, grappling with multiple thoughts as he tried to encapsulate an incredible life's journey.

He concluded by alluding to one of his favourite authors: 'It's more like I believe the Aldo Leopold approach about intelligent tinkering and our creative actions in a landscape. As Leopold says, "a thing is right when it tends to preserve the integrity, stability and beauty of the biotic community. It is wrong when it tends otherwise." Agroforestry by farmers is a good story. If designed well, it might be the "right way" we are looking for. So farm forestry is a good story – but a long one. We're still writing the early chapters.'

I now wish to conclude by telling the story of another exceptional change agent in regenerative agriculture. Some 2000 kilometres north of Rowan Reid resides Terry McCosker, at Yeppoon in north-central Queensland. With his wife, Pam, in 1985, he began a move into agricultural extension that eventually became Australia's leading broad-scale grazing and farming education company: Resource Consulting Services or RCS (Australia).

Because of years of a tough apprenticeship, and though taking a different pathway to Reid, Terry McCosker evolved a farmer-change organisation that has strong parallels to Reid's, in its focus on social learning, self-empowerment and self-transformation.

Terry was born in 1950 and grew up on a small dairy and mixed farm (peanuts, maize, lucerne, pigs and cream for a butter factory) near a small town, three hours north of Brisbane. 'We essentially grew up in poverty,' he told me, 'no flushing toilets, electricity or running water.' But he gained a love of the land from his mother and, because his father had survived as a soldier in the Second World War, 'I got a fair bit of humanity from him.' In addition, his parents instilled a high set of values in the boy, including an ethic to serve one's community: again similar to Rowan Reid. His father, however, counselled him against farming, telling the fourteen-year-old boy, 'the farm's not big enough. The next best thing, if you're really wrapped up in farming, is to go and become an advisor.' In a roundabout way, that is what happened, though Terry would radically break the mould of a traditional advisor.

Terry left school in Year Eleven and got a job in the Queensland Department of Primary Industries (DPI) as a cadet in 1967. Because of his knowledge of farming, he related well to dairy farmers and, by the age of nineteen, became 'a fully blown extension officer', while also doing a Diploma of Rural Technology.

In 1971, Terry was transferred to a research station at Innisfail in the wet tropics, where, while he finished his diploma, he became an experimental officer. He remained there for seven years, building up knowledge and contacts in the field of tropical agriculture. However, he found the work unchallenging and threw himself into wider community activities. Terry told me, 'money has no value to me. Helping people is probably the deepest thing, and it's been there from the beginning.' So his time was not wasted up north, and in the process in 1973 he made the best decision of his life in marrying his wife, Pam, whom he had met earlier at college in Brisbane.

In 1978, Terry was recommended by a colleague to apply for a job with the giant American multinational company W. R. Grace. They had bought what Terry described as 'rubbish country' in the Top End of the Northern Territory, where forty-four per cent of the 800,000-acre (324,000-hectare) station went under water in the wet. This station, Mount Bundey, was located a third of the way from Darwin to Katherine and ran 12,000 cattle. The company invested heavily in a southern Australian industrial-agriculture approach by attempting to sow annual legumes and apply superphosphate: methods completely inappropriate to the brittle savannah climate of the Top End. So they were looking for someone who understood pastures.

Terry told me that he took the job 'because they said something I've never forgotten: "We want somebody who has no preconceived ideas."'

After six months on a motorbike inspecting 150,000 acres (60,700 hectares), Terry found an utter mess. He delivered a report to senior executives who had flown in from the USA, and told them their strategies were failing and they should sell up. The Americans rejected this and instead backed Terry to enact a one-million-dollar research program over four years to solve their problems. 'Do what it takes,' they told him.

To cut a long story short, Terry threw himself into an intensive and unique research program – with the support of his boss, Texan Cliff Emerson. The strategy Terry put in place would revolutionise the northern cattle industry, which, up to that time, according to Terry, comprised 'a bunch of cowboys in a cattle hunting industry where you went out and caught what you could. No weaning, just harvesting once a year and sell what was saleable; just cowboy stuff, no fencing, no infrastructure, and the business survived because of low overheads.' What he found on Mount Bundey, however, was a 'high cost' but still same low productivity business.

Over the next four years, Terry called on all the outside expertise he could find (encompassing nutritionists, pasture ecologists, statisticians and others) and conducted extraordinarily detailed research and comparative trials, including with over 6000 plots of introduced pastures. He read enormous amounts of literature and changed the entire management strategy. In the end, he upended every conceivable paradigm, and in the process did the equivalent of three PhDs worth of work.

The results were dramatic. He took breeder mortality from fourteen to two per cent and weaning from forty-four to seventy-two per cent, while weaner growth rates nearly doubled. From Terry's point of view, he had maximised a unique opportunity, and this experience – in deductive research with a whole-of-systems approach – 'just set me up for what I do today'.

However, just as Terry's efforts were ready to bear fruit, the Americans sold out. By then, his ex-boss Cliff Emerson was with the Northern Territory Cattlemen's Association, and they realised the huge significance of Terry's work. He was funded for a year to write up his findings. Not only had Terry conducted the best practical research ever done in the Territory to that time, but he also now published eighteen papers in twelve months across eight different disciplines.

It was then, at a difficult moment in Terry's career in 1985, that private Darwin agricultural consultant David Hanlon took him on as a partner in his company RCS. This productive partnership would remain until 2005.

In the next few years, Terry gained experience in Nigeria working on desert reclamation, but then began a sequence of steps that would change his life and the path of Australian agriculture. It started when Cliff Emerson again stepped into Terry's life in 1985 on returning to Australia. Emerson was full of praise for the benefits of a new grazing management system he had seen and researched: then known as 'cell grazing'. Terry read Emerson's material and told him, 'Cliff, this is the greatest pile of bullshit I've ever seen in my life. Just forget it.' But Emerson returned yet again in 1988, with more material on Stan Parsons' and Allan Savory's work. Again, he got the same response from Terry. But the next year, Stan Parsons and his wife, Hazel, came to Australia on holidays, and Emerson persuaded them to go to Darwin so Parsons could address the Northern Territory Cattlemen's Association. He then put Terry and Parsons together in a room. 'Everything made sense, except this grazing stuff,' said Terry.

However, what eventuated was that the Secretary of the Department of Agriculture Peter Plumber offered to underwrite the budget to get Parsons back. Thus began the entry of ideas on holistic grazing regimes into Australia: arguably the single greatest broad-acre regenerative agriculture movement to turn around land degradation and profitability in farming landscapes in this nation.

Upon splitting with Allan Savory in 1983, Stan Parsons had developed the Ranching for Profit School, which he taught in the USA and South Africa. The key to introducing holistic grazing ideas into Australia was that, from October 1989, Terry was co-opted to organise four one-day workshops in Rockhampton, Charters Towers, Alice Springs and Katherine. In January 1990, Terry, who was still sceptical about the cell-grazing elements of Parsons' teaching, got Parsons back for more workshops. He and Stan then put on the first seven-day 'Ranching for Profit' course in Rockhampton in March 1990. After this, the course was renamed 'Grazing for Profit', and then a second course was run in July 1990, when thirty-three more farmers turned up. It was then that Terry said to himself, 'There's something going on here.'

There was also something else going on. Terry ran into a government official one day who said, 'I hear you're bringing Parsons and Savory to Australia?' 'Just Parsons,' said Terry. 'Well,' said the official, 'we've got a pile of literature four inches deep on where they've destroyed country everywhere they've been in the world.' And, said Terry, 'he told me, "We're going to run them out of Australia, and if you get mixed up with them, you're going to go down with them."' The problem, said a shaken Terry McCosker, was, 'I was broke. I had no way to fight back.'

This was classic but vicious paradigm defence. So Terry got hold of some DPI literature where they accused him and Parsons of being snake-oil salesmen. He gave them a solicitor's letter saying that if it happened again, they would go to court. The printed attacks stopped but negative telephone advice didn't. 'They very nearly succeeded in putting us out of business,' said Terry, 'because I had this massive opposition, 5000 of them and one of me. And it wasn't just them, it was CSIRO, it was everything.' Later on, as Terry's RCS expanded into other states, the established scientific and extension power bases in each state reacted the same way.

To sort out the issue of both conflicting evidence and scientific and extension opposition, Terry applied for and was awarded a Churchill Fellowship in 1991. 'I decided the only way to work this out was to go around the world and go everywhere Parsons and Savory had ever been,' said Terry, 'to visit all the properties.' This he did in the USA, South Africa, Namibia, Zimbabwe and Botswana. Along the way, he did a holistic management course at Savory's HQ in New Mexico and attended a Savory workshop in Namibia. It was thorough homework, talking to everyone involved with Savory. One of these, for example, was Bob Vaughn-Evans, an extension officer in the Kwekwe region of Zimbabwe who ground-proofed and extended a lot of Savory's evolving thinking.

Says Terry: 'Stan Parsons became recognised as one of the best teachers in the world, and I had the privilege of being an understudy to him for five years.' Each of them then went on to develop the Executive Link and the Graduate Link components in response to the demand for support that could not be met by individual consulting.

After being exposed to Holistic Resource Management in 1991, Terry made the conscious decision to stick with the Parsons approach. 'There were several reasons for this', said Terry. 'Firstly, Stan's approach was to save the planet, one farmer at a time – i.e. bottom up. The HRM approach was more top-down. Secondly, Parsons' approach was more holistic with an even balance between economic, land, people and production systems. Thirdly, the production and economic approach was principle-based. And finally, we jointly worked on developing an effective support system'.

Nevertheless, says Terry, 'the world owes a lot to that combination of intellects of Savory and Parsons, who worked together for thirteen years'.

One further lesson Terry brought back from his exhaustive Churchill world trip was another core philosophy. 'I decided from the moment I arrived back,' he told me in 2015, 'that I would only talk to people who wanted to listen, and I would only work with farmers who wanted to work to improve things. So that's what I did. And that strategy I came back with in 1991 after the Churchill Fellowship,' he concluded, 'gave me the stories, slides and ideas to be able to launch this thing in Australia.' Therefore, soon after returning he undertook a series of one-day workshops through Queensland, Tasmania and Victoria. He also did a large run in New South Wales, organised by then leading Hassall Associate's officer Greg Brennan from their Dubbo office, who saw the potential of the Savory–Parsons system.

So, while Terry returned to Rockhampton and a new home in Yeppoon, in many ways 1991 marks the real start of RCS's revolutionary teaching in what Terry calls 'cell grazing' – as it was originally named by Parsons and Savory on the Liebigs' ranch in Zimbabwe. But RCS quickly evolved into a holistic teaching and support enterprise for farmers undergoing world-view change in business management and regenerative agriculture. This was a totally new form of a holistic extension service for broad-acre Australian farmers. In my opinion, today it remains a world leader in the field.

A major reason for this is the extraordinarily strong partnership between Terry and his wife, Pam, and the latter's level-headed loyalty.

Three key experiences served to reinforce the high-value system that marks the McCoskers and RCS. The first was a harrowing event in both Terry's and Pam's lives that prepared them for the tough road ahead. This occurred in their first year at the Mount Bundey station. On the day their second child, Kelly, was due to be born, with Pam largely incapacitated and Terry out on the station, their first child – two-year-old Shaun – drowned in a nearby river. Understandably, Terry told me with considerable emotion, 'When he died, I vowed that his life would not be in vain.'

The second tough experience was when Terry and Pam temporarily ended up nearly broke. 'We arrived in Yeppoon with the clothes we stood up in, a car worth a thousand dollars, and a $70,000 debt. Had to start again at the age of forty,' said Terry. That was one way to get your values realigned, he confirmed, of getting money in perspective.

The third experience was when RCS was nearly destroyed by an aggressive internal takeover attempt by once-trusted employees, which in turn led to a spin-off organisation that now competes in the same field.

It was these three experiences in particular that led to the McCoskers' business being built on values and not a relentless quest for money; to placing the highest priority on family and people and not things and commerce; but also to Terry putting too much pressure on himself – and sometimes on others.

RCS initially licensed Stan Parsons for a few years to establish their teaching system. This quickly grew into their renowned seven-day 'Grazing for Profit' courses. The company burgeoned, building the Grazing for Profit course into a life-changing experience that covered ecology, holistic grazing management, animal nutrition, business management and life decisions. Continuous improvement was their motto.

The ongoing evolvement or emergence of a new mind among clients meant developing new systems to enhance the adoption rate following their courses. What emerged were small groups of 'Farmer Boards' made up of course attendees from different districts who met at each other's farms through the year, with facilitation and further education by the company's officers. These became the famous Graduate Link and then Executive Link (and later Growth Link) boards. They were built around continuing education and the philosophy that the journey was about personal growth and continuous learning. Combined with repeat courses and evolving new ones, the package led to a vast increase in farmers' adoption rates.

What these boards entailed was the same as what Rowan Reid had instigated with the Otway Agroforestry Network: a sophisticated and powerful form of social learning or 'communities of practice' to support and mentor those undergoing change, and to provide a safe community in which to change and continue growing. In the process, RCS sowed ideas of change by subtly teaching different forms of ecological literacy. They also thrived on challenging entrenched paradigms.

Part of the reason for the success of the company was due to a major lesson Terry received when attending a course on adult education in Brisbane. Here, he was told about the four stages of adult learning: unconscious incompetence; conscious incompetence; conscious competence; and unconscious competence. 'For me then,' he said, 'a light bulb went on.' 'That's it!' he exclaimed, 'and so that became my business model.' This was because, as he told me, 'I knew from my previous experience as a Government Advisor and then getting on the management side of the fence and listening to my own inadequate advice, that the traditional extension model was broken. But it took me years to know how to build an alternate approach.'

The RCS Rural Profit System is structured in different stages based on this model. Results in triggering change have been remarkable. Incompetent farmers, managers and their spouses have become new families, building healthy farms and solid off-farm investments. Some RCS farm boards have continued for over two decades; marriages and farming enterprises have been saved. From nil decision-making around succession planning, farming families have moved to previously unimagined solutions.

Not every case is a success. There are dropouts because paradigm change or family obstacles proves too difficult. In some cases, impatience kills change during sometimes depressed production and income levels as a farm's landscape ecology shifts from a simplified monocultural industrial state to an agroecological system. And in other cases normal human flaws, droughts, commodity crashes and other factors prevent people making the regenerative leap, and so 'creep-back' occurs. As Terry summed it, the largest contributor to creep-back is peer pressure. Peers and neighbours do not want others to step outside the norm, and it takes character to overcome that pressure. But transformative change is more the rule than the exception.

As we talked that day in 2015, we sat on the spacious deck of Terry and Pam's house just outside Yeppoon, looking out from the steep hillside over his small farm of tropical grasses and plantations of rainforest timbers. Terry gazed across to the ocean with tears filming his eyes. 'Yeah,' he reflected, 'to see couples that were tearing each other apart three years ago when they walked onto the farmer-board, to be sitting there holding hands now; to be trying with each other; to be just saying how happy they are now: that is my greatest reward.'

When I look back on the rise of regenerative agriculture in Australia from the 1980s onwards, I see at the forefront Terry and Pam McCosker and their RCS organisation. It was Terry who initially introduced holistic grazing regimes to Australia through Stan Parsons in 1989. Only later did Allan Savory bring his holistic management courses to Australia and set up his holistic management chapter. In the meantime, Terry kept introducing innovative, sometimes radical change agents. In 1999, he brought out Bud and Eunice Williams from the USA to run what evolved, through Queensland stockman and manager Jim Lindsay, into low-stress stock-handling schools in Australia. This in turn, via Jim Lindsay, Rod Knight and Graham Reece, evolved into the KLR marketing schools and methodologies. In 2000, Terry invited Dr Elaine Ingham to Australia, who triggered substantial change in biological agriculture.

Terry also brought out leading 'healthy soil–healthy human' advocate Dr Arden Andersen, and in 2002 he invited subtle-energy pioneer Dr Phil Wheeler, who introduced the skills of radionics and dowsing more widely to Australian agriculture. Then, in 2005, Terry, through his interest in kinesiology and via an Executive Link board, enabled Peter and Anne Downy to present and subsequently evolve the Dawson human health program. And in 2010, at his company's twenty-fifth-year anniversary conference in Brisbane, Terry introduced subtle-energy practitioner Dr Patrick McManaway to Australia.

Over the years, more than 7000 farmers have now been trained by RCS in Australia and South Africa. Among many accolades at the anniversary conference in July 2010, American counterpart David Pratt wrote:

> Terry and his team have helped people see that economics, ecology, finance, production and sociology are not separate

topics but part of one whole ... How many millions of tons of soils were saved? How many millions of gallons of herbicide were not sprayed? How much more effectively are minerals cycling, water cycling and energy flowing? Whether measured in time, energy or money, how much more life do their clients have? They don't make calculators that can count that high.[18]

Today, Terry has moved into the next big thing. He wants to set up a carbon trading scheme for regenerative farmers before he retires, and he has formed a company, Carbon Link Pty Ltd, to do so.

Thousands of RCS farmer clients are now running with the ball, constantly evolving, learning, growing and innovating in what appear to be self-organising, Emergent systems of ongoing growth and increasing resilience. I personally can vouch for their seminal influence. Terry's work (and that of many leading regenerative farming innovators such as Rowan Reid and the Savory Holistic Management group) reveals that transformative change to a better agriculture is possible. The urgency now, in the face of a mounting Anthropocene challenge, is to accelerate the pace of that change: not just in agriculture but more widely, throughout society. I believe the examples in this book provide the inspirational stories to do so. This is because what has been instigated is a powerful pathway to transition from a post-Mechanical mind to that of an Emergent mind.

Thereby, a regenerative agriculture future is based on a different relationship with the land. In turn, this applies to everyone in society and their wider relationships. Such a two-way partnership flows on to all human relationships: with each other, with the landscape and with the planet.

CHAPTER 20

Healing Earth

Inspiration is not garnered from the recitation of what is flawed; it resides, rather, in humanity's willingness to restore, redress, reform, rebuild, recover, reimagine, and reconsider ... Healing the wounds of the earth and its people does not require saintliness or a political party, only gumption and persistence. It is not a liberal or conservative activity; it is a sacred act.

PAUL HAWKEN, *Blessed Unrest*[1]

On a windy morning in late June a few years back, I left our homestead soon after dawn to go fencing. I was headed for a paddock we call the 'Chandler'. It was our last remaining large paddock (some 200 acres or eighty hectares) and too hard to manage for regenerative holistic grazing. By dividing it into ten (and according to what the landscape, soils and vegetation dictated), I could then inject proper energy and rest into our grazing system. I would also have ten better lambing paddocks rather than one poor one.

The Chandler paddock is named after some long-forgotten early settler. Like most parts of our farm, it carries layered memories: of rabbiting as a child; a heavy fall from a bolting horse; difficult musters and 'breakaway' lambs at lamb-marking time; and a hot January day in the early 1990s when I was fire captain and a dangerous bushfire was lit by a fire-bug.

After packing my fencing gear this particular morning, I drove north along a two-wheel track that follows the spine of a ridge. I went slowly to take in the morning light. A golden-pink efflorescence shone on the leathery, upturned leaves of snow gums. Beside me, an ethereally beautiful light ran along fence lines, catching tall-grass seed heads bending to the dawn breeze and highlighting the shiny green serrated edges of speargrasses that glinted as they too bobbed and swayed in the north-west wind.

Further along, I noticed three fat, cryptically coloured stubble-quail run beside poa tussocks, then disappear behind the chiselled faces of basalt rocks whose angular planes also glinted in the early light. The sun is life, and I knew in the next hour this life (even in winter) would awaken.

As I fenced through the day this late June, I frequently paused just to soak in the landscape and imbibe the changing rhythms of life around me. Flame robins flitted along fence lines, atop boulders and beneath snow gums, tinkling their evocative calls. Crimson rosellas would swoop by, their breasts as shockingly vivid as arterial blood. And already, around me in the tall candlebark gums, wood ducks were paired and occupying nest sites in nearly every tree: the male and female bonding to the nest site and each other with a constant chattering refrain (a soft 'kuk-kuk-kuk-kuk') and a restless to-ing and fro-ing that went on for much of the day – until they would repair to an adjacent dam for a paddle and preen.

This fencing I was doing was doubly significant. It was not just that it would enable me to undertake regenerative grazing but also that it was the first time since white settlement that there would be an intensive focus on land regeneration in this paddock.

This was vividly brought home to me when landscape ecologist Richard Thackway recently spent three days on our property, much of it focused on the Chandler paddock. Richard looks a bit like an archetypical Santa Claus: white beard and whiskers, round, jolly face and a twinkle in his penetrating blue eyes. But behind the twinkle is a forensically sharp brain, and he has serious credibility as an ecologist. He was a key architect of the delineation of Australia's eighty-nine bioregions when working for the Federal Department of the Environment. Subsequently, he worked all over Australia, refining an applied new ecological analysis system he has termed 'VAST' (Vegetation, Assets, States and Transitions).

Using his knowledge of how landscapes function, Richard has devised a standardised diagnostic system that can be applied to any landscape in Australia: whether a severely modified grazing terrain like our Chandler paddock, a disused mine site or a remnant pocket of rainforest in northern Australia. Richard thus spends a lot of time trying to understand the previous history of the site and, in the case of our Chandler paddock, what it was like in 1820 under Aboriginal care and management before white settlers arrived. Using nearby natural, healthy benchmark sites as the comparative standard of a healthy historical landscape, he then assesses the structure of the existing vegetation, the capacity of the site to regenerate (i.e. its resilience) and the composition of the vegetation (its species diversity). From this, Richard then uses various measured characteristics within ten main landscape function indicators to assess the health of a particular site or landscape.

Richard's vegetation-focused analysis effectively comprises a landscape ecologist's toolkit equivalent to that of a medical doctor's stethoscope and blood-pressure kit. It therefore came as a shock when he delivered his preliminary diagnosis on our Chandler paddock: severely modified (i.e. suffering a major illness) and permanently compromised in some areas (i.e. loss of some vegetation structure and biodiversity, with erosion effects that would be very hard to restore to full health), but still with some regenerative capacity. That is, with the right management and nurturing, the Chandler paddock could be returned to a higher level of health and productivity, but never fully restored to its pre-1830s state (even if that were possible). Were the Chandler a human patient, 'Doctor' Thackway was advising me that to raise its life expectancy it had to get off the booze, smokes and junk food and undertake a rigorous exercise regime.

Being forced to examine and research the land-use history post–white settlement in our region was sobering, as it made me acutely aware of what had been lost. Most alarming was the fact that ecological decline had begun almost from the arrival of the first European shepherds. Only now are we getting scientific information to confirm what virtually all us landholders on the Monaro deny or are ignorant of: that European settlement was devastating to landscape function from the start. For example, in 2005, research into 7000-year-old lake sediments on the Snowy River flood plain where it runs out to the sea in Victoria revealed a doubling in sedimentation levels (i.e. erosion impact) from the immediate arrival of Europeans. These sediments then increased thirty-eight-fold by the late 1950s.

This latter level is linked to ongoing land clearing and especially overgrazing due to both rabbits and a loss of mountain grazing leases.

Concerning the latter, many Monaro families (including our own) had summer grazing leases in the Snowy Mountains, which were cancelled upon the creation of the Kosciuszko National Park. Loss of these cheap leases meant increased grazing pressure on the home farms. And there was also increased tillage at this time, in combination with the damming of the once beautiful Snowy River. Biotic water quality declined accordingly. These changes, the scientists concluded, were as pronounced as any experienced through the whole of the preceding millennia of the Holocene. It was sobering for me to reflect that all our farm – including the Chandler paddock – drains into the Snowy River, which is but five kilometres away.[2]

But there was an upside, because all was not lost for the Chandler paddock. There was still resilience and regenerative capacity left, and in all cases from around the year 2000, we had begun lifting virtually all of Richard's key indicators. The reason for hope and positive action was that I knew Mother Nature (up to a point) was both forgiving and resilient, and that love and care could largely regenerate the Chandler and let that landscape self-organise again.

In an essay in 1983, the contemporary agrarian philosopher Wendell Berry made the wise observation that 'our crying need is for an agriculture in which the typical farm would be farmed by the third generation of the family'. This, he said, would therefore involve 'the lengthening of memory'. By this, he meant a longevity involving an intergenerational view of land care, love and regeneration where '[p]revious mistakes, failures, and successes would be remembered': that is, where '[t]he land would not have to pay the cost of a trial-and-error education for every new owner', and where the present state of health of the farm 'could be measured against its own past – something exceedingly difficult outside of living memory'.[3]

In applying his VAST system to our land and Chandler paddock, Richard Thackway was giving my family the tools to go well beyond the third generation in memory, and to learn from our and others' mistakes and successes in order to continue regenerating our land. The full disclosure of what had really gone on and what was lost had to be the very first step in changing what still exists as strong collective denial. On reading the Snowy River sediment story, I was forced to reflect that there seemed much truth in Wes Jackson's observation, 'Conquerors are seldom interested in a thorough-going discovery of where they really are.'[4]

My fencing of the Chandler that beautiful winter morning a few years back was the continuation of an extensive plan for both regenerative grazing and biodiversity plantings in wide corridors and large mosaic blocks and patches. But it was long, solo days among the candlebarks and the 'kuk-kuking' wood ducks that led me to contemplate the significance of Richard's analysis of the Chandler paddock. This was because his report was a microcosm of what had happened to the entire Australian continent from 1788.

However, given our understanding of the processes of self-organisation and emergence, I know also that with concentrated management I could to a large extent turn around and revitalise the paddock's health. For it had become clear to me that if we could not only understand what had happened to Australia after 1788 but especially why, then we would have the understanding to begin addressing and turning around (at least in part) the ecological destruction of this continent. Nevertheless, as I soberly reflected, even more challenging was how to ameliorate the horrific degradation of our continent's vast, complex and ancient Indigenous heritage.

I remember as a young undergraduate in 1974 purchasing a slim paperback that had become a bestseller but which, unsurprisingly, was ignored (or attacked) by the world's leading decision-makers. *The Limits to Growth* was published in 1972 by a group of young systems-thinkers using computer models. It stands as a metaphor for present times because it raised the issue of what the late Australian scientist Tony McMichael called 'planetary overload'. Today, our 'global footprint' reveals that humanity is already consuming over one-third greater the resources than Earth can replenish. Within a few decades, we will be consuming and trashing over two planets' worth. Those in denial (who constitute the leaders of the 'thinking' and driving organisations of modern capitalist society plus the majority of our politicians) choose to ignore these fundamental facts. They choose instead to cling to a denialist 'cornucopian enchantment': a belief by scientific, business and political leaders that technology and human ingenuity will get us out of the mess. Well, reality has now caught up – notwithstanding unconscionable denial by such leaders. That is the significance of the powerful new metaphor of humanity tipping planet Earth into the Anthropocene era.[5]

We have previously seen that the Anthropocene epoch is the fruit of the shift from the Organic to the Mechanical mind. But while the causal factors are due to this *mechanistic* world view, its progeny constitute the following beliefs, which are serving to take us deeper into the Anthropocene. These comprise: *Materialism* (human needs being met via consumption of goods and services); *Anthropocentrism* (that nature belongs to humans, not humans to nature); *Contempocentrism* (a focus on the present and a discounting of the future); and *Technocentrism* (a belief that more-complex, energy-intensive technologies, aligned with humankind's inventiveness, will solve all problems and will deliver all human needs). But these are the beliefs of a fool's paradise.

What is so confounding and frustrating as we and our planet accelerate headlong towards disaster is the complacent, almost schizophrenic, delusional nature of our ruling belief system and shibboleths: that, as John Ralston Saul points out, 'The texture of our reigning mythologies is so thick that no one can see what is actually happening behind this intellectual and emotional camouflage.' Even more frightening, as Saul affirms, '[d]eprived of direction, we are determined to go there fast.'[6]

Chillingly, evidence is mounting that we are 'going there' far too fast. Recently, a group of leading scientists led by Will Steffen of the ANU stated, 'We are the first generation with widespread knowledge of how our activities influence the Earth System, and thus the first generation with the power and the responsibility to change our relationship with the planet.' The onus of this first-time 'widespread knowledge' is awesome, because, as Steffen and company concluded, 'effective planetary stewardship must be achieved quickly, as the momentum of the Anthropocene threatens to tip the complex Earth System out of the cyclic glacial-interglacial pattern during which *Homo sapiens* has evolved and developed'.[7] That is, if we reach one or more tipping points in any of Earth's key systems and trigger runaway positive-feedback events, then who knows what might happen? What is truly alarming is that it is almost certain the result would be disastrous for humanity and most other life forms. Surely it isn't a risk worth taking, otherwise, as Martin Luther King so presciently stated, it might be '[o]ver the bleached bones and jumbled residue' of our own civilisation that 'are written the pathetic words: "Too late"'.[8]

Crucially, when I examined the extensive literature on the Anthropocene, most sobering for me was the finding that a prime causal factor in the crossing or threatened crossings of the safe operating limits of Earth's systems is industrial agriculture. But instead of causing despair, I believe that in this information resides hope and potential salvation. This is because I know that it is a regenerative agriculture that can also provide many of the solutions to this global crisis: that agriculture can play a huge role in pulling us back from the brink.

Intuitively and presciently back in 1911, pioneering conservationist and naturalist John Muir had perceived what modern science is now telling us (as referred to in previous chapters): that 'when we try to pick out anything by itself, we find it is hitched to everything else in the universe'. Indeed, the more we dig, the more we find just how 'hitched' we are – and in vastly more extraordinary ways than anyone could have dreamt. So, it is this hitching – if we empower a broad-scale shift to regenerative agriculture – that I believe gives great cause for optimism in the vital mission of healing Earth. Let me explain why.

In giving examples of regenerative agriculture in previous chapters, I attempted to demonstrate that what returned health (and profitability) to ecosystems was restoring health to the key landscape functions. Moreover, as we also saw, regenerative agriculture is the quickest and most efficient way of pulling carbon dioxide out of the atmosphere and storing carbon in the soil. This can only be sustainably done by the healthy regeneration and interdependent synergy of all five landscape functions.

A hint of the power of regenerative landscape management (in conjunction with major natural events) to affect large global systems was revealed in 2016 when a group of geophysical scientists and climatologists finished crunching the numbers on the year 2010–11: an extremely wet year for the southern hemisphere. It was so wet that (seemingly counterintuitively) the world's ocean levels, in bucking a long-term trend, fell by six millimetres. And Australia's so-called 'dead heart' bloomed, its grasses and other vegetation pumping millions of tonnes of carbon into the ground. In fact, in that wet year Australia took out of the atmosphere and squirrelled in the ground one-quarter of all carbon produced globally through the annual burning of fossil fuels.

But there is a catch. In an article evocatively titled 'Rain Falls, Mulga Rises, Carbon Sinks', journalist Matt Cawood explained that in 2010–11, 'millions of square kilometres of mulga and spinifex grasses flourishing in the inland drew down massive amounts of atmospheric carbon' but 'only to return much of it in 2012–13 when rainfall over much of the semi-arid zones was half the long-term average.'[9] Associated with this drying factor was extensive overgrazing by domestic livestock and feral animals. The key message here is that natural processes – if creatively and empathically managed – have enormous potential to address issues such as global warming and climate disruption.

Corroborating the untapped potential of regenerative agriculture to impact large climate cycles, in 2009 the leading and highly respected group of Australian scientists the Wentworth Group wrote a paper in the midst of the political debate over carbon emissions. This paper was called 'Optimising Carbon in the Australian Landscape'. They pointed out two glaringly obvious facts, but missed a third key one. First, that vegetation clearing and land degradation alone are responsible for twenty per cent of annual global carbon emissions. So reducing land clearing and poor agricultural practices is a no-brainer. Second, carbon stored in the world's landscapes (vegetation and soil) is three times greater than the amount held in the atmosphere. Therefore, increasing not just sequestration but particularly sequestration of long-lasting carbon is the second no-brainer. But the third and critical point, missed entirely, is that in regenerative agriculture and across broad landscapes we have a ready, effective and proven means of fixing enormous amounts of long-term carbon in the soil (or what is called soil organic matter).

The background to this is the simple equation that for every twenty-seven tonnes of carbon sequestered in the soil via biological processes, one hundred tonnes of carbon dioxide is removed from the atmosphere. Conservative estimates reveal that twenty-five per cent of Australia's annual carbon dioxide emissions could easily be sequestered if just fifty per cent of Australia's cropping land swung over to biological and other regenerative farming systems: and this without instigating other non-cropping regenerative agriculture practices such as increased ground cover via holistic grazing regimes and agroforestry. A basic principle involved here is that 'the quantity of carbon contained in soils is directly related to the diversity and health of soil biota'.[10]

As the Wentworth Group concluded, 'At a global scale, a fifteen per cent increase in the world's terrestrial carbon stock would remove the equivalent of all the carbon pollution emitted from fossil fuels since the beginning of the industrial revolution.' Unfortunately, they just hadn't yet seen that the solution – regenerative agriculture and the enhancement of landscape function – sits right under their noses.[11] Moreover, this solution would do more than redress the balance, as was demonstrated in that 2010–11 wet year (as described earlier).

There is an important nuance attendant here. We know that soil organic carbon exists in four different carbon pools. The four pools are graded in terms of ease of decomposition. These comprise (1) the 'labile' (particulate) pool (easiest and quickest to degrade, with a half-life of weeks or months); (2) 'humus' (that is biologically generated in situ in rootzones and which drives soil fertility via humate products, and which can disappear quickly); (3) 'less labile' (recalcitrant in decomposition, with a half-life taking years and up to decades to decompose); and (4) the 'inert' pool, such as charcoal and biochar. This latter carbon pool is crucial in addressing long-term climate change as its half-life decomposition can involve centuries to millennia. So again, addressing climate change by burying long-lasting, slow-degrading carbon derived from regenerative agriculture would appear a no-brainer – as we saw earlier in Chapter 9, where the symbiotic roles of both arbuscular and ectomycorrhizal fungi deliver resilient, long-lasting carbon compounds like glomalin and chitin.

Leading USA-based Australian soil scientist Greg Retallack (mentioned in Chapter 14) supports this view. Citing fossil records of soils, Retallack wrote in 2013 that due to plant carbon sequestration, the evolution of early land plants after 444 million years ago, and then the evolution of trees after 390 million years ago, on both occasions caused atmospheric CO_2 levels to fall, and the result was global cooling. Similarly, he says, the 'global expansion of grasslands and their newly evolved, carbon-rich soils (Mollisols) over the last 40 million years may have induced global cooling and ushered in Pleistocene glaciation.'[12]

Indeed, with the rise of more modern, more efficient carbon-fixing C_4 grasslands in the last twenty or so million years, the cumulative effect of total grassland evolution and higher organic soils and their spread 'from 0 per cent to 40 per cent of the world's land area', says Retallack, explains 'long-term downward trajectories in temperature, precipitation, and CO_2 seen in paleoclimatic records'.[13]

That is, Retallack and colleagues see grassland evolution 'as a biological force for global change through coevolution of grasses and grazers', and because 'modern grassland ecosystems are a potential carbon sink already under intensive human management, and carbon farming techniques may be useful in curbing anthropogenic global warming.'[14] Conversely, says Retallack, the great work of natural grasslands in the last forty million years had been undone 'by human exploitation of soils and fossil fuels.'[15]

Significantly, using healthy plant ecosystems to sequester carbon is far more complex than just locking up ground carbon, as Retallack further explains. 'Plant creation of reduced organic compounds and export of bicarbonate and nutrient cations in solution', he says, 'are forces for global cooling, because soils, lakes, and oceans store carbon that may otherwise be liberated to the atmosphere as the greenhouse gases CO_2 and CH_4.'[16] Grasslands therefore play a big role here, and not just via increased soil carbon sequestration but through higher surface albedo [in this case the fraction of solar and/or other electromagnetic radiation reflected off earth into the atmosphere] and reduced evapotranspiration.[17]

If broad-acre-agriculture continents such as Australia, Africa, North and South America and the steppe lands of Asia and Europe swung to regenerative agriculture, then the first safe boundary already crossed of the Anthropocene – climate change – would begin to be dramatically addressed. But then so would the other two crossed boundaries: biodiversity loss and degradation of the nitrogen cycle. Equally significantly, by regenerating landscape function and thus fixing more soil carbon, most of the other endangered operating limits of planet Earth would also be addressed, including: unsustainable and polluted freshwater use; the degradation of the phosphorus and other biogeochemical cycles; and deleterious land systems change, along with chemical pollution.

What hasn't even been considered because of the Intergovernmental Panel on Climate Change and general scientific focus on carbon and climate is the fact that available moisture and hydrology play a huge role in the climate story. We have seen that the water cycle is one of our key landscape functions. Well, voluminous scientific work confirms that, in the words of Walter Jehne, 'water and its hydrological and heat transfer processes are the dominant factor governing some ninety-five per cent of the heat dynamics and climate of the blue planet'. The simple fact is that regenerative agriculture can restore the water cycle to health, retain more moisture in the soil and earth systems, and thus play a key role in climate amelioration.[18]

Restoring adequate water for cooling the climate via regenerating landscapes means, says Jehne, restoring former nature 'soil carbon sponges and in-soil reservoirs globally'. Such a carbon bio-sequestration process to rehabilitate these carbon sponges, continues Jehne, is governed by plants and soil microbes, particularly fungi, and how effectively they can:

1. Maximise the photosynthetic fixation of sunlight, CO_2 and water into plant matter.
2. Minimise the oxidation of that plant matter back to CO_2, particularly by fires.
3. Maximise the fungal polymerisation of this plant matter into stable soil carbon.

    As such, the soil carbon sponge is largely fixed solar energy that is stored in a stable carbon matrix.

This sponge, he says, 'enables plants and bio-systems to operate as a positive feedback process, progressively increasing the draw-down of carbon to extend healthy terrestrial bio-systems over most of the land'.

We have seen in earlier chapters that the secret is how carbon-rich soil absorbs vast amounts of water. Clearly, as Jehne concludes, these carbon sponges 'are thus critical in capturing, retaining and sustaining the supply of water: of which one crucial function is that it enables us to restore the natural hydrological cooling processes and climate balances'. As we have seen, regenerative agriculture can do this.[19]

Much scientific work now confirms that degradation of all key landscape functions has led to an inability of upper-soil levels to store water. However, while these levels are drying out, we have the paradox that groundwater levels in many areas are increasing (leading, for example, to rampant salinity problems). This is due to degraded soil profiles allowing (via a variety of degraded mechanisms) more rapid and deeper recharge of water than is the case in a healthy landscape situation.[20]

Leading expert on the global water and associated energy and nutrient cycles, limnologist Wilhelm Ripl of Berlin is adamant that modern industrial agriculture is playing a major negative role in destabilising Earth's operating systems. These systems are the result of aeons of self-organising evolution, whereby water plays a crucial role in safely dissipating the vast majority of life-giving energy – and doing so in an orderly, non-random, closed-loop manner. But the modern use of fossil fuels (which vastly increases heat loads) and the wide application of damaging industrial farming techniques (which has disrupted self-organised cooling and nutrient recycling and distribution patterns) have, says Ripl, grossly distorted 'the global cooling system through interference with the water cycle and vegetation cover'. This is because, he says, '[s]ociety has hitherto failed to realise that the most crucial stabilising process is the short-circuited water cycle – between evapo-transpiration and precipitation.'[21]

Through extensive study, Ripl shows that:

> the conversion of local evapo-transpiration processes into the passage of water through the soil [the loss of valuable nutrients and crucial soil cations, which are the equivalent to irreplaceable plasma in our bloodstream, for example], along with the increased use of groundwater – both being responsible for the loss of nutrients and minerals – have proved to be adverse to sustainable life processes.

That is, the water, solar, soil-mineral and dynamic-ecosystem community landscape functions (which coevolved to synergistically work in stable, non-randomised ways in both local and global Earth and natural systems) have been hugely disrupted by industrial agriculture. The resulting run-down of key recyclable nutrients Ripl calls 'landscape ageing'. Ripl says of Germany, with its youthful, rich soils and humid climate, 'Since the introduction of non-renewable energy [that is, fossil fuels], the irreversible material flow has increased, per hectare, for the whole of Germany, by a factor of more than fifty and even up to one hundred times.'[22]

On dry continents such as Australia and much of the rest of the world, accelerating and more obvious land degradation and desertification (that is, accelerated earth-ageing) reveals just how rapidly we are degrading these long-evolved functioning mega-cycles that have hitherto, in self-organising fashion, regulated a sustainable and liveable environment on

Earth. As Ripl said, 'our collective intelligence dealing with the process nature seems highly underdeveloped'.[23]

The good news is that, given industrial agriculture is a key destabiliser of Earth's self-organising systems, a shift to regenerative practices will have powerful 'turnaround' impacts – and this will result in 'the local remediation of fundamental water-based ecological functions'. Yes, 'water and matter cycles have been degraded in unprecedented fashion', but the many examples and basic principles of regenerative agriculture disclosed in this book reveal just how vital it is, in Ripl's words, that 'Man has to be reintegrated in ecosystems as the most suitable intelligent controller of water cycles and matter cycles' – that is, bar Nature herself.[24]

Similar arguments as discussed above concerning the water cycle also apply to another big Anthropocene boundary-crossing disaster: global biodiversity loss. There are numerous examples in this book of how, through holistic grazing regimes, new cropping regimes, farmer-driven agroforestry, planned biodiversity plantings and regeneration, and other innovative ways to retain water in landscapes and to fix more carbon and so on, that biodiversity is consequently also restored. Halting all vegetation clearing is clearly an urgent imperative.

The same applies to the third crossed planetary-system boundary: human destabilisation of the nitrogen cycle. Injudicious capture and use of synthetic, fossil-fuel-derived nitrogen fertilisers, combined with ongoing land degradation and lack of soil-fixing nitrogen bacteria (with the concomitant release of nitrous and ammonium oxide), have resulted from poor farming techniques.

A number of biologists and agricultural, soil and landscape scientists are now challenging the power base of industrial agriculture and ringing the bell on the alarming dangers of synthetic nitrogen. In short, long-running trial data now reveals that high synthetic nitrogen inputs deplete soil carbon, impair water-holding capacity and, as a result, end up being obtusely counterproductive by leading to depleted soil nitrogen. A key component of this is that synthetic nitrogen grossly simplifies the soil food web – especially of microorganisms and valuable mycorrhizal fungi. This is because an imbalanced level of microbes that devour organic matter are favoured. This hugely negates all the benefits of soil organic matter, valuable humates, and long-lived carbon in regard to water-holding capacity, the enhancement of soil structure, the retention and recycling of nitrogen, and so on.

In the view of Illinois soil scientists Richard Mulvaney, Saeed Khan and Tim Ellsworth, synthetic nitrogen use creates a kind of treadmill effect. As organic matter dissipates, the soil's ability to store organic nitrogen declines and so a large amount of nitrogen simply leaches away. As Mulvaney observed in an interview in 2010, 'the soil is bleeding'[25], which is exactly what Wilhelm Ripl said about the distorted water cycle.

That regenerative farming practices actually work in addressing climate change and other Anthropocene boundary issues has recently been quantified on a farm near Yass in southern New South Wales owned by John and Robyn Ive.

John was previously a researcher at CSIRO but was summarily ditched in one of its many restructures. Still a scientist at heart, he sought solace by going into farming in 1980 when he and Robyn bought the 605-acre (245-hectare) farm Talaheni. What he found, in one of the most saline-affected areas in New South Wales, was a grazing farm that had been savagely cleared of vegetation since the late 1800s. As a result, there was a rising water table, widespread dryland salinity and soil acidity, and increasing sheet and gully erosion, plus continuing loss of native vegetation – sadly a common story across vast areas of Australia. All this was accompanied by declining agricultural productivity and an increasing loss of biodiversity.

However, in a planned and systematic way, John and Robyn set about rectifying what they called a 'basket case'. Following extensive fencing off of woodlands, the planting of over 250,000 trees and shrubs in long corridors and blocks, plus paddock subdivision, extensive scientific monitoring of landscape function, the sowing of perennial grasses, and much more, the Ives have transformed their land. Dryland salting has markedly declined, as has overall soil degradation, and erosion has been halted and then healed. Biodiversity at most levels has increased dramatically, and despite taking out of production thirty per cent of the farm, there has been an overall increase in stocking rate of 2.5 dry sheep equivalents per acre in thirty-six years – or more than a doubling of the original level (but this on just seventy per cent of the original area). At the same time, higher-quality superfine wool, beef and timber are being produced.

Then, in what is an Australian first in examining rejuvenated farmland and its carbon sequestration, Melbourne University researcher Natalie Doran-Browne recently undertook a comprehensive carbon emissions and sequestration study of Talaheni. This revealed some astounding results. Setting aside tree planting, since 1980 the Ives' regenerative farming approach has sequestered enough soil carbon to offset all the farm's live-stock enterprise and farm-running emissions plus that of other activities utilising fossil-fuel energy. Putting this in context, at a global level, non-regeneratively grazed livestock emissions are a huge source of anthropogenic greenhouse gas emissions. Moreover, from an Australian perspective industrial agriculture is responsible for fifteen per cent of our national greenhouse gas emissions.

So, from the work of John and Robyn Ive, when revegetation is included, in total Talaheni has sequestered around eleven times more carbon in soils and trees than total farm emissions over thirty years. This includes impres-sively lifting soil carbon levels from one per cent to four per cent. It is a dramatic indicator of the potential of regenerative agriculture to address – in integrated fashion – a whole group of critical Anthropocene boundary-crossing areas, not the least of them climate change.[26]

John and Robyn Ive are just another of many examples in this book illus-trating an emerging shift to a new regenerative agriculture. As such, they provide a touchstone to the great conundrum that underlies this entire book: that, as the alleged most sapient of all species, we have precipitated Earth into the Anthropocene, and that this is because of a fatal lack of understanding of ourselves.

In Chapter 19, I discussed how this ignorance and blindness is exhib-ited: that the vast majority of us farmers (and indeed the general populace) are inculcated in the Mechanical mind and thereby unthinkingly commit to both industrial agriculture and a consumer society driven by the giants of global business. However, also discussed in Chapter 19 were examples of how it is possible to enact the process of transformation to change one's paradigms and shift to regenerative agriculture: in short, to begin to incul-cate an Emergent mind.

The exciting thing is that, once regenerative farmers move down a path to ecological agriculture in a two-way, push–pull process of personal transformation, there is invariably a multiple shift in one's mind: not just cognitively but ethically, empathically, philosophically and often spiritually also. In short, regenerative farmers enact some form of transformation in themselves and begin to move from a Mechanical towards an Emergent mind: to truly beginning to understand themselves and their place in a Gaian Earth and her systems.

This preparedness and ability to diagnose what is wrong with the ruling world paradigm and thus agriculture and everything else about our industrial society also constitutes a step towards radicalisation. And it will be this change that is needed to extricate ourselves from our Anthropocene dilemma. This is because we need a wider and more rapid shift to an Emergent mind than is just occurring at the cutting edge of regenerative and urban agriculture. In turn, this shift means that teaching ecological literacy to all levels of society becomes an urgent imperative: that from the start we all need to be exposed to more sensitive, humble, holistic and Earth-empathic thinking and feeling.

Only this teaching will allow us to truly 'know ourselves' and begin the process of attaining a close intimacy with Earth and her systems. Therefore, if we do this and at the same time inculcate widespread Earth-friendly living, farming and consumption, then we can all play a pivotal role in pulling us back from the brink.

Moreover, there is an additional and quite extraordinary bonus in shifting to regenerative agriculture, which is that, in healing Earth, we begin a process of also healing ourselves – both physically and mentally. This I will address in the next chapter.

CHAPTER 21

Healing Ourselves

Like it or not, we are slightly fat, furless, bipedal primates who crave sugar, salt, fat, and starch, but we are still adapted to eating a diverse diet of fibrous fruits and vegetables, nuts, seeds, tubers, and lean meat. We enjoy rest and relaxation, but our bodies are still those of endurance athletes evolved to walk many miles a day and often run, as well as dig, climb, and carry.

DANIEL LIEBERMAN, *The Story of the Human Body*[1]

As a child, one lives in and for the moment. Our farm vegie garden and a variety of fruit trees were a happy foraging ground for an only child, and secret raids through the high netting garden fence were part of the fun. There, one could snatch sappy climbing peas and beans, uproot a young carrot, munch on a juicy tomato and, if game, bite into the tart stem of crimson rhubarb. In the mornings, we had fresh eggs and Jersey milk and separated cream, and our sheep-meat and occasional steer, like the milk, was off natural pasture and laden with taste: the crispy, nutrient-laden fat especially memorable.

I particularly recall carefree summer days in the 1950s spreadeagled up an ancient mulberry tree that dated to an earlier time in the nineteenth century. The berries were not giant and elongated like today's bland, highly bred, tasteless pap, but smallish and compact yet intense in taste: somehow semi-sweet yet powerfully tart at the same time. They stimulated saliva glands as one gorged while sharing the fruit among foraging birds. Crimson juice dribbled from mouth corners onto arms and clothes, staining hands. It was childhood joy, and I was completely oblivious to the fact these mental, physiological and stimulated behavioural responses went back to the very childhood of our species.

Time and again throughout this book I have reiterated the significance of processes such as coevolution and self-organisation, where species are embedded in, and indivisibly part of, their environment. For humans, the time of greatest significance in this coevolution process (with the most profound physical and mental implications for the future) was when our species evolved into its modern form in the last one million years. At that time, our complex cerebral and physical beings were rapidly changing, and simply because our brains and bodies were in a coevolutionary process of adaptation and change with our environment so that we could best survive in it. This coevolutionary behaviour never stops. A more recent example (some 8000 or so years ago) is of humans developing lactose tolerance once they had begun to domesticate animals and use their milk products.

This conjoined physical and cultural coevolution with our landscape and natural world particularly applies to the unimaginably various and complex physiological adaptations in our bodies in keeping with our hunter-gatherer diet. The fact we are adapted to this hunter-gatherer lifestyle has two important health consequences that I wish to highlight in the context of regenerative agriculture returning us to health in this industrial world.

Evolutionary biologist Daniel Lieberman sets in context the underlying causes behind the conundrum of disease in contemporary industrial society: ostensibly the best fed and most advanced, energy-rich human cohort there has ever been. A key part of this conundrum is the fact that the industrial agricultural and food system is not just poisoning us but is also, confoundingly, making us obese while starving us at the same time.

Our modern industrial society, Lieberman explains, is in disjunction with our evolutionary bodies – bodies that, in his words, have an evolutionary story 'that matters intensely'. Says Lieberman, 'We get sick from chronic diseases by doing what we evolved to do but under conditions for which our bodies are poorly adapted, and we then pass on these same conditions to our children, who also then get sick.'[2] The result is that modern industrial society increasingly has lower mortality but higher morbidity (ill health).

The key issue is that we have outsmarted our evolutionary biology through our cultural evolution. For the first time in the history of any species, our cultural evolution (or adaptation) is occurring faster than our biological evolution, and is causing what Lieberman calls both a gross evolutionary *mismatch* between the two and thus also *dys-evolution* (a form of cultural evolution through which 'we pass on the behaviors and environments that promote mismatch diseases').[3]

By 'mismatch diseases', Lieberman means 'diseases that result from our Paleolithic bodies being poorly or inadequately adapted to certain modern behaviors and conditions'.[4] Resultant modern mismatch diseases now gobble up much of healthcare spending across the world. A clear example is the massive obesity problem in modern society. That is, as a hunter-gatherer it was a clever adaptation to have a craving for energy-rich foods coupled with a predisposition to lay down fat (the most efficient way of storing energy) for inevitable lean times and hard exercise. Having spare fat/energy didn't just enable individual physical survival, it also increased fertility – or species survival. But today, as Lieberman explains, 'our bodies are inadequately adapted to cope with relentless supplies of excess energy'.[5]

The latter situation is due to the way we have prostituted modern industrial food, substituting raw and diverse nutrient quality with denatured nutrients in combination with vast quantities of salt and processed fats and sugars. Spring-boarding off our maladapted obesity are many of the modern killers and debilitators: type 2 diabetes, cardiovascular disease, cancer and a host of others.

The second form of mismatch I wish to highlight is the destruction of nutrient quality and density in foods by modern industrial agriculture and the connected manufacturing, distribution and retailing of industrial foods.

As passionate nutritionist the late Jerry Brunetti has pointed out, human hunter-gatherers ate a vast diversity of foods. Our bodies evolved to detect, savour, digest, biochemically process, convert (into various chemicals, fats and proteins), store and/or excrete them and their derivatives. These foods were packed, first, with primary nutrients. But they were also packed with micronutrients. These came to early humans not just via gathered plants but also through the meat they hunted. And their bodies had evolved the most extraordinary mechanisms to ingest and process these myriad compounds and chemicals – including handling toxins.

So when one has the most extraordinary, ready-made, self-organising system for health and survival, then it seems obvious that only fools would ignore it, let alone tamper with it. But this – as with the energy–fat issue above – is what has progressively occurred in Western society (and increasingly now globally).

The oldest and nearest example to ourselves as ancient hunter-gatherers are the !Kung bushmen of the Kalahari Desert. Today, they use over seventy-five different plants in their diet and the nutrients of tens of thousands more through their hunted game. Modern diseases such as cancers are virtually unknown in their society. Anthropologists and ecologists have worked out that, before agriculture, hunter-gatherer humans across the varied seasons of a calendar year utilised tens of thousands of different types of foods and tens of thousands more food nutrients through the game they ate.

Studies of nutrients in the unadulterated traditional agrarian and hunter-gatherer diets of the past, when compared with a post-1940 Western diet, for example, reveal astounding differences. The amount of nutrients in the original diets we are coevolved for was staggeringly greater. As we saw in the Introduction, for key nutrients such as calcium, phosphorus, magnesium, iron and the fat-soluble vitamins, the levels were seven to ten times more in the original diets. Even well before 1900, our original rich and varied diets had already become simplified compared with pre-agricultural times. As 'Wild Food' advocate Jo Robinson concludes, 'we will not experience optimum health until we recover a wealth of nutrients that we have squandered over ten thousand years of agriculture, not just the last one hundred or two hundred years'.[6]

But today, what of the modern American or Australian diet gleaned from a supermarket or junk-food outlet? Answer: only five to ten plants and two main sorts of meat are predominantly used. With this goes a massive decline in food nutrient density. For example, we know that a typical apple in the USA in 1914 supplied twenty-six times more iron than a modern apple that has been bred for production and cosmetics and not taste under a high industrial production system (taste, as with my mulberries as a kid, being a good indicator of nutrient density and variety). This drastic decline extends to most nutrients. Across the board in standard vegetables and also in our bulk grains (wheat, oats, buckwheat and white rice), there are significant drops in such key minerals as calcium, iron, potassium and magnesium. Consequently, various studies since 1948 confirm that there is a dramatic relationship between declining minerals in our food and an opposite escalation in mineral-deficiency and related diseases. And this without looking at the decline of more complicated molecules that play huge roles in disease prevention.[7]

The most frightening aspect of the huge decrease in healthy, natural food diversity and the destruction of nutrient density and variety today is that it is directly related not just to mineral-deficiency diseases but also to the huge increase in human cancers and other modern Western diseases.

While we know that a sedentary lifestyle plus smoking and alcohol are strongly involved, it is now becoming increasingly clear that a nutritionally compromised food supply is a major factor in this chronic disease epidemic – and especially when linked to the energy–fat mismatch issue discussed earlier. Obesity, type 2 diabetes mellitus, cardiovascular disease, hypertension and stroke, autism, allergies, Alzheimer's disease and many types of cancers and other diseases are leading to hugely escalated volumes of premature death and disability, plus associated chronic illnesses and mental diseases. These in turn are swamping the medical and social systems of developed and, increasingly, less developed nations. Even such establishment bodies as the Food and Agriculture Organization and World Health Organization are talking about nutritionally compromised food supplies, and that such global under-nutrition comprises a non-communicable epidemic that in fact constitutes a global pandemic.[8]

The tragedy is that virtually all these modern diseases are preventable.

But why are high nutrient density and variety important? Well, it would take a hundred textbooks to explain the importance of a hugely rich and varied complement of nutrients in our diet, but suffice to say our coevolved bodies and their mechanisms are unbelievably complex and comprise an extraordinarily fine-tuned, interconnecting array of different systems. All, however, are dependent on adequate nutrition. Just to give a brief 'taste' of the importance of nutrients, we know this: they support biological processes without interfering with them, and they are required for every biological process, for either the energy or the enzymes they supply. As Dr Arline McDonald of the Feinberg School of Medicine in the USA explains, 'Nutrients are the raw materials that support physiologic and metabolic functions needed for maintenance of normal cellular activity.' Crucially, this also includes our immune systems.[9]

What is rarely appreciated in this area is the huge role of microorganisms in both our general and gut health. In short, we coevolved to eat a wide variety of nutrient-dense foods to feed not just ourselves but also the vast zoo of bugs in our body on which we depend. Very few of us appreciate the size, extent and key role of our gut ecosystem or microbiome – especially the estimated thirty-nine trillion or so microbes in our gut, which comprise one thousand or more species and work 24/7 to break down our food and toxins of all kinds.

Our gut, says Associate Professor Andrew Holmes of Sydney University, 'is the largest immune organ, the largest endocrine organ, and the second largest neural organ – second only to the brain' (indeed, the approximate two kilograms of microbes in our bodies are nearly the weight of the brain). Crucially, he concludes that the gut is 'where you have this – more or less – seamless connection from diet to microbes to metabolic health to immune health.'[10]

And this is where our modern diet causes problems. First, hunter-gatherers and others who ate traditional diets ingested from 60 to 120 grams of 'fibre' a day ('fibre' meaning a range of materials left over after the small intestine had absorbed most of the readily available sugar, fat and protein). This 'fibre' is then broken down in the large intestine or colon, where most of our microbiota – especially bacteria – live. Here, the bugs turn the fibre into short-chain fatty acids, which are so vital to our metabolism, health and immune function, and behaviour. Moreover, this 'fibre' level – one that we and our gut-flora evolved for – maintains a high diversity of microbes. The problem is modern Western diets only deliver less than twenty per cent of this crucial fibre, as modern processed foods are largely digested in the small intestine, leaving the mega zoo in our large intestine to either starve or be vastly simplified – with huge impact on our health and immune system. Then, directly concerning the chemical glyphosate in our food as we will see, man-made poisons get into our food and thence our gut, where they wreak havoc on the microbiome.

In short, we have become utterly divorced from the very world that sustains our health and to which we are coevolved. And this without calculating the concomitant mental-health issues involved in our species' disassociation from our natural environment. This predicament is nothing short of suicidal, for it is a contravention of the basic laws of natural systems. As Wendell Berry states, 'There is in fact no distinction between the fate of the land and the fate of the people. When one is abused, the other suffers.'[11]

The evidence is overwhelming, for those who want to hear, that a key pathway to returning to health is eating healthier and more diverse foods off healthy, regenerative landscapes, and also reconnecting to nature, along with becoming more physically active again.

Even back in 2001, as leading specialist medico and nutritionist Rima Laibow outlined, 'preventable chronic disease accounted for more than sixty per cent of all global deaths (nearly thirty-four million deaths) and approximately forty-six per cent of the global disease burden'. Is it any wonder, therefore, as Laibow states in regard to the USA, that 'we have more cardiovascular disease, cancer, diabetes, autism and other neurological disorders, macular degeneration, arthritis, osteoporosis, MS and a host of other diseases than any population known in history'.[12] Australia is now rapidly emulating this predicament.

Laibow says, 'The US provides a sad but illustrative example of a country spending enormous sums on everything except the powerful combination of preventive nutrition and natural medicine focussed on optimal health.' The USA's efforts, concludes Laibow exasperatedly, 'are toward slamming the high-tech barn door after the nutritional horse is well away'. Instead, says Laibow, 'Nutritional status and immune competence are the cornerstones upon which the house of health rests.'

Just to finish what is a rapid gallop across an immense field, there is one area of great relevance to this story on regenerative agriculture, where time and again I have emphasised (in the wake of Sir Albert Howard and others) the huge importance of having grazing animals in an agricultural system. Hitherto, I have outlined sound ecological reasons for this. But in addition, having animals genetically adapted to healthy landscapes and natural, human management regimes has huge implications for human health. This is because such animals on healthy landscapes provide vital nutrients that a plant only–based diet cannot.

There is now much evidence to show that non-industrial, non-high-growth herbivorous animals that eat grass and shrubs off healthy landscapes have amazingly desirable ratios of beneficial nutrients. This includes high ratios of long-chain omega-3 to omega-6 polyunsaturated fatty acids (known as antioxidant cancer-prevention markers) off grass-fed beef, lamb, pork, poultry, milk and eggs (and shown to be passed on to humans); increased vitamins E, A and beta-carotene in similar healthy grazing situations in a range of products; and on and on. Similar results are recently shown for the fatty acid CLA (conjugated linoleic acid) – thought to play a major role in disease prevention.

As with all the above areas concerning human health, this specific question of animal products versus vegan or vegetarianism is complicated and caught up in vested interests and opposing worldviews. Epitomising this is the debate concerning the dangers of saturated fats (saturated fatty acids, or SFAs) in 'unhealthy' diets and their role in leading to elevated LDL-cholesterol levels and thus an increase in cardiovascular disease. It turns out this connection now underpins a whole health business.[13]

However, this theory and practice only applies in the modern Western food system, with its high levels of saturated fats and its poverty (even total absence) of available 'health-giving' nutrients. As Daniel Lieberman explained, humans have coevolved to have moderate and even high intakes of saturated fats, but only when they are in balance with an active lifestyle and the optimal intake of vitamins, minerals and other cofactor nutrients that can only be found in a healthy natural diet. When in a balanced situation, we know such fats are integral to healthy human function in a large number of ways. Confirming this is the knowledge that the longest-lived communities of humans enjoy traditional diets that are rich in animal fats.

There is a clear message here: if we change our lifestyles and the ancient diet to a modern Western diet full of sugar, fats and over-processed foods that are bereft of most of the goodies off healthy landscapes and soils, then saturated fats become deadly disease-causers: and particularly when things such as antioxidants are missing. This same upside-down demonisation or ignoring of 'good' nutrients in a proper healthy situation applies to lots of other nutrients.

As the classical founder of the honourable tradition of medicine Hippocrates stated way back in 400 BC, 'Let food be thy medicine, and medicine be thy food.' Or as Michael Pollan playfully put it in a twist on conventional wisdom, 'You are what you eat eats.'[14]

Concerning this frightening and suicidal predicament that modern humanity finds itself in, it is not as though we weren't warned. The American naturopath Henry Lindlahr (1862–1924) was banging on about healthy food and natural forms of healing back in the early twentieth century, as were the two British medicos Sir Robert McCarrison and Guy Wrench – each as forerunners to Sir Albert Howard, having been profoundly influenced in India by the connection between their ancient organic agriculture, and healthy soil and its relationship to diet and health. Another medico, the gastrointestinal specialist Charles Northern, also linked human diseases to poor nutrition from the early 1930s. But then in the 1930s there emerged an extraordinary giant in the field: the Canadian-born Cleveland dentist Weston A. Price (1870–1948). He is best remembered for his groundbreaking work (beginning in 1894) linking diet to tooth decay and facial deformity, and thus pioneering holistic dentistry. But his second huge contribution linking diet to overall human health came when he investigated a wide variety of cultures, especially indigenous and agrarian people (groups, Price reasoned, who had diets similar to our ancestors thousands of years ago), and then compared these to modern Western people.

Price's results linking health, diet and nutrients were profound. He discovered that traditional foods have vastly greater levels of key minerals, vitamins and other essential nutrients when compared with the contemporary American diet. He also found that disease levels were in opposite proportions. In turn, Price was followed and supported by other scientists and writers who pointed out the powerful connections between diverse foods off healthy landscapes and human wellness, along with their converse. Leaders of this group included many of the principal figures of the emerging 'organic' tradition.

Following the Second World War, and in response to the rise of industrial agriculture, a new crop of thinkers and writers appeared in this field, which continues to grow to this day.

So the relevance of all this to regenerative agriculture? Absolutely enormous, and for five key reasons. First, due to the rise of industrial agriculture, we have a widespread escalation in the demineralisation and nutrient stripping of our agricultural soils and thus foods. This is caused by poor cultivation processes, but especially by the application of synthetic fertilisers, herbicides and pesticides, which kill off both the soil biology and other biotic diversity. This (as we saw in Chapter 9) means crucial organisms such as mycorrhizal fungi are not sourcing a wide variety of good nutrients and not screening out toxins and pathogens.

Second, due to the widespread industrial practice of what Michael Pollan exposed as concentrated animal feedlot operations (CAFOs) – plus animals grazing on non-diverse grasslands or chemically treated soils and crops – much of the beef, lamb and chicken (and even fish from fish farms) eaten today across the world is lacking in key health-enhancing nutrients. At the same time, this approach concentrates deleterious nutrients, chemical additives, antibiotics, synthetic hormones and the like.

Third, as part of the productivist efficiency drive of modern agriculture, research goals have delivered fast-growing, high-production, cosmetically pleasing plant and animal strains geared to high industrial input, CAFO, and/or processing and marketing systems. This, however, means such artificial plants and animals have lost the ability to both source and concentrate desirable nutrients for maintaining and enhancing their own and human health (i.e. many have become drug addicts, dependent on human inputs). The bulk of these when they reach the human table are nearly empty of essential nutrients.

Fourth, due to the globalisation of agriculture and the domination of large, multinational food-growing, distribution, processing, packaging, transport and trading conglomerates (what Paul Hawken aptly dubs 'corporate kleptocracies'), already poor nutrient density in foods is either exacerbated in further processing or degraded further with the increased time taken in processing plus transport to retail.[15]

And fifth and finally (as we will see directly concerning glyphosate), the prostitution of modern agriculture by not just the excessive use but the use at all of synthetic fertilisers, herbicides, pesticides and pharmaceutical products in food production means significant amounts of man-made chemicals are known to get into our foods. Not only are many of these toxins directly harmful to human health, but in addition this toxic load imposes on our bodies and immune systems a further need for nutrients that is far in excess of that experienced by our hunter-gatherer and agrarian ancestors. This is now doubly exacerbated because modern foods are increasingly bereft of such urgently needed nutrients.

In short, what a diabolical mess! And yet the solution seems so simple. Along with addressing our increasingly sedentary lifestyles, we just need to shift back to eating food that comes off healthy soils and diverse grazing landscapes, and which has a short time span between harvest and consumption. As Thomas Berry states, 'A healing of the Earth is a prereq-uisite for the healing of the human.'[16] Sadly, this simple solution is difficult to implement in a widespread way because of that crucial fifth factor of landscape functions: the human–social. This is manifested by modern society's commitment to an economic-rationalist, growth-based capitalist globalisation of greed and a discounting of the future.

In an article I came across the other day, I was powerfully reminded of just how captured modern society is by its ruling institutions and mindsets, and by the behemoths of global business. The article appeared in the *MIT Technology Review Business Report*. But first I had to skip past an article titled 'Creating a Better Apple', which highlighted in bold the grab-line 'Non-browning apples could be pre-sliced and left on shelves for a couple of weeks, perhaps creating a new market for kids' snacks.' Moreover, this article opened with, 'When the world's first biotech apple was approved from U.S. and Canadian regulators this year, it demonstrated that with plenty of patience even a farmer can develop a GM [genetically modified] plant.' Clearly, the technology must be simple if a farmer can handle it!

Just how low farmers stood in the food chain was then revealed in the feature article that had caught my eye. This was titled 'The Nestlé Health Offensive' and concerned how 'the world's largest food company tries to overcome technical challenges and popular tastes to make its food health-ier'. In a breathless paean to big business and the apparent power of science and technology, the author, Corby Kummer, first asked rhetorically, 'Could the food industry engage in, as the public-health community would put it, "harm reduction"?' Kummer then answered, 'Certainly big industry has the technical and marketing expertise to do so – skills that far surpass those of any farmer, produce consortium, or artisanal business.'

Nestlé is the world's largest food company by revenue, and at its 'Research Centre' just outside Lausanne, Switzerland, it employs 600 peo-ple (250 with PhDs) who collaborate with over fifty universities 'to conduct research into food composition, physiology, taste perception, and health'. So, after Nestlé's people looked closely at 2005 World Health Organization recommendations on 'healthy' food, Kummer said the food manufacturer had recently 'branded itself as a nutrition, health, and wellness company'. This, reported Kummer, meant implementing 'an initiative to reduce sugar, sodium, and saturated fat across its entire product line by the end of

2016'. In 2014 alone, continued Kummer, Nestlé 'claimed to have reformulated – "renovated", in its own parlance – 10,812 of the products that its
2,000 separate brands made in more than 442 factories in 86 countries'.

As Kummer revealed, the intriguingly named Nestlé Nutritional
Foundation 'set a target in 2013 to reduce levels of these "nutrients" by 10
per cent between 2014 and 2016'. Other gobsmacking 'healthy nutrition'
moves, said the reporter, included such things as 'reducing the size of Kit
Kat bars to bring down the calories and fat in each portion'. In this slavish
kowtowing to the big end of town, Kummer concluded:

> Many, many smaller and boutique companies are staking
> their new product lines on claims of fresher, less-processed,
> lower-calorie food, of course. But it is the often quiet changes
> made at big companies that can have more of an effect on
> people's health. More people are eating their foods.[17]

Appropriately, it was Vandana Shiva who stated that 'the common citizen
is politically orphaned in a world shaped by corporate rule – farmers' rights
and people's rights to food are extinguished'; that, in short, 'consumerism
lubricates the war against the earth'.[18]

However, there is also lots of good news. This book reveals many examples of regenerative farmers returning their soils and landscapes to health.
Also highlighted are the many holistic-thinking soil scientists, ecologists,
medicos, nutritionists and others, such as the macro-thinking Fred
Provenza, who are beginning to part the veil for us on how healthy landscapes work and why returning them to a state of functional health is so
important. Ultimately, however, the solutions rest in the hands of caring,
ethically driven regenerative farmers, urban gardeners and consumers,
who are increasingly flowering.

This is just as well because agriculture's turn to the industrial has other
quite frightening implications for human health over and above the
stripping-out of food nutrient density and variety. As mentioned in the
Introduction, these include, first, alarming new information on the dangers of the world's most widely used herbicide, glyphosate, and, second,
the implications of industrial agriculture's destabilisation and destruction
of human health through the increasingly understood role of epigenetics.

While I touched on the epigenetics story in Chapter 15, this unfolding phenomenon is now assuming greater significance in regard to human health. This is because our anciently coevolved bodies now find themselves encountering situations and environments they were not adapted for – such as in terms of what humans eat and also of other environmental influences.

Modern Western consumer society (particularly following the Second World War and the rise of the petrochemical, plastics, pharmaceutical and associated industries, not to mention pollution-crammed modern cities) has filled our world with alien chemicals that our ancient genome isn't equipped to handle.

These influences, both benign and deleterious, can have immediate heritable yet hidden impacts on succeeding generations via epigenetics. As discussed earlier, factors such as low nutrient density, man-made chemicals, refined sugars, saturated fats, recreational and medicinal drugs, and then in addition elements including oxidative stress, mental stress and unnatural shocks, all constitute a minefield of unintended consequences.

The alarming issue is that these factors can have transgenerational repercussions. As illustrated by the work of Fred Provenza, if ever there is a cautionary tale about what we do to our environment and ourselves, then it resides here in this sphere of epigenetics. This is because this emerging knowledge field puts front and centre the importance of the interface between us (or our plants and animals) and our genome: that an environmental impact can directly influence our genome and be passed on to the next generation. Therefore, epigenetic impacts are hugely relevant to the food we eat and its quality and nutrient integrity, and also to poisons such as glyphosate in our food and indeed the very unnatural environments we live in, as these can impact our gene expression.

In Chapter 15 when discussing epigenetics in the context of animal breeding and management and what I call adaptive landscape genomics, I pointed out that not only is the vast array of life microbial, but that we ourselves (like our animals) are enveloped in a microbial world: that the cells of the microbiota on and in us far outnumber our own cells, and that this is especially true of our gut flora – our enterobiome. Arguably, therefore, it is our gut – the interface with our ingested food and other external stimuli (such as stress-related hormones) – that is a key site of epigenetic activity. Thus, the microbiome is the front line of interaction with our environment because over fifty per cent of the genetic interface with our environment exists here in the microsphere.

To put this in context, a number of leading thinkers in this area now talk about the 'hologenome', defined 'as the sum of the genetic information of the host and its microbiota'.[19] The 'host' can be a human, plant or animal, for example, and the combination of this 'host' and its associated microorganisms is called a 'holobiont'. Because of the symbiotic, parasitic and/or commensal relationships between hosts and microorganisms, such thinkers (including my friend Professor Richard Jefferson who first described these concepts in a talk in 1994) now see the holobiont (animal/plant and microbiota) as the real unit of selection in evolution.[20] This in turn means the 'holobiont' is key also to epigenetic adaptation (or maladaptation) and thus adaptability or fitness and evolution.

It doesn't take much imagination therefore to anticipate what may happen to the human organism if it is repeatedly exposed to either harmful environmental influences or, for example, insidious poisons in our food – as we will see directly concerning the globally dominant herbicide glyphosate.

To grasp the significance of epigenetics, we can view our genome as the hardware in our 'system', but it is impotent without the software of epigenetics. Another helpful analogy used by American geneticist Jill Escher is that a musical keyboard has eighty-eight keys but it can be played in many different ways. Thus, our 'epigenome' is how our genome is played and in what order. Most important of all, it is our environment that is playing the keyboard.[21] What then extrapolates the complexity of such epigenetic behaviour is the fact our genome is a vast collective, in the form of the holobiont.

A brief example of epigenetic impacts is that mothers who are already obese or become badly overweight in pregnancy will predispose their children to becoming obese. Part of this is that a mother's diet during pregnancy and early in foetal life also influences the child's future food preferences. That is, influences in life can have enduring impacts on form, function and behaviour of humans and animals: factors inherited by the next generations and also through the next multiple generations of the microbiome.

What is even more sobering is that the molecular phase of the life cycle of sperm and egg development takes place well before fertilisation – in fact, years before. Thus, for example, a twenty-five-year-old pregnant woman can have her epigenetic expression derived from fifty or more years ago, when the germ plasm in her mother (now the grandmother of the foetus, but then a twenty-something mother) was preparing to be allocated to her gonads during the process of foetal differentiation. There are also other periods of our life that render us particularly vulnerable to environmental-epigenetic influences, such as early childhood, males' pre-puberty, females' peri-ovulation, peri-conception and so on.

A recent but well-studied example of such multigenerational impacts of epigenetics involves the first and second generations of children descended from Dutch mothers who were pregnant during the few months of November 1944 to April 1945. This was a period known as the 'Dutch Hunger Winter', when, due to Nazi occupation in a bitterly cold winter and a catastrophic cut in food availability, the Dutch population attempted to survive on around thirty per cent of normal daily calorie intake – including by eating grass and tulip bulbs.

While the children of starved mothers exhibited issues including higher obesity rates and a greater proportion of other health problems (including mental diseases), the surprise was that the grandchildren of those mothers who were starved in the first three months of pregnancy also exhibited higher proportions of ill health than children whose grandmothers had not experienced that terrible winter.[22]

There are thus enormous epigenetic implications for human health. However, they don't relate just to starvation such as occurred in Holland during the Second World War but also, as discussed earlier, to the situation where there have been serious declines in both primary and secondary (micro) nutrients in our industrial food, and particularly following the post-War explosion of industrial agriculture. For example, recent research into modern spring wheat cultivars, and especially in the industrial soft white wheat market, reveals moderate to major decreases in such crucial elements as copper, iron, magnesium, manganese, phosphorus, selenium and zinc. Researchers Kevin Murphy, Phillip Reeves and Stephen Jones go on to conclude that '[t]he large increases in the percentages of people suffering from micronutrient malnutrition over the last four decades coincide with the global expansion of high-yielding, input-responsive cereal cultivars.' Consequently, they say, 'global food systems are not providing sufficient micronutrients, resulting in an increased prevalence of micronutrient deficiencies.'[23]

To put this in context, Murphy and colleagues state that we now know '[t]he diet of approximately three billion people worldwide is nutrient' and 'micronutrient deficient (a condition referred to as "Hidden Hunger").' This is the cruellest form of what is called 'slow violence', and furthermore, it can have alarming transgenerational impacts.

Author Rob Nixon used the term 'slow violence' to describe our industrial, over-processed, nutrient-bereft food that is strongly implicated in a huge spectrum of modern diseases. The escalating predominance of a false, Mechanically minded world view that destroys people and community along with Mother Earth is also a form of slow violence: 'a violence that occurs out of sight, a violence of delayed destruction that is dispersed across time and space, an attritional violence that is typically not viewed as violence at all'.[24]

From the epigenetic research mentioned above, and from this wheat and similar work mentioned earlier across most modern grains, vegetables, fruit and meat, it appears almost certain that much of the huge escalation in issues such as autism, ADHD, asthma, Alzheimer's disease and other factors to do with our immune systems is due to these epigenetic effects induced by either extraordinary circumstances like the Dutch Hunger Winter or such influences as our poisonous, nutrient-deficient, out-of-kilter modern lifestyles. Autism, for example, has increased by over 3000 per cent since the early 1980s in Western society (as evidenced in California, the state with the most comprehensive records in the USA).[25]

There is an old saying that 'You are what you eat.' However, a knowledge of epigenetics now behoves us to add the rider: 'You are also what your mothers and your grandmothers ate.'

Another alarming factor in industrial agriculture's impact on human health relates to an exponentially increasing use of chemicals in our environment.

In the course of pursuing this story on regenerative agriculture, I found myself one winter morning in 2010 in the south-east of South Australia, some 225 kilometres from Adelaide near the little town of Keith. This 'new town' had rapidly grown from the late 1950s after CSIRO and other scientists discovered that the Naracoorte Sand Plain on which it resided was deficient in a number of trace elements. By correcting this through additives to fertiliser, what was once considered 'useless' scrubby sand country (despite its remarkable natural diversity) suddenly gained value as potential cropping and grazing land. Keith also happens to be the home town of my wife, Fiona, for her parents were some of the early settlers and developers who took up land following this 'trace element revolution'.

The reason I was in Keith (with Fiona tagging along for a nostalgia trip) was to visit two farmers, Peter and Pam Cooks, just on the outskirts of town. In the early 2000s, the Cooks' lives and farming practices – at the time high-input standard industrial agriculture – were completely upended when their entire farming environment and Peter's health began to fall to bits. Fortuitously, in casting about for a better solution they came across one of the leading innovators in today's regenerative soil agriculture: Dr Maarten Stapper. As a result, they were able to turn around their farm and personal health (the latter temporarily), their finances and their lives.

I have alluded to Maarten previously, but in brief: he was a Dutch immigrant cum ex–CSIRO scientist. It was not by accident that Maarten, like Dr Christine Jones, had become another square peg who didn't fit in the round holes of an increasingly reductionist and ideologically driven CSIRO. In Maarten's case, he came from the Division of Plant Industry, and the more it went down the path of industrial agriculture and genetically modified plants under its then chief, the more Stapper's conviction grew that something was wrong. He was eventually forced out in 2007.

Maarten – today a passionate and single-minded young sixty-year-old – had trained at Holland's famous agricultural university of Wageningen. However, his make-up and background from the start seemed to predestine him to become a maverick who sought a better agriculture for the world. Through serendipitous circumstances, Australia became his home. Along with others including Adrian Lawrie, Christine Jones, Graeme Sait, Helen and Hugo Disler, Walter Jehne and visiting American biological agriculture proponents such as Dr Elaine Ingham and Dr Arden Andersen, Maarten became one of the leading evangelists of the new 'biological agriculture'. This, however, followed a slow journey of awakening.

Maarten believes that to kick-start a regeneration in agriculture and transition from chemical dominance, what is needed is the 'reintroduction and enhancement of humic and soil biological activity'. He is a pragmatist in this, believing that biological agriculture can allow 'for minimal use of the most microbe-friendly fertilisers and herbicides with humic additives and molasses or sugars to enhance effectiveness and reduce damage to microbes'.[26]

Why the Cooks turned to Maarten (after hearing him at a field day when he described both their situation and also its solution) was that, by the 1990s, they were having major health issues with the sheep and goats on their farm. Animals increasingly were not thriving nor performing well and were getting lots of cancers, facial eczema and dermatitis in the wool. Also, the animals seemed more and more susceptible to intestinal worms, requiring increased drenching and vaccinating to keep them alive. In short, their immune systems were shot. But that was only half the story, because the Cooks' crops and soils were also deteriorating. Their cropping included grain and irrigated lucerne (alfalfa). 'This used to be a productive farm,' Peter told me, 'but by the early 2000s the soil was dead. Problem was, we used fertilisers and herbicides galore.' His wife, Pam, agreed. 'We were just trapped into this chemical merry-go-round,' she said.

'We employed an agronomist,' said Peter. 'He'd come in here; he'd go around and look at things, write out his docket. "Right, this is what's got to be done!" Boom, boom, boom, boom. Off he'd go. It was all chemically based, and he'd just print it off his computer.' On the list, said Pam, 'there were pre-emergents, selective hormone herbicides, glyphosates, various fungicides, just to name a few.' And that was without a toxic mix of acidic or salt-based and often heavy metal–laden fertilisers.

The Cooks gave me this agronomist's last printed list of chemical applications for the cropping season, which I hold before me as I write. There are a gobsmacking thirty-eight different recommended chemical treatments from May to November 2003 (with such names as 'Staccato', 'Clincher', 'Roundup Powermax', 'Striker', 'Hammer', 'Kwikin', 'Archer', 'Broadstrike', 'Victory', 'Tigrex', 'Gladiator', and so on): an aggressive nomenclature reeking of warfare and killing.

'Well, he'd leave a $15,000 bill of chemicals just to put on the paddocks,' said Peter. 'But the more chemicals went out, the worse it got. The agronomist would come in and he'd say, "Okay, now that paddock we need to use this, this and this," so there'd be three chemicals. Then he said, "Now, it's fine, mix them all together and put them on. That won't have any reaction." But the problem was we were getting the cocktail effect on our land and it was killing everything. That agronomist killed the paddocks, the soil biology. Absolutely wrecked it.'

The soil became more compacted, the stubbles wouldn't break down, the yields declined and there was escalating plant disease; the sheep were more unhealthy; massive pasture-snail plagues hit, souring the grass; the birds and insects disappeared; and on and on. Plus, Peter noticed his own health rapidly worsening. This whole working environment in time led to their son Michael and young family leaving to seek a safer environment. Sadly, Peter has died since I visited them in 2010.

So, they acted and sacked the agronomist. Initially, they used biological-farming consultant and product supplier Adrian Lawrie, and then Maarten Stapper from 2005. As a result, the Cooks experienced a rapid and spectacular turnaround in the entire health and productivity of the farm, and also, for a time, in Peter's health. Inevitably, as part of this regenerative transformation, the Cooks also went on a massive journey of knowledge discovery and world-view change via extensive reading and attendance at 'alternative' field days. This included radical changes to their own diet.

However, as I finished my visit, the Cooks gave me one cautionary example of how hard it is to fully overthrow the powerful industrial-agriculture system. They told me, 'Our neighbours all think we're mad.' Yet such neighbours, said the Cooks, were also trapped in the 'madness' of an industrial system that can't cater anymore for healthy, nutrient-dense food. Peter recalled driving a truck full of his harvested, biologically grown lentils to drop at the local bunker for storage and shipment. 'There was a huge difference in colour,' he said. 'You could see our lentils from quarter of a mile away ... bright orange. All the other growers' lentils were a dull charcoal. Same variety, but ours was biological and no fungicides. Then they just bulked it in for shipping with the other shit.'

In 2010, twenty-odd years into the second major agricultural chemical revolution, Wes Jackson noted, 'Since the minimum-till and no-till techniques were developed, made possible by the use of herbicides, we effectively were acknowledging that we have to poison our soils to save them.'[27]

Concerning the modern agricultural chemical revolution connected to no-till and the cynical use of the term 'conservation agriculture', here's the rub: the entire no-till chemical agriculture revolution largely rests on

Monsanto chemists' invention of glyphosate (as we have seen, commonly known as Roundup). For decades, the industrial establishment has assured the world that glyphosate is safe: some agronomists and product suppliers even stating 'you can safely drink it.' This establishment (glyphosate is now off-patent and so is sold by other companies) continues to insist that glyphosate has no residual effect but is totally broken down and does not enter the food system.

Today, glyphosate is the most widely used chemical in the world, with over 825,000 tonnes used in 2014 (a fifteen-fold increase in twenty years, and now annnually approaching a billion tonnes), and value of sales worth tens of billions (more than the value of all other herbicides combined).[28] Moreover, usage has continued to sharply increase: one key reason being that genetically modified crops are predicated on its increased use.

But now the global industrial agricultural system is facing a crisis point, though not yet admitted or acknowledged. This is because: (a) we now know glyphosate is residual and does enter the food system[29]; and (b) once in the human body, it has huge impact, and so potentially appears to be one of the worst chemicals ever to be introduced into the environment.

Furthermore, the cat is out of the bag. Increasingly voluminous medical and other research is now exposing the canard that glyphosate is safe. The first batch of evidence, unsurprisingly, relates to negative impacts on soils and farming systems. Leading soil microbiologist with the USDA Dr Robert Kremer outlined why glyphosate is damaging. First, it kills beneficial soil microbes that keep pathogens in check, while at the same time promoting 'bad' bugs. Second, it binds up all the nutrients in a plant so that any farm animals and their progeny ingesting this as fodder end up suffering from nutrient deficiencies. In addition, their gut microbial ecology is effectively destroyed. This in turn leads to other diseases. Third, the plants can't defend themselves because they have lost control mechanisms. Fourth, glyphosate accumulates in the new growth points of plants (roots and buds). Fifth, glyphosate's breakdown product AMPA is more toxic than its parent, and deadly on soil microbes. And sixth, at least forty documented crop diseases increase as a direct result of glyphosate use.

This is also part of the modern multi-paradox of industrial agriculture's drive to simplify nature: that the more weedicides and pesticides you put on, the worse you make things. For example, we now have weedicide resistance in scores of 'targeted' weeds (necessitating even stronger poisons). While pesticides destroy predatory, controlling insects and other bugs, this enables ever-bigger 'pest' outbreaks and devastation.[30]

But, in agronomy, glyphosate is at the forefront of this simplification and amplification of the very problems that are targeted. Glyphosate – the active ingredient in Roundup – works as a chelator: that is, it grabs hold of key elements in the soil (such as calcium, iron, manganese, zinc and so on) and locks them out from use. In effect, it starves the plant and makes it susceptible to disease. Its particular impact is to work on inhibiting specific enzymes that play a vital role in the detoxification of outside biotic elements. So, look out if it gets into the human body! This non-selective, systemic herbicide was patented in 1964, and Monsanto registered it in 1974. We now know that around twenty per cent of glyphosate that hits the foliage of plants is exuded through a plant's root system directly to the soil – where it remains.

What finally helped blow the lid on the insidious danger of glyphosate was a lengthy paper by two brave biomedical American scientists, Anthony Samsel and Stephanie Seneff. They reviewed 286 diverse papers and other material in the medical and bioscience fields to piece the story together. Their findings are of seismic proportions, but the entire agribusiness world is not even acknowledging a register on any seismic instrument.

Concerning glyphosate, Samsel and Seneff concluded, 'The industry asserts it is minimally toxic to humans, but we argue otherwise. Residues are found in the main foods of the Western diet.' Then comes a summary of the alarming and diverse evidence. 'Glyphosate enhances the damaging effects of other food-borne chemical residues and environmental toxins.' The result? 'Negative impact on the body is insidious and manifests slowly over time as inflammation damages cellular systems throughout the body.'

Alarmingly, Monsanto and others claim that glyphosate can't be harmful to humans and other mammals because it only works on plants by disrupting what is called the shikimate pathway, which is involved with the synthesis of essential amino acids. However, what they neglected or at best conveniently ignored is that, in Samsel and Seneff's words, 'this pathway *is* present in gut bacteria, which play an important and heretofore largely overlooked role in human physiology through an integrated biosemiotics relationship with the human host'. That is, what Samsel and Seneff are referring to is the wider context of the holobiont – in this case a human in combination with its associated microorganisms as a key unit of selection in evolution.

Evidence on the central importance of gut bacteria to human health and well-being is mounting by the minute. In addition to aiding digestion, we now know our gut microbiota synthesise vitamins and detoxify xenobiotics (i.e. harmful outside chemicals), and are crucial to our entire immune system, to homeostasis and to our gastrointestinal tract's permeability. Therefore, stated the brave authors, just interference alone with the vital CYP enzymes involved has multiple knock-on effects in 'disruption of aromatic amino acids by gut bacteria, as well as impairment in serum sulphate transport'. It is disruption of these elements that triggers the other disease factors.[31]

Samsel and Seneff concluded their landmark paper by stating that the consequences of this massive biological disruption are 'most of the diseases and conditions associated with a Western diet'. This involves a long and comprehensive list of all the modern terminal and chronically debilitating diseases, such as: gastrointestinal disorders, obesity, diabetes, heart disease, depression, autism, infertility, cancer and Alzheimer's disease. In their work, they add many more to the list, such as liver diseases, various other brain-related disorders such as dementia and Parkinson's disease, cachexia (muscle wasting), developmental problems and multiple myeloma.

Samsel and Seneff pointed out that all this is because 'the Western diet is a delivery system for toxic chemicals used in industrial agriculture. The diet consists primarily of processed foods based on corn, wheat, soy and sugar, consumed in high quantities. Chemical residues of insecticides, fungicides and herbicides like glyphosate contaminate the entire diet' – with the highest levels of glyphosate being found in grain and sugar crops. And this alarming situation exists without cataloguing similar impacts on bees and other farm animals and organisms crucial to a healthy functioning agro-ecosystem and indeed the entire ecosystem.

Modern society at the moment, therefore, appears skewered on a massive, multi-horned dilemma. First, Monsanto, brother chemical companies and the entire agri-business sector (including most industrial farmers) deny there is a problem. Second, almost all of the industrial agricultural system is predicated on using massive and ever-escalating amounts of glyphosate to produce food needed for an ever-increasing global population. But third, as Samsel and Seneff concluded, '[g]lyphosate is likely to be pervasive in our food supply, and, contrary to being essentially non toxic, it may in fact be the most biologically disruptive chemical in our environment.'[32]

President F. D. Roosevelt may have said in 1935 'a nation that destroys its soils destroys itself'. But on the threshold of the modern chemical revolution in industrial agriculture, there was no way, even in his wildest imagination, that he could have anticipated the full import of those words. Were we to have an equivalent of Roosevelt today, she/he would need to add this rider: 'A nation that poisons its soils also poisons its own people and their natural life-support systems, not to mention those of future generations!'

There is only one sustainable solution to this emerging disaster: that industrial agriculture, as much as is practically possible, weans itself off practices predicated on genetically modified crops, synthetic fertilisers, pesticides, herbicides, antibiotics and other pharmacological crap and swings over to proven regenerative and agroecological practices outlined in this book and elsewhere. Given how some sixty to seventy per cent of the globe's food is produced by a small-scale peasant, ecologically-based agriculture on farms averaging only five-acres or less (along with the fact of huge food wastage in the industrial system), we already know that not only can this work, but also an organic/regenerative agriculture can feed the world.

The problem, of course, is the big end of town: that raft of many of the world's biggest transnational companies and its attendant financial infrastructure in food manufacture, marketing, distribution, retail, and in industrial agricultural products. The evidence would suggest that this coterie of big chemical, big pharma, big retail, big food processors, big traders care only about shareholder value and profit, and not the well-being, and certainly not the physical and mental health, of customers, society and our sustaining Earth. Given their strategic manipulation of food products and additives (to name just salt, sugar and fat for starters), and given their knowledge of the impacts of their products on human health, these 'corporate kleptocracies' could well be also accused of multiple and hidden forms of genocide and mass human suffering.

So this won't be a top-down change. As I therefore argue throughout this book, the change needs to be bottom-up: from the consumer, the urban and small food grower, and especially linked to regenerative farmers. It needs to be an underground insurgency. As Lady Eve Balfour (one of the founders of the modern organic agriculture movement) stated in 1943,

'If the nation's health depends on the way its food is grown, then agriculture must be looked upon as one of the health services, in fact the primary health service.'[33]

This brings me back to my grandson Hamish's profound question cited earlier, as to why people have to kill things to grow things. As I reflected at the time, I could not yet tell him this glyphosate story. Far better, I decided, not to burden Hamish or his older sister Alexis with this knowledge, but instead to let them taste the intense variety of saliva-stimulating, nutrient-dense home-grown foods and the rich pleasures of mulberry juice dribbling down one's chin; appreciate bush orchid and flowering gum or the clap of wings and then straight-line flight of a bronze-wing pigeon; or chase a skink sunbaking on a rock. For it is this that helps generate a healthy body, soul and mind.

It is to 'mind' that I finally turn, and to our mental health: that other, invariably overlooked but indivisible component of a holistic health picture. This is because anyone living on the land (let alone in a nature-divorced city) has invariably encountered this 'other' element: whether in the face of drought, financial stress, or some family travail along life's journey. For me, the encounter came early.

One August day when I was four-and-a-half years old, my father took me in our old Land Rover for the three-kilometre drive to our mailbox. It was mail day, and I always looked forward to what might be in the locked blue-canvas mailbag. Plus, it was often accompanied by fresh-smelling bread and other tantalising groceries.

Back at the house, we can only assume that, on hearing the Land Rover depart, my mother immediately went to the gun cupboard and selected a shell and the 4.10 shotgun: a gun renowned as a light 'ladies' gun' and a good snake-killer. We don't know how long she took, but soon after there was a single gunshot.

To this day also, no one knows the full reason why she took her life. But we know she was sick. I have seen the letter she left, full of apology and pain; the writing like a child's: wandering and barely legible.

We know that mental ill health, in its various forms, is an often nebulous but nevertheless genuine illness. Unfortunately, research shows that in many industrial countries (including Australia) there are higher levels of mental ill health in the bush than in non-farming environments. Farming 'is associated with a unique set of characteristics that is potentially hazardous to mental health', concluded one academic study, while another stated, 'even farmers scoring positively on mental health or wellbeing measures appear more likely than non-farmers to feel hopeless about the future, have suicidal ideation or contemplate suicide ... apparent in several countries ...'[34]

There can be many reasons for this perplexing, corrosive and damaging situation: biochemical, hormonal, marital problems, debt and financial concerns, seemingly endless drought, loneliness and isolation, different forms of depression, a diagnosed serious or terminal illness, and so on. But why the incidence of increased rural ill health is so perplexing is that one would think that living out in nature and working with animals in clean air away from the noise and stresses in the city should be good for one's soul, heart and mind: after all, it is an environment we were coevolved for. Way back in 1984, E. O. Wilson popularised the word 'biophilia': 'the urge to affiliate with other forms of life'. What it described is our innate emotional affiliation to nature, which goes beyond nature's provision of basic needs to include our needs for aesthetic, intellectual, cognitive and spiritual meaning, satisfaction and well-being. Yet clearly, many rural people are not experiencing 'biophilia', or else are not having their deeper needs met by it.[35]

Increasingly now, I am beginning to feel that one factor in the cause of rural ill health is somehow connected to the way we farmers treat our landscapes. This is because, like most in the city, we have become largely divorced from intimate connections with nature and with local human community. Can this be because we are trapped within the confines of the Mechanical mind, and that our deep inner psyches are in revolt?

My mind was cast back to Peter and Pam Cooks in South Australia and their list of thirty-eight chemicals for the six-month cropping season, and the image of their consultant (whom they eventually nicknamed 'Dr Death') accelerating away in anger on his motorbike after he had been sacked. And again my grandson's question popped into my head: 'Grandpa, why do you have to kill things to grow things?' In other words, why do we perpetrate so much violence, even hate, on the Mother who sustains us? The answer could be simpler than we think: our deep embedding in the Mechanical mind. And somewhere, in our ancient Organic minds, I believe our psyches are rebelling. Subconsciously also, I believe we can never fully belong in our landscapes unless we acknowledge, and become reconciled with, the previous Indigenous managers of our land, whom our culture dispossessed.

I will not attempt here to address the huge and complex issue of mental health, except to touch on our divorce from nature as being one piece of the puzzle.

At home, our family regularly takes a walk that follows a circuit of around four kilometres or so. It first traverses granite boulder country interspersed through grassland and revegetation areas, and gives views of fifty kilometres or more to the west and the main range in its tiered ranks of fading blue. Then it meanders through ribbon gum, peppermint and candlebark forest, passes our 1850s woolshed and blacksmith complex, and thence returns via open basalt grasslands and distant views to the smoke-blue coastal range in the east.

The walk provides a chance to talk, laugh and plan, and to show my grandchildren Alexis and Hamish the signs of animal visitations of that day: to identify bird calls, speculate on whether the micro-prints beside a dusty bowl were pipit or quail, and to point out wildflower, native fruit and seed and other endless bequests left by nature.

Last winter, we regularly saw on this circuit a particularly quiet but large, dark and solid wallaroo, who spent much time on the green barley grass around the woolshed yards. For some inexplicable reason, the children named him 'Mr Kev'. There were frequent discussions as to his solitary nature, and speculation on his friends, family and habits. But somehow his 'christening' had given him substance and personality, and the children now saw him as a distinct individual, with his own habits, personality, extended family and existential presence.

The sad thing is that few children have this privilege.

Recently, the not-for-profit organisation Planet Ark Environmental Foundation conducted public surveys into Australian children's contact with nature. The findings are disturbing. In Planet Ark's words, they highlight 'a dramatic shift in childhood activity from outdoor play to indoor activity in the space of one generation'.[36] That is, interaction with nature has plummeted. The survey data reveals that more than one in four children (27 per cent) have never climbed a tree; that nearly one in three (31 per cent) have never planted or cared for trees or shrubs; and that only one in ten children play outside once a week or more. Similar findings relate to participation in bushwalking, camping and so on.

The reasons are many. Like the USA, Australia now boasts the largest homes in the world, and this is closely tied to an increasing trend in suburbia of having small backyards, and, universally, to the fact that 'our love affair with television and the internet shows no sign of abating'. 'For every hour we spend on outdoor recreation,' concludes Planet Ark, 'we spend just over seven hours in front of screens watching television or accessing the internet.' As the Planet Ark authors quip, 'we can't see the forest for the screens'.[37]

These trends fly in the face of worldwide evidence that spending time outdoors and in nature is good for our mental and physical health – and in so many different ways. In his powerful book *Last Child in the Woods: Saving Our Children from Nature-Deficit Disorder*, chairman of the Children and Nature Network Richard Louv makes a compelling case for what he calls 'nature-deficit disorder' among modern-day children. Louv states, 'The children and nature movement is fuelled by this fundamental idea: the child in nature is an endangered species, and the health of children and the health of the Earth are inseparable.'

In talking about the modern situation of 'de-natured childhood' and a 'generational break from nature', '[n]ature-deficit disorder,' says Louv, 'describes the human costs of alienation from nature, among them: diminished use of the senses, attention difficulties, and higher rates of physical and emotional illnesses. The disorder can be detected in individuals, families and communities.' Moreover, he says in summarising a huge amount of powerful evidence, 'long-standing studies show a relationship between the absence, or inaccessibility, of parks and open space with high crime rates, depression, and other urban maladies'.[38]

And so it is absolutely vital we heed Richard Louv and his warning that our children are suffering from a deficiency of 'Vitamin N' (Nature).[39]

Furthermore, our divorce from nature has implications for, and causal factors contributing to, our ill health. Previously, I have discussed the huge, hitherto non-considered role of epigenetics. We evolved in and for nature, so severance from it must have major epigenetic and developmental consequences. We have seen evidence that supports how our visual and sensual environment profoundly affects our physical and mental well-being. This must have transgenerational (let alone immediate) impacts. But such deprivation can also be due to longer time frames. That is, linked to epi-genetic implications is what some researchers call 'ghosts': those evolutionary remnants of past experiences that are hard-wired into a spe-cies' nervous system. This undoubtedly links to our coevolution in Africa but, in combination with epigenetic factors, can be more recent in human history. For example, psychologist Michael Gurian believes that 'our brains are set up for an agrarian, nature-oriented existence that came into focus five-thousand years ago. Neurologically, human beings haven't caught up with today's over-stimulating environment.'[40]

There is a simple lesson here: become impoverished by stepping away from nature and the world one coevolved in, then expect negative conse-quences. For example, there are now recognised physical and behavioural impacts to overexposure to electronics in children. On the physical side is interference in developmental pathways of the plastic immature mind at cru-cial stages of early growth, and among the behavioural pitfalls are addictions and an increasing narcissistic use. This is without this pervasive technology being the overwhelming cause of seducing children away from nature.

It was back in March 1946 that, in a speech to Westminster College in Fulton, Missouri, Winston Churchill publicly called out communism and defined the inception of the Cold War. This came in his famous metaphor-ical prognostication, 'From Stettin in the Baltic to Trieste in the Adriatic an iron curtain has descended across the continent of Europe.' Well, given the Anthropocene crisis for humanity and the fact that computers, game consoles, social media, selfies and the rest are overwhelmingly diverting not just our children but also adults from engaging with nature, it could be said that an electronic curtain is descending across the Earth. This curtain is now serving to divide us not just from each other but also from the natu-ral world that bore us and sustains us. It constitutes but one more example of a Mechanical mind gone rogue.

In addition to the obvious impact of electronic seduction, there are other emerging physical consequences of becoming divorced from the natural environment we coevolved in. This, for Daniel Lieberman, is one more form of 'mismatch': which includes a 'dys-evolution' whereby our culture is now passing on disease-inducing environmental conditions. For example, there is an alarming increase in short-sightedness in children today: they don't receive enough natural light or exercise enough distance vision to fully develop eye function. Again, the hypnotic electronic screen is the great seducer.

But the story is wider than this and need just be touched on here. The key issue – due to the dys-evolution of culture and biology – is, in Lieberman's summation, 'disuse'. By this, he means, 'To grow properly, almost every part of the body needs to be stressed appropriately by interactions with the outside world'. Diseases and health problems can 'result from too little of a formerly common stimulus'.[41]

In the resultant category of mismatch diseases caused by dys-evolution, and in addition to short-sightedness, Lieberman cites such issues as osteoporosis, dementia, ill-developed jaws and teeth structure, and the modern epidemic of immune diseases and immune-related issues including ADHD, allergies, autism and asthma – and largely because our immune systems weren't sufficiently challenged at the appropriate time in early development by germs and other factors. As Lieberman quips, 'a little dirt never hurt', and 'growing up needs to be stressful' because 'we did evolve to "use it or lose it"'.[42]

In short, we coevolved in and with our environment to healthily grow and develop by interacting with our environment. And this also includes an immersion in the natural world.

Back in 1956, Rachel Carson said, 'Help your child to wonder ... It is not half so important to know as to feel' when introducing a young child to the natural world.[43] She was right. Young children do need to experience 'wonder' and to 'feel' the natural world. As Richard Louv says, 'a sense of wonder and joy in nature should be at the very centre of ecological literacy' – because only then can education be a full 'portal to the outside world'.[44]

The process of our modern alienation from nature is slow and insidious. In 1997, Susan George (at that time a high-profile anti-war activist, exposer of corporate greed and strong advocate for global social justice) wrote about 'the age of exclusion': where our so-called free market excludes people from meaningful participation in a life we were designed for.[45] Another way to put this is that much of our population is experiencing 'slow violence'.

The beginnings of solutions to this crazy dilemma seem obvious. We know the ancient coevolved mechanisms of developing children's plastic brains are being hijacked by hours of immersion in electronic gadgetry; we know that opposite immersion in nature has multiple health, developmental and social benefits; we know dozens of other indicators point to our need to re-engage with nature and meaningful biophilic, non-exclusionist, non-slow-violent lifestyles; we know from recent studies that both farmers and urban people are healthier when they engage with nature.

So it is time to throw off our 'ecophobia'; time, as academic food writers Allen and Sachs pinpointed, to subvert and change the 'current political organization of agriculture, the purpose of which is primarily to produce profits, incidentally to produce food'; time to overthrow the 'money code' as providing our only set of values; time to build community – rural and urban – and the 'civil commons'.[46]

We must return to what our bodies and minds were coevolved for: a biophilic immersion in nature, in healthy regenerative farming, in urban community and school gardens, home vegie patches and good urban and rural design. In short, a 'return to Eden', to allow our mental and physical systems to re-engage with Mother Earth. We need to shift to the Emergent mind and acknowledge that perhaps Nature does know best: because, after all, she has had a few billion years of practice to work out her systems.

As John Muir said in 1915, 'When we contemplate the whole globe as one great dewdrop, striped and dotted with continents and islands, flying through space with other stars all singing and shining together as one, the whole universe appears as an infinite storm of beauty.'[47]

So, isn't it time to gaze at the stars and seek 'to live life in radical amazement'?[48]

The last story about amazing farmers I wish to tell affirms, in a number of dimensions, Wendell Berry's insightful observation, 'To be healed we must come with all the other creatures to the feast of Creation ... the wilderness of Creation where we must go to be reborn.'[49]

This story concerns a knockabout Western Australian wheat cocky and shearer. After much travail, he had finally succeeded in living a life in 'radical amazement'. Through this and previously being stripped bare, his life had become a poignant tale of how the natural world can heal brokenness and give a reason and meaning to living.

A few years back, I was given access to a collection of frank interviews conducted in 2002 by Greening Australia. Among many outstanding pieces, one story stood out. It was that of a sixty-year-old farmer (who will remain nameless – so I will use the pseudonym 'Joe') in the Mallee wheat country of Western Australia. He grew up on his parents' wheat block and now lives on a small farm that he took up in 1958. To survive as a kid, he caught rabbits to sell, and birds' eggs also, and 'we used to … shoot parrots with a shanghai,' he said. 'I went shearing when I was sixteen,' he continued, 'and I built a little camp on it … There wasn't much cleared on it … no roads or anything like that.'

Life was tough for Joe, with endless days of hard work and then back-breaking clearing of mallee with an old tractor and roller on the weekends on his small block. 'Them days we had to clear a certain amount because they [the state government] would have taken the blocks off us,' recalled Joe. He then got married and had a family, but circumstances turned sour. 'Things have been pretty rough at times,' he concluded. He became an alcoholic, but, 'I had a good wife and she worked hard and everything … In the end, she left. She got with another fella … I still talk to her … and then I got with AA, Alcoholics Anonymous … that's a spiritual thing … that's how I live now … I mean I believe now that I don't have much control over my life, I just leave what I call my higher powers to look after most things … That's why everything's positive from here on … I do have a God. I believe that I'm part of the universe … part of the bush and everything … That's helped me with the healing … Like I'm really free. I'm a caretaker, see? … I'm just here to be part of it … You imagine you're looking after it for future generations … someone who hasn't been born.'

Joe lived alone, but what helped turn him around and find peace was his engagement with the environment he had grown up in. In only six or so years prior to his interview with Greening Australia, he took to active land and bush regeneration. This involved planting thousands of trees every year (many from seeds he collected), and he got to know his native grasses. He also placed covenants on his land (which eventually comprised thirty-seven per cent of 'bush'). 'My world is mostly the environment now,' he said. 'I'm trying to rebuild what we spoiled … I let nature work for me … because nature sorts it out … I feel like I'm part of the bush, so I suppose that means indigenous, yeah.'

However, it was the critically endangered mallee fowl that particularly touched Joe's heart. 'Like I don't control the Mallee fowl,' he said. 'That's part of the pattern … they like being here. I've seen them here, like every morning. I sleep outside under a verandah, and every morning one or two of them wander past and have a bit of a scratch. They live here. There's always half a dozen outside … I control the foxes and dogs … Some people might say I'm giving them a chance, but they're probably giving me a chance … There's something that we can't see. And that's what causes it, those things that we don't see.'

At the end of the interview, clearly in reflective mood, Joe concluded, 'And I get woken up by the Mallee fowls every morning … You can go all your life and never ever smell a flower, and then all of a sudden you realise that they're there and the whole world's full of them.'

CHAPTER 22

Towards an Emergent Future

One bird sitting quietly on a leafless tree branch
Sang the world into a premature spring,
Because he gave all of his strength to his art, and fretted not
Of the yield he would bring.

LULU CURME BRETNALL, *c.*1922

It is May 2016 – the last stage of this book journey. In the pre-dawn dark, I walk again to my office. Frost is in the air and glistens on the ground. Dry elm and gum leaves crunch underfoot, for we have missed our autumn rains. May has been more like August: day after day of cold, drying winds as fronts sweep up from the Southern Ocean and Antarctica beyond. They deposit rain on the cropping lands west of the Great Divide in Victoria and New South Wales, and early snow on the mountains. This means relentless, drying winds across our Monaro tablelands. It also means our growing season has closed. When the winds drop at night, in the morning hard frosts turn any green remnant to brown and black.

Consequently, I have spent hours calculating available dry matter of feed (expressed as available stock-days per acre). This is the equivalent of a factory inventory, except it has to last beyond mid-September, when growth begins again. The result is that, after numerous calculations, I sell down 1000 dry sheep equivalents – while the animals are still fat and markets still buoyant. This enables our holistic grazing plan to work comfortably through to October.

But life has not shut down, despite the absence of green across our subdued, variegated brown and khaki landscape. After the huge 2003 bushfires in the mountains, yellow-tailed black cockatoos began visiting regularly. This season, a small group have returned to feed on green pine cones. They float down like lazy kites, large spinnaker tails and pinions buffeted as they utter high, companionable cries. In the walnut tree, their cousins – sulphur-crested cockatoos and demure crimson rosellas – quietly clean up the last of the nuts. While across paddocks and in timber, flame robins flit like aerial jetsam, choughs squabble interminably with magpies, and 'blow-away grass' seed heads do just that: stacking halfway up fences.

However, as I reach my office this early morning, it is not just my familiar companions I look to: not just the Milky Way, Southern Cross and Seven Ice Sisters; nor the whispering microbats, the early yellow robin and the noisy kookaburras welcoming the day.

No, as I shut the door to my writing and thinking space, I sit down to gather my thoughts in an attempt to put this story in context. Inevitably, my eyes travel to the bookshelves and their rows of old friends: a global brains-trust across time, landscape, continent and distance, each with wisdom and fertile seeds potentiated for sprouting. Most of them in one way or another have both sung praises to Mother Earth and also been the context-shapers, the urgers and prompters. And yet it is not they who have been the real inspiration for this book. The real galvanisers for this story have been the many dozens of free-thinking regenerative farmers and like-minded urban associates who have transformed their land and/or their communities and themselves. It was they who forced me to confront my own parallel but delayed, bumbling journey.

Moreover, in their positive, alternative journeys, they had exposed the negative: that why we trash instead of nurturing and regenerating our farms and landscapes comes down to the dominant Mechanical mind. Our entrenchment in its belief systems stands in the way of becoming ecologically literate, of developing an understanding of how our landscapes function. It is this that determines whether our farming activities end up being destructive or constructively creative.

The Mechanical mind as expressed in industrial agriculture has led to our lack of empathy and care for a living, diverse world in this timeless Gondwanan continent. It is why we continue acting as aliens in a perceived foreign and insensitive environment, and why there is still so much racial prejudice against our ancient Indigenous peoples.

In contrast, the leaders of this insurgent regenerative movement have come to realise the fact (as my friend, Indigenous leader Kerry Arabena articulated) that 'as we are all indigenous to the Universe, we are deeply and profoundly connected', and this 'requires us to acknowledge the natural and Universe laws in which our lives are lived'.[1] As I and many others have found, a powerful way to come to empathically, intuitively and rationally comprehend the depth of these 'natural and Universe laws' is to have a sincere and deep dialogue with Indigenous or First Nation peoples. In this context, Kerry Arabena states, 'We have to be terrified by what we have done, but not without hope ... To care for country is all our business, it is the necessary and transformative element of a reconciliation agenda in Australia, and the world'.[2]

But above all, and whether we like it or not, we humans still live in a self-organising, self-making universe. For self-making is the essential characteristic that distinguishes living from non-living systems: systems (whether they be ecological or other systems or organisms, or even the human brain and Earth itself) that continuously recreate themselves within a boundary of their own making.

This then leads to a recognition that such systems – and in this case humans and their society – are indivisibly connected to their environments (what leading thinkers in this field Humberto Maturana and Francisco Varela call 'structural coupling'). What this means is that the internal structure and driving force of self-organisation within a system or organism are what creatively drive the changes, adjustments and adaptation.[3]

Crucially, the evolutionary process that has led to such self-making systems, and is indivisible from them, is neither random nor determined but creative. This reality, therefore, is a total rebuttal of the nihilistic mechanism and meaninglessness of 'chance and necessity' as popularised by thinkers from Jacques Monod to Richard Dawkins. This creative, self-making process is also one that gives higher meaning, values and a spiritual dimension to our lives: a reality I found in my investigations of regenerative farmers who had undergone a mind-shift from the Mechanical to a newly Emergent mind.

Therefore, part of an accompanying post-Mechanical, post-ancient Organic mind – the Emergent mind – comprises a realisation that we humans are indivisibly part of all life and Earth's living systems. The story of our own emergent evolution reveals this: that our 'descent' was and is not simply linear. As Lynn Margulis and Dorion Sagan concluded, 'Life did not take the globe by combat, but by networking. Life forms multiplied and complexified by co-opting others, not just by killing them.'[4] It thus appears the Emergent mind is the next stage in our cultural-biological evolution.

But the irony is that certain elements of the emergent properties of the Mechanical mind have now mutated and become malignant, and that these now-dominant aspects of the Mechanical mind have become closed. Such a Mechanical mind, therefore, is no longer truly 'emergent' – given that the entire process of self-organisation and its use of novelty and emergent properties is to re-equilibrate after disturbances in order to maintain an organism's or system's survival and enhance resilience. That is why it is clear to me that the Mechanical mind has gone rogue. Moreover, the alarming reality is that if we don't address our key role in precipitating the Anthropocene, then this once novel emergent property will be seen as the greatest of all failed experiments: remembering that life's history is littered with extinct, ultimately non-adapted species.

In his groundbreaking book, *Design with Nature*, back in 1971 the provocative pioneer of ecological and urban design Ian McHarg concluded with a searching question. He stated:

> ... nature is our creation and we shall dominate and subjugate
> it, for that is our divine destiny. We relinquished integration
> when we found consciousness and in rejection we move to
> disintegration. But the prospect offers us several choices – the
> quickest is annihilation as anthropocentric man produces the
> holocaust ... The land will be raped and creatures extirpated
> because the insistencies will be so loud; who can plan for the
> long term when survival is today? The cities will grow as they
> have, enlarging the pathology of their hearts, growing into
> necropoles [sic]. What other prospect can you see?[5]

In answer to this rhetorical question concerning both planning and acting for the long term in order to create positive prospects, my book provides a positive answer and a pathway to a bottom-up revolution. For the multiple solutions presented here – driven by a new way of thinking about our landscapes, food, society, personal relationships and a new way of open-ended, Earth-empathic thinking – allow for the positive and creative enablement of self-organising systems to regenerate farming, urban, social and global ecosystems.

Therefore, the urgent challenge and question for humankind is 'Can we rapidly enough shift our modern, industrial, economic-rationalist, selfish, consumeristic, non-biophilic world view to one that nurtures instead of destroys?' That is, as Thomas Berry articulated in 1999, 'any recovery of the natural world in its full splendour will require not only a new economic system but a conversion experience deep in the psychic structure of the human'.[6]

There are two main reasons why I am optimistic about humankind making this conversion. First, it is clear to me that, as evidenced by not just regenerative farmers but also a host of Earth- and nature-empathic thinkers, the third great stage of the human mind is spontaneously already present (albeit as a small and marginalised presence, and not yet fully revealed). Moreover, unlike the previous Organic and Mechanical minds, which were also the result of emergent properties (both in biological and cultural evolutionary terms), this new and third mind – the Emergent mind – seems to lie in another dimension.

It involves a change of consciousness for our epoch. This change is embedded in a readiness to step back and let self-organisation processes be expressed (that is, to allow emergent properties to appear). In other words, in those who have transformed there appears to be a deep biophilic capacity to implicitly see ourselves as in unity with Earth and its systems and with all other creatures (and not separate to them or dominator over them). As with the appearance of the Organic and Mechanical minds, this shift to an Emergent mind therefore creates a different relationship between people and their landscapes and wider natural systems. This encompasses more than just functional ecology and landscape literacy, because it includes social elements such as food safety, public health and wider issues to do with power, justice and equity.

Therefore, an Emergent mind combines elements of the previous Organic and Mechanical minds, but its true difference is an openness to the ongoing processes of emergence and self-organisation. From this, as regenerative farmer David Marsh expressed to me recently, 'watching self-organisation occur (in even as short a time as seventeen years in our case), indicates it is nature's/Earth's store of ecological knowledge from the evolutionary past that is key to unleashing its power'.[7]

Moreover, by the very nature of such processes, we need to appreciate that, unlike the fixedness of the Mechanical mind, which was all too prone to going rogue, the newly appearing Emergent mind must therefore be an interim mind, ever on the way to becoming something else; something constantly capable of adapting to changing circumstances but always in keeping with the necessity of sustaining/regenerating life's and Earth's support systems. In this respect, such a mind represents a new step in evolution, for it entails a new way of thinking, feeling and reacting that is empathically embedded in life's evolved systems, rather than simply changing from one set of constructs to another (as in the previous two minds).

Therefore, in common with the Organic mind, an Emergent mind has a capacity for apprehending the 'sacred', or the spiritual/transcendent. With this and a deep biophilia goes a high level of moral consciousness. And finally, making such a new, nature-evolved Emergent mind more difficult to grasp, it can only emerge through a process and so can't be defined but only described and part-explained. Making this even harder to grasp still is that an Emergent mind involves us humans developing the capacity to learn from life's lessons over evolutionary time. In contrast to our modern mechanical culture, which seeks a technological 'fix', this will often mean stopping doing things, which is doubly a challenge to the reigning Mechanical mind.

But in addition to what I see as the spontaneous appearance ('emergence') of an Emergent mind, there is a second reason why I believe this new mind will continue to appear and evolve. This reason is that enormous crisis usually also brings opportunity and confrontation with the status quo, which can stimulate change for the future. As we've seen, in Australian bush vernacular, 'the thinking that got us in the shit won't get us out of it'. Having got ourselves into the Anthropocene, only the best of our evolving modern brains and a totally new and different thinking will get us out of it – through a truly Emergent mind.

This is revealed in the numerous examples in this book, which comprise a plethora of solutions 'from the ground up' via a wide application of various forms of rapidly evolving agricultural practices and thinking. And this lead in agriculture is occurring across every continent. It is bringing post-Mechanical, Emergent thinking to the spheres of healthy eating and meaningful design. It is being embraced in urban centres and in movements such as the drive for Fair Food, and in organisations such as La Via Campesina. Thus, an Emergent mind is fuelling the global peasant-farmer-driven revolt against the Mechanical mind and its dominant globalised, degrading powers.

Significantly, the new regenerative agriculture is not just being practiced on small backyard plots or half-acre (quarter-hectare) allotments (for these too are part of the solution, and some of the practices apply and are indeed being practised there). No, the approach is also being applied on tens of millions of acres, while elements of it have the potential to be applied equally broadly worldwide, as I have described in this book. These include holistic grazing management, grassland grains, pasture cropping and natural intelligence agriculture, a new agroforestry, farmer-managed natural regeneration, biological agriculture and many others.

Therefore, this revolution has the potential to make massive inroads into addressing the key Anthropocene challenges of climate change, biodiversity loss, distortions to the nitrogen, phosphorus and other biogeochemical cycles, land degradation, malfunctioning of the water cycle, and so on. In addition, regenerative agriculture has astounding potential to address human ill health; to therefore address the huge burden of modern health costs on society; to make farmers more profitable and free of the clutches of banks and large multinational chemical and energy suppliers; and to free farmers and urban cousins alike from capture by the merciless behemoths of agribusiness, commodity trading, financing, retailing, food processing and the like.

Consequently, we need to be aware that a new regenerative agriculture is subversive: but in a truly constructive way. That is why I call it an 'underground insurgency', for it is a bottom-up revolution instigated by a new mind.

The process of transformation does not occur overnight. Nor is it without deep self-examination and considerable emotion. This is for two reasons. First, we need to confront the great conundrum of humankind: our lack of understanding of ourselves. And second, as outlined in Chapter 19, it is because the shift to an Emergent mind involves a major cognitive and affective (emotions and feelings) leap. For the metaphor switch from the

Mechanical to an Emergent mind involves dismantling and then rebuilding the entire superstructure of one's belief system and world view.

The release of energy in this transformation process fuels active searching for learning about alternative practices and also stirs the emotions. It is these that become the crucial motivator and energiser of change. Also triggered is a strong ethical and moral element and often an openness to the spiritual dimension. Such transformational change, therefore, is a true 'emancipatory learning': a true 'symmetry-breaking' experience. Almost invariably, the end result of this shift to a new Emergent mind is the development of a greater interdependent relationship with, and thus a compassion for, society and nature – aligned with a distinctively holistic approach to knowledge.

David Orr – outstanding proponent of the need for greater ecological literacy – in a 2003 university seminar took this thinking to the next step. He pointed out, 'The transition to sustainability will require learning how to recognise and resolve divergent problems, which is to say a higher level of spiritual awareness.' The term 'divergent problems' came from E. F. Schumacher in his famous book *Small Is Beautiful* of 1973, who said such problems could not be solved by rational means alone. Therefore, said Orr, 'By whatever name, something akin to spiritual renewal is the *sine qua non* of the transition to sustainability.' But he also pointed out that 'the spiritual acumen necessary to solve divergent problems posed by the transition to sustainability cannot be just a return to some simplistic religious faith of an earlier time. It must be founded on a higher order of awareness that honours mystery, science, life and death.'[8]

People who have already attained a deeper spiritual state of mind (and for various reasons) can be more readily open to recognise or discover emergent properties that may aid the regeneration of nature and society. This feature is exemplified by many regenerative farmers discussed in this book.

As one such person, the leading American regenerative farmer, scientist, Christian minister and philosopher Fred Kirschenmann says, 'Since Descartes insisted that facts and values must be kept in separate worlds, both scientists and farmers have been reluctant to use the word "spirituality" in connection with anything having to do with science or agriculture.' But he and many others – including Aldo Leopold, Wendell Berry, Wes Jackson and the biophilic theological giant Thomas Berry – do instead call for an embracement of the spiritual. Morris Berman, in his powerful book *The*

Reenchantment of the World, is unabashed about such a heart-and-mind shift that 'would embrace value without sacrificing fact'. Says Kirschenmann, 'the "spiritual" is simply a way of understanding our world that acknowledges the connection and relationship to the rest of the expanding universe'. And so he concludes, 'Doing agriculture within the context of spirituality will lead us to pay attention to all of the relationships in which farming is involved.'[9] In short, this is another element of our 'structural coupling' with the environment that sustains us.

For me, therefore, a key trait of an Emergent mind is the attainment of a new, higher level of moral consciousness, which requires a re-examination of the 'sacred'. It also requires unconditional love.

I believe that love is *the* essential ingredient in human and human–Earth relationships. It is clear that pouring herbicide on the Earth is not an act of love, nor is aggressive ploughing, clear-felling healthy forests or the simplification into monocultures of complex creative systems, nor locking up animals in confined cages and feedlots and stuffing them with food and additives they were not coevolved for. Instead, these activities represent a culture of death. Moreover, an absence of love is seen in the ongoing colonial psyche in Australia and its lack of remorse for our Indigenous people's loss of sacred country. It is also seen in an ongoing deep racial prejudice and lack of empathy (which means a lack of reflection about the past and thus a lack of recognition and mutual respect). Until there is true reconciliation and appropriate reparation and acknowledgement of the past injustices with Indigenous people, we modern non-Indigenous Australians won't be able to belong here; to truly set deep roots and our psychic selves in this ancient country; to truly become structurally coupled with the Australian earth.[10]

Being of an Emergent mind means un-learning many things. Thus love in creative landscape management ironically entails stepping back and allowing self-organisation and creativity to function. It means relinquishing the need to control that is inculcated by a Mechanical mind, and instead embracing the deeper connections to Earth and its universe (even, for some, pursuing subtle energies and ancient pathways), and to the spiritual.

In 1977, Lady Eve Balfour told an organic-farming conference, 'We cannot escape from the ethical and spiritual values of life for they are part of wholeness.' If we follow through the new thinking on self-organisation and emergentism, then this, in a biological context, also opens up the idea of love and spirituality being an imperative for our very survival.

So, how do we create change? As quoted in the frontispiece of this book, J. L. Thompson exhorts us to exercise 'considerable imagination, critical thinking, subversion and undutiful behaviour' so as 'to destabilize and de-construct the authority of the inevitable'.

For me, therefore, one fact is clear. As David Korten wrote, 'The Great Turning … requires reframing the cultural stories by which we define our human nature, purpose and possibilities.' That is, as he explains, 'Change begins with a story of the wonder and beauty of life and the cosmos', stories 'so compelling as to displace in the public mind the story of Sacred Money and Markets'.[11]

However, change can't be forced either. Surprisingly, I find myself agreeing with one of the 'fathers' of modern economic rationalism: Milton Friedman. In his signature book *Capitalism and Freedom*, he observed, 'Only a crisis – actual or perceived – produces real change.' He then pointed out, 'When that crisis occurs, the actions that are taken depend on the ideas that are lying around. That, I believe, is our basic function: to develop alternatives to existing policies, to keep them alive and available until the politically impossible becomes politically inevitable.'[12]

As we can't force change, then instead we have to make sure that regenerative agriculture readily presents itself as a viable alternative for whenever people are ready for change. I believe the key to this is telling stories or narratives that are meaningful, substantive and relevant depictions of a new reality. And if ever there was a time for the citizens of this Earth to be galvanised by *story* as we hurtle headlong into the Anthropocene crisis, then that time is now. For undoubtedly the four horsemen of the apocalypse have been let loose and are already moving from a canter into a gallop.

Significantly, as the Anthropocene crisis continues to express itself it may, paradoxically, be the very factor that will trigger wide-scale change. Past 'axial' ages of change in human history (such as the periods of agricultural domestication; the eighth and third centuries BC in Greece, Palestine, Persia, India and China; and the Renaissance in Europe, followed by the scientific, industrial and capitalist revolutions that led to the entrenchment of the Mechanical Mind) were all moments of massive change or collapse:

shifts in time and mind that led to humanity seeing the world in a different way and thus with a different 'mind'.

Such a moment is now, with the greatest crisis and potential impending collapse of modern civilisation that humanity has ever seen – and, if it occurs, almost certainly the last it will ever witness. And as Paul Kingsnorth (nature writer, poet and co-founder of the Dark Mountain Project) states in what is an echo of Martin Luther King cited in my Introduction, entering the Anthropocene means, 'It is too late now to plan for the future or to issue warnings about it. The future is here. We are living in it'. And this is because '[w]e tell a story that we can mold the world to the needs of the self, rather than molding the self to the needs of the world.'[13]

'Story' is thus the touchstone for our hearts and minds because it is fundamental to our cognitive, metaphorical, symbolic mind that emerged some 250,000 or so years ago. Intervening at this seminal point of our psyches is therefore key. As the courageous, pioneering systems-thinker and environmentalist Donella Meadows said in terms of complex creative systems, 'People who manage to intervene in systems at the level of paradigms hit a leverage point that totally transforms systems.'[14]

So, central to this entire process of transformation is that humans are made for 'story'. Paul Kingsnorth says that the story that will save us – the one 'by which our species will live or die', the one that will come through 'a radical alteration in people's lived experience' due to 'a crisis that forces people up against the consequences of what we have done' – is a form of an old organic connection to Earth. Moreover, it is one on the spiritual–ethical level that may well be 'an echo-location from the Earth'. But he also says, concerning this 'oldest of tales', that 'we don't know ... what to do with it', that 'we don't know how we might even start to live it again.'[15]

Well, I beg to differ. As I said earlier, the multiple stories in this book, expressed by increasing numbers of Emergent minds, reveal the story is being taken on-board – and deeply – and this is what is leading to transformative change and regeneration.

In the case of regenerative agriculture and a rural-urban-interface transformation, instead of a story about plunder, poison, productivity, meaningless consumption and 'growth', those in the process of change can instead imbibe stories about regenerating landscapes, beauty and natural health, sustainable livelihoods, community, family and human health.

Ultimately, therefore, it is a story about Mother Earth and her systems and our deep co-dependency with these.

What storylines do for farmers and others making the regenerative shift to a new Emergent mind and to a new Earth and its biotic and human communities, is provide them with a set of symbolic references that suggest a common understanding and 'the creation of coalitions'. Moreover, storylines provide a shared language and so are instrumental in developing a shared discourse and building a community of discourse.

In many respects, that is what this book is about. For it presents a plethora of examples that aren't just 'lying around' but instead were painstakingly developed, are being constantly refined, and are proven to be working so as to profoundly regenerate Earth and its systems. I believe it is in this that our hope lies: that crisis will galvanise us to seek this Great Story of our salvation.

Thomas Berry stated in his book *The Dream of the Earth*:

> For peoples, generally, their story of the universe and the human role in the universe is their primary source of intelligibility and value ... The deepest crises experienced by any society are those moments of change when the current story becomes inadequate for meeting the survival demands of a present situation.

Utterly relevant to this story about a new agriculture, Berry points out, 'Our story not only interprets the past, it also guides and inspires our shaping of the future.'[16]

The future that Berry envisions is of humanity and its planet moving into a new 'geobiological' era of Earth: what he calls the Ecozoic era. From the evidence presented in this book and elsewhere, it is my belief that a newly Emergent mind can enable us to move from the Anthropocene to the Ecozoic era by 'furthering our relationship' with Mother Earth and its self-organising systems.

Berry concludes in his definitive book *The Great Work: Our Way into the Future*, 'Earth as a biospiritual planet must become for us the basic referent in identifying our own future.' It is clear to me that this means us collectively shifting to an Emergent mind for the next stage of our existence.[17]

Pondering all these stories and the way they have coalesced into one large Emergent story, just after dawn I have walked up a hill above our farm to sit beside the bole of a favourite old snow gum. It is a still autumn morning, late May 2016. My position provides a view across the farm below and wider region beyond. Frogs are calling across the landscape, which glistens with dew.

Beneath me, beyond the homestead and frosted, dun-brown native grasses, is Buckley's Lake – or what a few local Ngarigo people still know as Lake Bundawindirri. This morning, it reflects the blue sky. My attention is drawn to a large mob of around a hundred cawing crows who have taken off from their grub hunting. They gravitate up high and their blackness becomes framed against the light-blue lake and sky, the deep-blue hills and the creamy grass heads and variegated brown landscape: a chiaroscuro of colour and movement.

But what my eye finally turns to is a land bedecked in silk: a million glistening spiderwebs, virtually one to each grass clump. It is like a prime-val scaffold that cloaks and entwines the landscape, a silken songline connecting the Earth and sky, horizon to horizon. And, close by, I marvel at the brilliant construction of a web. I then remember some research I read the other day about how some spiders can customise their webs to ensure they get the diet they need (bigger catching area, smaller mesh size for small flies; bigger mesh size, stronger and stickier for grasshoppers and crickets). This put me in mind of Les Murray's 'tankstand spider', which 'adds a spittle thread to her portrait of her soul'.[18]

However, what are the spiders really telling me? They are an indicator that our grassland is healthier because of our holistic grazing regime. To sustain millions of spiders, there must be a corresponding diversity in the food chain, and healthy landscape function above and below ground. The fact we no longer get regular devastating wingless-grasshopper plagues is just another indicator of increasing health.

Having full grass cover and floral diversity that is never eaten into the ground has other bonuses. With every dew or frost, there is a huge increase in 3D collective 'roof area' for moisture to condense on and then run to earth: perhaps an extra millimetre or two of moisture each night. Below ground, in a burgeoning mass of life and activity that is tenfold that above ground, fungi, bacteria and other organisms have begun to create and sustain an entirely different, living, absorbent soil structure: the very heart and essence of healthy farming and landscape function. And so, when the big rains come – even five to six inches (125 to 150 millimetres) with an intense low off the east coast – we get little run-off owing to absorbent soils.

The spiders, birds and absorbent soils are our reward. So too is the energy and life I can feel and see across our landscape. As a kid, we had no swamp wallabies nor wallaroos. Now, there are scores of them. With their cute, anvil faces and little paws and ears, the wallabies placidly examine me each morning. Only the other day, we had the excitement of discovering a dunnart (a small, hopping marsupial mouse), a native bush rat (on a wild-life camera), three new frog species, and two previously unrecorded woodland birds on our farm. And after collecting blackthorn seeds the other week, when I opened the plastic bag at home it was crawling with at least eight different micro-spider species, of weird colours and patterns. I was in awe. Since giving it a chance, just how much diversity does our land-scape now hold that I am completely unaware of? But then I thought, 'So what?' It doesn't have to be all catalogued. Knowing it is there and func-tioning is terrific in itself, as is growing healthy, chemical-free, nutrient-dense food and fibre off our sun-driven landscape.

So I feel I am getting there – more closely attuned (or 'structurally coupled') to the healthy functioning of our landscape and the entirety of our surrounding natural world despite being a slow learner. My regret now is that I am running out of time with so much to do. I'm in need of another lifetime now that I'm only just graduating from kindergarten.

Sitting beside the snow gum, feeling its cool, knobbly bole, I reflect on our modified landscape, which I know is slowly returning to health. My shift in management thinking to begin catalysing the turnaround was largely due to the host of wonderful regenerative farmers I was privileged to learn from. They enabled me to begin to grasp how landscapes function and thus to gain some ecological literacy. Yet this is a literacy still not taught to farmers, and certainly not in traditional courses at university or agricultural college.

I also realised that two other factors were involved in me starting to regenerate our landscape. The first of these was a heart-and-mind issue. My innate biophilia was enabled to be expressed so that in time I saw my role as an 'enabler' and not a 'dominator'. And so I am learning to further love and bless our landscape, not treat it as an unfeeling, inanimate substrate. The final factor was my development of friendships with some remarkable Indigenous people. This allowed me to gain some understanding of their Organic approach to, and nurturing of, country: a concept that anthropologist Deborah Bird Rose so poignantly defined as 'a nourishing terrain'.

I think back in particular to my friend Rod Mason (Ngarigo senior lawman) and his embracing of our kurrajong tree, accompanied by his gentle, drip-fed lessons about country and cool-burning and his people's alive and present spirit world and energies in the landscape. He and his people were and still are indivisibly structurally coupled to country – to the substrate and ether from which they are derived and which sustains them, and for which they are ethically, morally and spiritually responsible. And so, as I've said, until we come to acknowledge our Indigenous people's dispossession and collaborate with them to jointly regenerate and care for country, we can never achieve proper reconciliation between people and with country. In conjunction with the Mechanical mind with its subterranean racial prejudice, this remains one of the great blockers of a societal shift to an Emergent, Earth-loving mind in our nation: a blocker to us collectively addressing the Anthropocene and thus to any chance of surviving and moving into an Emergent Ecozoic era.

I now realise also that we cannot return our landscapes to a state that existed before the invasion of the Europeans. Self-organising systems don't work like that. If allowed to regenerate following gross disturbance, landscapes will invariably give rise to different and novel species and properties. The secret, therefore, is to simply restore healthy landscape function and allow nature to do the rest.

In Chapter 1, I discussed the unique 'writing' of the Australian landscape: its different and ancient patterns; its nuances; its startling beauty; and its natural innovations and adaptations. I then floated the idea of a natural 'palimpsest': the term for a manuscript on which the original writing has been effaced to make room for a second writing. In this context, I alluded to the fact that 50,000-plus years of Indigenous management of Australia's 'Great Estate' had imposed a human organic palimpsest; that this had then largely been erased; and on the fragile Earth instead had been imprinted a brutal writing of the Mechanical-industrial mind. In turn, I was laying the groundwork to suggest that there are the beginnings – through the wonderful examples of regenerative agriculture in this book

– of a new, post-Mechanical, Emergent-mind palimpsest that involves restoring health to the landscape.

But now I realise 'palimpsest' is an inappropriate metaphor. If we as regenerative landscape managers enable self-organisation, we don't write the inscription. Nature does that, through throwing up emergent properties (and via such iterative, creative functions as epigenetics and other mechanisms), which allow for health-giving self-organisation. No one can exactly predict what will happen. 'Inscribing' (even in a healthy manner on landscapes) still smacks of a controlling Mechanical hubris, an attempt to impose a new 'writing' on a substrate without understanding its ancient, coevolved, self-organising 'imprint' and functions. Instead, we need to realise that Mother Nature is in charge. Her work can't be pre-scripted, and at best we can be empathic, collaborative 'enablers' as part of an indivisible partnership in the creation of a new 'nature-writing'.

In this respect, much of the lead is coming from regenerative farmers who are the subject of this book, and who, through undergoing personal transformation, have experienced profound and life-changing transitions. Stimulated often by crisis, these farmers experienced a sort of psychological 'punctuated equilibrium', a tipping point that precipitated a phase-shift in consciousness to a newly Emergent mind.

The overwhelming issue, therefore, is about managing ourselves: about being able to change and transform; about truly coming to 'know ourselves'. If we can do this at this moment of our greatest crisis, then, as Thomas Berry says, we can 'be present on the earth as a beneficial and not destructive species'.

So on this late autumn morning as I sit by the old snow gum, watching mist rise off hydrated country and the glistening silk of spiderwebs, and as the crows disperse while currawongs silently loop down towards the homestead for a day of noisy berry scrounging, my mind inevitably wanders.

It drifts to stories of the past and a hard-to-imagine fully hydrated Monaro. To a story Rod Mason told me of his great-grandfather in the 1860s or thereabouts spearing a jabiru stork on Lake Bundawindirri. And in my mind's eye, not only can I see small bands of past Ngarigo people travelling through the grasslands and woodlands, pausing to fire-stick a patch here and there, but I can also imagine down by the waters of a brimming Lake Bundawindirri a scene of vibrant activity: of jabiru storks, magpie geese and brolgas, egrets and herons; huge flocks of blue and green budgerigars; swans; duck of many kinds; and waterhen and coot, pelican, cormorant and water rats. And I can even glimpse the dark shapes of enormous two-tonne diprotodonts, wading out to forage on rushes and other succulent water plants.

This landscape below me would have changed so many times over the vast reaches of millennia since the last lava flowed some forty-odd million years ago: warm and hot, then ice cold and characterised, in turn, by drought, trees and shrubs, grasslands and short-tussock and sedge – a kaleidoscope of different fauna, flora, microbiota and changing colours, scents and sounds.

Such changes are never-ending, each one a new creative expression – a new nature-writing ineffably evoked on and through the landscape by varied temperature, levels of moisture, day lengths, soil structure, and the varied activity of diverse organisms, from invisible microbes to ancient megafauna. But also by fire, and by humans of the Organic, the Mechanical and the newly Emergent minds.

And then suddenly, out of the reed beds of Lake Bundawindirri, I hear a bird calling. Swaying on the rustling stems, it seems she is giving her heart as she '[s]ings the world into a new spring, giving all her strength to her art and fretting not of the yield she would bring'.

For it is the beautiful, piercing song of a reed warbler. But is it a song of joy on returning again, or a song for a mate, of loneliness or for what was lost? I like to think it is a song for the Earth – a song of possibilities, an exultation that, if allowed, the Earth and its life forms and creatures can be regenerated again. Yes, she is calling us in a poignant, heartfelt cry for all creation – a metaphor for us humans to once more become the enablers, the nurturers, the lovers of Earth.

And this call of the reed warbler resonates deeply, penetrating my soul.

ACKNOWLEDGEMENTS

The old truism that it takes a village to raise a child has equal validity concerning the creation of a book such as this. For indeed it did take a 'village' of collaborators to craft and deliver this book. It is that collective of generous people and their combined wisdom and commitment of precious time and effort whom I would like to thank and acknowledge.

First to my family. To my wife, Fiona, for her love, patience, tolerance and never-ending support despite the mess of paper, strange hours, my absences on road-trips and the sometimes chaos left behind at home. I deeply thank her for her steadfastness and support, and invaluable input – you are such an inspiration, Fi! To our daughters Alison and Tanya: thank you for your insightful and valued editorial input and encouragement, and Tans, for some crucial feedback at the end; and to daughter Emily and son-in-law Andrew for their equal support, patience and holding the fort during my various absences. Not least also, I wish to mention the inspiration of my grandchildren, Alexis and Hamish, for their open-eyed wonder at the world and pain at its travails, but also their reminding presence that it is theirs and the next generations who will inherit the Earth that our cohort leaves behind.

Second, besides the heroic subjects of this book (the landscape managers and other regenerative thinkers in Australia and globally), the book was 'seeded', 'nurtured', 'watered' (via good coffee) and 'fed' in the Fenner School of Environment and Society at the Australian National University (ANU). I thank its then-director Professor Steve Dovers for his support along the way, and the federal government for a PhD scholarship – without which I couldn't have undertaken the initial stages of this project.

However, it is Professor Valerie Brown to whom I owe the largest debt for the creation of this book. She was my tutor in Human Ecology at ANU when I was an undergraduate, a wise mentor in between, and then thirty-six years later courageously backed me to gain entry to do a PhD at Fenner in the Human Ecology sector. As my subsequent PhD supervisor (that is, gentle 'handler', sounding-board, counsellor and 'prompter'), she patiently oversaw the somewhat serendipitous growth of my PhD thesis: an experience I liken to being sent as a jackaroo into a twenty-thousand acre paddock, thick with mallee, to muster a mob of stragglers (stray sheep), but sans dog or compass. By the time I emerged from the mallee with some sheep, the thesis had morphed into a book and Val had become close friend, collaborator, and an inspiration on my life's journey. It is only fitting she contributed the book's Australian foreword.

During my time at the Fenner School, there also arose two key groups of people, each creatively dealing with the big issues of the Anthropocene (and again, instigated by Val Brown). Individuals from both groups have played an integral support role in this book. To those members of ARLASH (the Alliance for Regenerating Landscapes and Social Health), I want to thank Dr John Harris, David Marsh and Dr David Freudenberger in particular, for our brain-storming sessions but also for their critique of the manuscript, and also Professor Kerry Arabena, Professor John Finnigan, Walter Jehne and Dr Richard Thackway for their similar wise counsel, friendship and support, and the many discussions on this, the biggest of all contemporary issues.

To the members of the Fenner writing group, fuelled as ever by more good coffee and our uniquely creative and productive discussions (which included constructive criticism of various chapters of my book), I also wish to express my indebtedness: especially to Dr Kate Andrews, Craig Ashhurst, Dr Liz Clarke, Dianne Dibley, Dr Dana Kelly, Dr Julia Wolfson and cross-members of ARLASH. These two groups, in bringing wide expertise across many different disciplines, epitomised the best of us collaboratively growing through a uniquely joint enterprise in transdisciplinary discussion, dialogue and social learning. Deep friendships were forged along the way.

Before I thank the many farmers / landscape managers at the heart of this story, I also wish to thank one other group of collaborators who devoted large chunks of time and energy as readers of this book. In addition to those mentioned above, I wish to thank Charlie Maslin, Bruce Maynard and Dr Dean Revell. Huge thanks go also to long-time friend Trish Dixon for her outstanding cover photograph.

Acknowledgements

My thanks also go to others in the scientific, academic and research area who were of great assistance, including: Professor Richard Bawden, Kerry Cochrane, Federico Davida, Dr Rob Dyball, Dr David Eastburn, Dr Josephine Flood, Dr Bill Foley, Professor Bill Gammage, Dr Malcolm Gill, Dr Stuart Hill, Dr Lynette Liddle, Professor Fred Provenza and Dr Nick Rose.

This now brings me to that other large group of 'village' members: the regenerative agriculture community. In particular, I thank the following for their willing cooperation in allowing me to feature their family's story in the book, and for their constructive criticism of relevant draft chapters. They include: Geoff Brown, Pam Cooks and the late Peter Cooks, Graham and Cathy Finlayson, Di and Ian Haggerty, Carol and Mervyn Hardie, David and Su Holmgren, Professor Steve Hughes, John and Robyn Ive, Walter Jehne, Beatrice and Tobias Koenig, Norman and Jennie Kroon, Terri and Dayle Lloyd, Kate and Peter Marshall, Dianna Outhred, Dr Terry McCosker, Rowan Reid, Rob and Caroline Rex, Bronwyn Rix and Rowan Wright, Allan Savory and Jody Butterfield, Colin Seis, Andrew Stewart, Jon and Vicki Taylor, David and Jane Vincent, Ron and Sue Watkins, Richard and Jenny Weatherly, John and Jan Weatherstone, and Tim Wright.

In addition, during the course of researching my PhD and then book, I interviewed and had discussions with many other collaborators in the regenerative agriculture and other fields. These, too, I wish to thank and acknowledge for their help, including: Warwick Anderson; Charlie Arnott; Sarah and Stephen Barrington; Matt Barton; Mike Bennell; Julian von Bibra; Jane and Andrew Bond; Mick Boyce; Robert Bradley; Wendy Bradshaw; Greg Brennan; John and Fiona Cameron; Jock Campbell; Phil and Nicole Chalmers; Mike Chilvers; Darryl Cluff; Tony Coote; Anna and Michael Coughlan; Ian Dickenson; Dr Saskia von Diest; Darren Doherty; Anne and Peter Downey; Richard and Janet Doyle; David Dumaresq; Tom Dumaresq; Cynthia and Tom Dunbabin; Sally Duncan; Trish Dunnicliff and the late John Dunnicliff, and Jane Armstrong; Lauren van Dyke; Judi Earl; Tom Edgell; Dr Jason Emms; John Feehan; Cecily and John Fenton; Craig Forsyth; the late John Gallard; Bruce Gardnier; Rich Gilmore, Dave Hinchly and Fergus McDonald from The Nature Conservancy; Paul Griffiths; Richard Groom; Erica Gundry and family and the late George Gundry; Graham Hand; Vince Heffernan; Rob, Cathy and Mike Henry; Dave Hewlett; Bill Hill; Tammi Jonas; Shane Joyce; Nigel Kerin; Louisa and Michael Kiely; Brett Kissell; Moira Lanzarin and Mike and Clair O'Brien; Michael Lloyd; Angus and Ric Maurice; Gary McDouall; Bob McFarlane and the late Erroly McFarlane; Dr Patrick McManaway; George Mills; Richard and Victoria Moffat; John and Jill Neal; Dr Patrice Newell; Tom Nicholas; Millie Nichols; Biz Nicholson; Sue Ogilvy; Mike Parish; Kevin and Catriona Pearce; Malcolm and Kerrie Plum; Dave and Kay Quinton; Adrian Richardson; Dick Richardson; Pennie Scott; Robin Sparke; Bill Spence; Dr Maarten Stapper; Nicki Taws; George and Chad Taylor; Robbie Tole; Bill Twigg; Annabel Walsh; Rob and Jenny Warburton; Rob and Sandra Waterson; Gus and Kelly Whyte; Bob Wilson; Tim Woods; Mark Wootton; Dr Julia Wright; and Tony York.

My enduring thanks for his great patience, generosity and friendship also go to Rod Mason – senior Ngarigo law-man – who has begun the long process of tutoring a novice.

I would like to acknowledge the generous assistance of the Rockefeller Foundation for their Fellowship and thus a highly productive month in November 2015 at Bellagio, Lake Como, Italy, and the opportunity to engage with international minds all working to the same end.

My thanks go also to Margaret Connolly & Associates P/L for permission to use extracts from two of Les Murray's poems (and of course to Les Murray himself for his generosity), and to PN Review www.pnreview .co.uk, and also to Curtis Brown Australia for permission to use an extract from an A. D. Hope poem.

I also wish to thank Debbie McInnes and the support team at DMCPR Media for their sterling marketing efforts in the media area, and also to Paul Beaver – computer expert extraordinaire.

Finally, I would like to acknowledge two great teams of publishers. In Australia, first and foremost, huge thanks to UQP's non-fiction publisher Alexandra Payne, who saw the importance of the vision and patiently, stoically and courageously supported the book through all the vicissitudes of growth and production. Alex, this book would not exist without your vision and generosity, thank you.

I also would like to thank other members of the UQP production team, including Harriet Swatman Manasa and Kylie Rathborne for their work on marketing and publicity, and to Vanessa Pellatt for her patient, meticulous and efficient project management and editing when under pressure.

Acknowledgements

My thanks also to Sandy Cull for her excellent cover design. Huge thanks also go to the two outside editors who helped create order out of chaos: to Nikki Lusk for her clear, constructive and pragmatic early work, and especially Kevin O'Brien for his full commitment to the book's essence and his excellent approach and contribution to the project. There is now a wallaroo named in his honour by my grandchildren!

And then there is the wonderful team at Chelsea Green Publishing in Vermont, USA. I would like to express huge thanks to Brianne Goodspeed. Like Alex Payne at UQP, she has a vision for a future Earth and Society, and likewise immediately saw the importance of the field of regenerative agriculture and its wider importance to regenerating Earth and addressing the Anthropocene. So this American and wider edition would not exist but for her passion and leadership. Likewise also, my great thanks to the professional and committed production and marketing team at Chelsea Green: to Nick Kaye for his patient and meticulous work with the photos; to Alex Bullett for his professionalism in project management; to proofreader Angela Boyle; designer Melissa Jacobson; Christina Butt for strategy/media communication; and marketing director Sean Maher.

Finally, I would particularly like to thank Nicolette Hahn Niman for her insightful and kind foreword.

So, my sincere thanks to all the above: diverse and generous members of a unique agrarian and literary 'village'.

BIBLIOGRAPHY

Books

Acocks, J. P. H., *Veld Types of South Africa*, Botanical Research Institute, Department of Agricultural Technical Services, South Africa, 1975

Albrecht, W. A., *Soil Fertility and Animal Health: The Albrecht Papers Vol. 2*, C. Walters (ed.), Acres USA, Austin, Texas, 1975

—— *Soil Fertility and Animal Health: The Albrecht Papers Vol. 3*, C. Walters (ed.), Acres USA, Austin, Texas, 2013

Alexander, C., *The Nature of Order: An Essay on the Art of Building and the Nature of the Universe, Vol. 1: The Phenomenon of Life*, Center for Environmental Structure, Berkeley, California, 1981

Anderson, J., *An Enquiry into the Nature of the Corn Laws: With a View to the New Corn Bill*, Edinburgh, Mrs Mundell, 1777

—— *Essays Relating to Agriculture and Rural Affairs*, William Creach & T. Cadell, Edinburgh, 1797

Andrews, P., *Back from the Brink: How Australia's Landscape Can Be Saved*, ABC Books, Sydney, 2001

Arabena, K., *Becoming Indigenous to the Universe: Reflections on Living Systems, Indigeneity and Citizenship*, Australian Scholarly Publishing, Melbourne, 2015

Balfour, E., *The Living Soil: Evidence of the Importance to Human Health of Soil Vitality, with Special Reference to Post-War Planning*, Faber & Faber, London, 1943

Bandura, A., *Social Learning Theory*, Prentice Hall, Englewood Cliffs, New Jersey, 1977

Barber, D., *The Third Plate: Field Notes on the Future of Food*, Abacus, London, 2014

Bateson, G., *Steps to an Ecology of Mind: Collected Essays in Anthropology, Psychiatry, Evolution, and Epistemology*, Granada, St Albans, 1973

Bateson, P. and P. Gluckman, *Plasticity, Robustness, Development and Evolution*, Cambridge University Press, Cambridge, 2011

Becker, E., *The Denial of Death*, Free Press, New York, 1973

Bennett, H. H., *Soil Conservation*, McGraw-Hill, New York, 1939

Benyus, J. M., *Biomimicry: Innovation Inspired by Nature*, Harper Perennial, New York, 1997

Berman, M., *The Reenchantment of the World*, Cornell University Press, Ithaca, New York, 1981

Berry, T., *The Dream of the Earth*, Sierra Club, San Francisco, 1988

—— *The Great Work: Our Way into the Future*, Bell Tower, New York, 1999

Bibliography

Berry, W., *The Unsettling of America: Culture and Agriculture*, Sierra Club Books, San Francisco, 1977

Blewett, R. (ed.), *Shaping a Nation: A Geology of Australia*, Geoscience Australia/ANU, Paragon Printers, Canberra, 2012

Bohm, D., *On Dialogue*, Routledge, New York, 1996

Boyd, R. D. (ed.), *Personal Transformations in Small Groups: A Jungian Perspective*, Routledge, New York, 1991

Bride, T. F. (ed.), *Letters from Victorian Pioneers: A Series of Papers on the Early Occupation of the Colony, the Aborigines, Etc.*, Lloyd O'Neil, Melbourne, 1983

Brock, W., *Justus von Liebig: The Chemical Gatekeeper*, Cambridge University Press, Cambridge, 2002

Brown, V. and J. Harris, *The Human Capacity for Transformational Change: Harnessing the Collective Mind*, Routledge/Earthscan, Abingdon, Oxon, 2014

Brown, V., J. Harris and J. Y. Russell (eds), *Tackling Wicked Problems: Through the Transdisciplinary Imagination*, Earthscan, London, 2010

Cane, S., *First Footprints: The Epic Story of the First Australians*, Allen & Unwin, Sydney, 2013

Capra, F., *The Web of Life: A New Synthesis of Mind and Matter*, Flamingo, London, 1997

Capra, F. and P. L. Luisi, *The Systems View of Life: A Unifying Vision*, Cambridge University Press, Cambridge, 2014

Carey, N., *The Epigenetics Revolution: How Modern Biology is Rewriting Our Understanding of Genetics, Disease and Inheritance*, Columbia University Press, New York, 2012

Carson, R., *Silent Spring*, Penguin, Melbourne, 1972

Cathcart, M., *The Water Dreamers: The Remarkable History of Our Dry Continent*, Text Publishing, Melbourne, 2009

Christian, D., *Maps of Time: An Introduction to Big History*, University of California Press, Berkeley, 2005

Christman, R. F. and E. T. Gjessing, *Aquatic and Terrestrial Humic Materials*, Ann Arbor Science, The Butterworth Grove, Kent, England, 1983

Clarke, J. J., *The Self-Creating Universe: The Making of a Worldview*, XLibris, United Kingdom, 2013

Clarke, P., *Where the Ancestors Walked: Australia as an Aboriginal Landscape*, Allen & Unwin, Sydney, 2003

Cluff, D., *Farming without Farming*, Department of Land and Water Conservation, Canberra, 2003

Conford, P. (ed.), *The Organic Tradition: An Anthology of Writings on Organic Farming 1900–1950*, Green Books, Bideford, 1988

Connolly, B. et al. (eds), *Radical Learning for Liberation*, Maynooth Adult and Community Education, National University of Ireland, Maynooth, Kildare, 1996

Cook, J., *Captain Cook's Journal during his First Voyage Round the World: Made in H. M. Bark 'Endeavour', 1768–71*, Elliot Stock, London, 1893

Crosby, A. W., *Ecological Imperialism: The Biological Expansion of Europe, 900–1900*, Cambridge University Press, New York, 2004

Curr, E. M., *Recollections of Squatting in Victoria: Then Called the Port Phillip District, (from 1841 to 1851)*, Melbourne University Press, Melbourne, 1965

Cushman, S. A. and F. Huettmann (eds), *Spatial Complexity, Informatics, and Wildlife Conservation*, Springer, Japan, 2009

Dale, T. and V. G. Carter, *Topsoil and Civilization*, University Oklahoma Press, Norman, Oklahoma, 1974

Dampier, W., *A New Voyage Round the World: The Journal of an English Buccaneer*, M. Beken (ed.), Hummingbird Press, London, 1998

—— *Dampier's Voyages*, J. Masefield (ed.), Grant Richards, 1906

Darwin, C., *The Formation of Vegetable Mould, Through the Action of Worms: With Observations on Their Habits*, John Murray, London, 1881

—— *On the Origin of Species by Means of Natural Selection: or, the Preservation of Favoured Races in the Struggle for Life*, John Murray, London, 1859

Davie, T., *Fundamentals of Hydrology*, Routledge, Abingdon, Oxon, 2002

Dawborn, K. and C. Smith (eds), *Permaculture Pioneers: Stories from the New Frontier*, Melliodora, Hepburn, Victoria, 2011

Dawkins, R., *The Blind Watchmaker: Why the Evidence of Evolution Reveals a Universe without Design*, Norton & Co., New York, 1986

Delbanco, A. (ed.), *The Portable Abraham Lincoln*, Penguin, New York, 1993

Descartes, R., *Philosophical Works of Descartes*, E. S. Haldane and C. R. T. Ross (eds), 2 vols, Dover, New York, 1955

Diamond, J., *Collapse: How Societies Choose to Fail or Succeed*, Penguin, New York, 2005

—— *Guns, Germs and Steel: The Fates of Human Societies*, Jonathan Cape, London, 1997

Dovers, S. (ed.), *Australian Environmental History: Essays and Cases*, Oxford University Press, Melbourne, 1994

—— *Environmental History and Policy: Still Settling Australia*, Oxford University Press, Melbourne, 2000

El-Baz, F. and M. H. A. Hassan (eds), *The Physics of Desertification*, Martinus Nijhoff Publishers, Dordrecht, Netherlands, 2012

Eliade, M., *The Sacred and the Profane: The Nature of Religion*, Harcourt, Brace & World, New York, 1959

Éluard, P. (*neé* E. Grindel), *Oeuvres Completes*, Vol. 1, Gallimard, Paris, 1968

Endres, K. P., *Moon Rhythms in Nature: How Lunar Cycles Affect Living Organisms*, Edinburgh, Floris Books, 2002

Faulkner, E., *Ploughman's Folly*, Michael Joseph, London, 1945

Fenton, J., *The Untrained Environmentalist: How an Australian Grazier Brought His Barren Property Back to Life*, Allen & Unwin, Sydney, 2010

Fischer, T., D. Byerlee and G. Edmeades, *Crop Yields and Global Food Security: Will Yield Increase Continue to Feed the World?*, Australian Centre for International Agricultural Research, Canberra, 2014, http://aciar.gov.au/publication/mn158

Flader, S. and J. B. Callicott (eds), *The River of the Mother of God and other Essays by Aldo Leopold*, University of Wisconsin Press, Madison, Wisconsin, 1991

Flannery, T., *The Future Eaters: An Ecological History of the Australasian Lands and People*, Reed Books, Sydney, 1994

—— *Mammals of New Guinea*, Reed Books, Sydney, 1990

Flood, J., *The Original Australians: Story of the Aboriginal People*, Allen & Unwin, Sydney, 2006

Bibliography

Frankfort, H., *Before Philosophy: The Intellectual Adventure of Ancient Man; An Essay on Speculative Thought in the Ancient Near East*, Penguin, London, 1949

Friedman, M., *Capitalism and Freedom*, University of Chicago Press, Chicago, 1962

Gammage, B., *The Biggest Estate on Earth: How Aborigines Made Australia*, Allen & Unwin, Sydney, 2011

Geiger, R., *The Climate Near the Ground*, Harvard University Press, Cambridge, Massachusetts, 1957

Gerritsen, R., *Australia and the Origins of Agriculture*, Archaeopress, London, 2008

Gliessman, S. R., *Agroecology: The Ecology of Sustainable Food Systems (2nd edition)*, CRC Press, Boca Raton, Florida, 2007

Goodwin, B., *How the Leopard Changed Its Spots: The Evolution of Complexity*, Phoenix, London, 1994

Griffiths, T. and L. Robin (eds), *Ecology and Empire: Environmental History of Settler Societies*, Melbourne University Press, Melbourne, 1997

Gunderson, L. H. and C. S. Holling (eds), *Panarchy: Understanding Transformations in Human and Natural Systems*, Island Press, Washington DC, 2002

Hagedoorn, A. L., *Animal Breeding*, Lockwood, London, 1939

Hancock, W. K., *Australia*, Ernest Benn Ltd, London, 1930

Havel, V., *Disturbing the Peace*, Faber & Faber, London, 1990

Hawken, P., *Blessed Unrest: How the Largest Social Movement in History is Restoring Grace, Justice, and Beauty to the World*, Penguin, New York, 2007

—— *The Ecology of Commerce: A Declaration of Sustainability*, Harper Collins, New York, 1993

Heseltine, H. (ed.), *The Penguin Book of Australian Verse*, Penguin, Melbourne, 1972

Hillel, D., *Out of the Earth: Civilisation and the Life of the Soil*, New York Free Press, New York, 1991

Holmgren, D., *Essence of Permaculture*, Holmgren Design Services, Hepburn, Victoria, 2004

—— *Permaculture: Pathways and Principles Beyond Sustainability*, Holmgren Design Services, Hepburn Springs, Victoria, 2002

Howard, A., *An Agricultural Testament*, Oxford University Press, New York, 1943

—— *The Soil and Health: A Study of Organic Agriculture*, Schocken Books, New York, 1972

Hyams, E., *Soil and Civilisation*, Thames & Hudson, London, 1952

Jacks, G. V. and R. O. Whyte, *The Rape of the Earth: A World Survey of Soil Erosion*, Faber & Faber, London, 1939

Jackson, W., *Becoming Native to This Place: The Blazer Lectures for 1991*, University Press of Kentucky, Lexington (Kentucky), 1994

—— *Consulting the Genius of the Place: An Ecological Approach to a New Agriculture*, Counterpoint, Berkeley, 2010

—— *New Roots for Agriculture*, Friends of the Earth, San Francisco, 1980

Jackson, W. R., *Humic, Fulvic and Microbial Balance: Organic Soil Conditioning*, Jackson Research Center, Evergreen, Colorado, 1993

Jantsch, E. (ed.), *Evolutionary Vision: Toward a Unifying Paradigm of Physical, Biological, and Sociocultural Evolution*, Westview Press, Boulder, Colorado, 1981

Johnson, C. N., *Australia's Mammal Extinctions: A 50,000 Year History*, Cambridge University Press, Melbourne, 2006

Jones, R., *Green Harvest: A History of Organic Farming and Gardening in Australia*, CSIRO Publications, Melbourne, 2010

Jurskis, V., *Firestick Ecology: Fairdinkum Science in Plain English*, Connor Court Publishing, Ballarat, 2016

Kelly, G. A., *The Psychology of Personal Constructs*, W. W. Norton & Co., New York, 1955

Kelly, L., *The Memory Code: The Secrets of Stonehenge, Easter Island and Other Ancient Monuments*, Allen & Unwin, Sydney, 2016

Kiddle, M. L., *Men of Yesterday: A Social History of the Western District of Victoria, 1834–1890*, Melbourne University Press, Melbourne, 1962

King, F. H., *Farmers of Forty Centuries: Or, Permanent Agriculture in China, Korea and Japan*, Rodale Press, Emmaus, Pennsylvania, 1911

Kinsella, J., *The Silo: A Pastoral Symphony – Poems*, Fremantle Arts Centre Press, Fremantle, 1995

Kolb, D. A., *Experiential Learning: Experience as the Source of Learning and Development*, Prentice Hall, New Jersey, 1984

Kolisko, E. and L., *Agriculture of Tomorrow*, Acorn Press, Bournemouth, 1978

Laslett, P. (ed.), *Two Treatises of Government*, 2nd edition, Cambridge University Press, Cambridge, 1967

Latz, P., *The Flaming Desert: Arid Australia – A Fire Shaped Landscape*, N.T. Print Management, Alice Springs, 2007

Lave, J. and E. Wenger, *Situated Learning: Legitimate Peripheral Participation*, Cambridge University Press, Cambridge, 1991

Leach, J. A., *An Australian Bird Book (7th Edition)*, Whitcombe & Tombs Ltd., Melbourne, 1926

Lefroy, E. C. et al., *Biodiversity: Integrating Conservation and Production – Case Studies from Australian Farms, Forests and Fisheries*, CSIRO, Melbourne, 2008

Leopold, A., *Game Management*, Scribner & Sons, New York, 1933

—— *A Sand Country Almanac, and Sketches Here and There*, Oxford University Press, New York, 1975

Leopold, A., Callicott, J. B. and E. T. Freyfogle (eds), *For the Health of the Land – Previously Unpublished Essays and Other Writings*, Island Press–Shearwater Books, Washington, 1999

Lieberman, D., *The Story of the Human Body: Evolution, Health and Disease*, Allen Lane, London, 2013

Lindenmeyer, D., *On Borrowed Time: Australia's Environmental Crisis and What We Must Do About It*, CSIRO Press, Melbourne, 2007

Lines, W. J., *Taming the Great South Land: A History of the Conquest of Nature in Australia*, Allen & Unwin, Sydney, 1991

Lopez, B., *Crossing Open Ground*, Vintage, New York, 1989

Louv, R., *Last Child in the Woods: Saving Our Children from Nature-Deficit Disorder*, Algonquin Books, Chapel Hill, North Carolina, 2008

—— *The Nature Principle: Human Restoration and the End of Nature-Deficit Disorder*, Algonquin Books, Chapel Hill, North Carolina, 2011

Lovel, H., *A Biodynamic Farm for Growing Wholesome Food*, Acres USA, Austin, Texas, 1994

Lovelock, J. E., *Gaia: A New Look at Life on Earth*, Oxford University Press, Oxford, 1979

Low, T., *Where Song Began: Australia's Birds and How They Changed the World*, Viking, Melbourne, 2014

Lowenfels, J. and W. Lewis, *Teaming with Microbes: A Gardener's Guide to the Soil Food Web*, Timber Press, Portland, 2006

Mackay, S. M. et al., *Native Tree Dieback and Mortality on the New England Tablelands of New South Wales*, Forestry Commission of New South Wales, Sydney, 1984

Margulis, L., *Symbiotic Planet: A New Look at Evolution*, Basic Books, New York, 1998

—— *Symbiosis in Cell Evolution: Microbial Communities in the Archean and Proterozoic Eons*, W. H. Freeman & Co., New York, 1993

Margulis, L. and D. Sagan, *Microcosmos: Four Billion Years of Microbial Evolution*, University of California Press, Berkeley, 1997

Marsh, G. P., *Man and Nature, or Physical Geography as Modified by Human Action*, facsimile of first edition (1864), D. Lowenthal (ed.), University of Washington Press, Seattle and London, 2003

Marshall, A. J. (ed.), *The Great Extermination: A Guide to Anglo-Australian Cupidity, Wickedness and Waste*, Heinemann, Melbourne, 1966

Massy, C., *The Australian Merino*, Viking O'Neil/Penguin, Melbourne, 1990

—— *The Australian Merino: The Story of a Nation*, Random House Australia, Sydney, 2007

—— *Breaking the Sheep's Back: The Shocking True Story of the Decline and Fall of the Australian Wool Industry*, University of Queensland Press, Brisbane, 2011

Maturana, H. and F. Varela, *The Tree of Knowledge: The Biological Roots of Human Understanding*, Shambhala Publications, Boston, 1992

McHarg, I. L., *Design with Nature*, Doubleday/Natural History Press, New York, 1971

McKernan, M., *Drought: The Red Marauder*, Allen & Unwin, Sydney, 2005

McNeill, J., *Something New under the Sun: An Environmental History of the Twentieth-Century World*, W. W. Norton, New York, 2000

Meadows, D. H and D. L., J. Randers and W. W. Behrens III, *The Limits to Growth: A Report for the Club of Rome's Project on the Predicament of Mankind*, Universe Books, New York, 1972

Merchant, C., *The Death of Nature: Women, Ecology, and the Scientific Revolution*, HarperSanFrancisco, San Francisco, 1980

—— *Reinventing Eden: The Fate of Nature in Western Culture*, Routledge, New York, 2004

Meuret, M. and F. Provenza (eds), *The Art and Science of Shepherding: Tapping the Wisdom of French Herders*, Acres USA, Austin, Texas, 2014

Mezirow, J. and Associates, *Learning as Transformation: Critical Perspectives on a Theory in Progress*, Jossey-Bass, San Francisco, 2000

Mitchell, M., *Complexity: A Guided Tour*, Oxford University Press, New York, 2009

Mitchell, T., *Three Expeditions into the Interior of Eastern Australia* (two vols), Cambridge University Press, London, 2011

Mollison, B., *Permaculture: A Designer's Manual*, Tagari Publications, Tyalgum, New South Wales, 1988

Mollison, B. and D. Holmgren, *Permaculture One: A Perennial Agriculture for Human Settlement*, Tagari Publications, Sisters Creek, Tasmania, 1978

Monod, J., *Chance and Necessity: An Essay on the Natural Philosophy of Modern Biology*, Vintage Books, New York, 1971

Montgomery, D. R., *Dirt: The Erosion of Civilisations*, University of California Press, Berkeley, 2007

Muir, C., *The Broken Promise of Agricultural Progress: An Environmental History*, Routledge/Earthscan, Abingdon, Oxon, 2014

Muir, J., *My First Summer in the Sierra*, Riverside Press, Cambridge, Massachusetts, 1911
—— *Travels in Alaska*, Houghton Mifflin, New York, 1915

Mulligan, M. and S. Hill, *Ecological Pioneers: A Social History of Australian Ecological Thought and Action*, Cambridge University Press, Melbourne, 2001

Mulvaney, J. and J. Kamminga, *Prehistory of Australia*, Allen & Unwin, Sydney, 1999

Mumford, L., *Technics and Civilization*, Harcourt, Brace & World, New York, 1962

Murray, L., *Collected Poems*, Duffy & Snellgrove, Sydney, 2002
—— *The Ilex Tree*, University of Sydney Press, Sydney, 1965

Nelson, L. H., *Who Knows: From Quine to a Feminist Empiricism*, Temple University Press, Philadelphia, 1990

Norton, D. and N. Reid, *Nature and Farming: Sustaining Native Biodiversity in Agricultural Landscapes*, CSIRO Publishing, Canberra, 2013

Novacek, M., *Terra: Our 100-Million-Year-Old Ecosystem and the Threats that Now Put It at Risk*, Farrar, Straus & Giroux, New York, 2007

Nuberg, I., B. George and R. Reid (eds), *Agroforestry for Natural Resource Management*, CSIRO, Melbourne, 2009

Odum, E. P., *Fundamentals of Ecology (3rd Edition)*, W. B. Saunders Co., Philadelphia, 1971

Odum, H. T., *Environment, Power, and Society*, John Wiley & Sons, New York, 1970

Odum, H. T. and E. C., *A Prosperous Way Down: Principles and Policies*, University Press of Colorado, Boulder, Colorado, 2001

Ohlson, K., *The Soil Will Save Us: How Scientists, Farmers and Foodies are Healing the Soil to Save the Planet*, Rodale Books, New York, 2014

Orr, D., *Ecological Literacy: Education and the Transition to a Postmodern World*, State University of New York Press, Albany, 1992
—— *The Nature of Design: Ecology, Culture, and Human Intention*, Oxford University Press, Oxford, 2002

Pascoe, B., *Dark Emu: Black Seeds – Agriculture or Accident?*, Magabala Books, Broome, 2014

Pennick, N., *The Ancient Science of Geomancy: Man in Harmony with the Earth*, Thames & Hudson, London, 1979

Pfeiffer, E., *Pfeiffer's Introduction to Biodynamics*, Floris Books, Edinburgh, 2011

Plumwood, V., *Feminism and the Mastery of Nature*, Routledge, London, 1993

Podolinsky, A., *Biodynamic Agriculture: Introductory Lectures*, vol. 1, Gavemer Publishing, Sydney, 2004

Pohl, F. G. von, *Earth Currents: Causative Factor of Cancer and Other Diseases*, Fortschritt fuer alle-Verlag, Feucht, 1978

Pollan, M., *The Omnivore's Dilemma: The Search for a Perfect Meal in a Fast-Food World*, Bloomsbury, London, 2006

Popper, K. R. and J. C. Eccles, *The Self and Its Brain*, Springer-Verlag, Berlin and London, 1977

Porte, J. (ed.), *The Journals and Miscellaneous Notebooks of Ralph Waldo Emerson*, Belknap Press, Boston, 1982

Bibliography

Powell, J. M., *An Historical Geography of Modern Australia: The Restive Fringe*, Cambridge University Press, Melbourne, 1988

Pretty, J. N., *Agri-Culture: Reconnecting People, Land and Nature*, Earthscan, Sterling, Virginia, 2002

Rapoport, A. (ed.), *Australia as Human Setting*, Angus & Robertson, Sydney, 1994

Ratcliffe, F., *Flying Fox and Drifting Sand: The Adventures of a Biologist in Australia*, Chatto & Windus, London, 1938

Raupp, J. (ed.), *Main Effects of Various Organic and Mineral Fertilization on Soil Organic Matter Turnover and Plant Growth: EC-Concerted Action 'Fertilization Systems in Organic Farming'*, Publications of the Institute for Biodynamic Research, Darmstadt, Germany, Vol. 5, 1995

Reid, R. and P. Stephen, *The Farmer's Forest: Multipurpose Forestry for Australian Farmers*, Australian Master Tree Grower/Melbourne University, Melbourne, 2001

Reid, R. and G. Wilson, *Agroforestry in Australia and New Zealand*, Goddard & Dobson, Melbourne, 1985

Robin, L., *How a Continent Created a Nation*, UNSW Press, Sydney, 2007

Robinson, J., *Eating on the Wild Side: The Missing Link to Optimum Health*, Little, Brown and Co., New York, 2013

Rodale, J. I., *The Organic Front*, Rodale Press, Emmaus, Pennsylvania, 1984

—— *Pay Dirt: Farming and Gardening with Compost*, Rodale Press, Emmaus, Pennsylvania, 1946

Rodale, M., *Organic Manifesto: How Organic Farming Can Heal Our Planet, Feed the World, and Keep Us Safe*, Rodale Press, New York, 2010

Rolls, E., *Australia: A Biography*, Brisbane, UQP, 2000

—— *A Million Wild Acres: 200 Years of Man and an Australian Forest*, Nelson, Melbourne, 1981

—— *They All Ran Wild: The Story of Pests on the Land in Australia*, Angus & Robertson, Sydney, 1969

Rorty, R., *Philosophy and Social Hope*, Penguin, London, 1999

Rose, D., *Nourishing Terrains: Australian Aboriginal Views of Landscape and Wilderness*, Australian Heritage Commission, Canberra, 1996

Rose, N. (ed.), *Fair Food: Stories from a Movement Changing the World*, UQP, Brisbane, 2015

Rosen, J., *The Life of the Skies: Birding at the End of Nature*, Farrar, Straus and Giroux, New York, 2008

Russell, D. B. and R. L. Ison, *Agricultural Extension and Rural Development: Breaking out of Knowledge Transfer Traditions*, Cambridge University Press, Cambridge, 2000

Russell-Smith, J. et al., *Culture, Ecology and Economy of Fire Management in North Australian Savannas: Rekindling the Wurrk Tradition*, CSIRO Publishing, Tropical Savannas CRC, 2009

Saul, J. R., *Voltaire's Bastards: The Dictatorship of Reason in the West*, Penguin, London, 1993

Savory, A., *Holistic Resource Management*, Island Press, Washington DC, 1988

Savory, A. and J. Butterfield, *Holistic Management: A New Framework for Decision Making (2nd Edition)*, Island Press, Washington DC, 1999

Scherr, S. J. and J. A. McNeely (eds), *Farming with Nature: The Science and Practice of Ecoagriculture*, Island Press, Washington DC, 2007

Schilthuis, W., *Biodynamic Agriculture*, Floris Books, Edinburgh, 2003

Schumacher, E. F., *A Guide for the Perplexed*, Harper & Row, New York, 1977

—— *Small Is Beautiful: A Study of Economics as if People Mattered*, Abacus, London, 1973

Schwartz, J. D., *Cows Save the Planet: And Other Improbable Ways of Restoring Soil to Heal the Earth*, Chelsea Green Publishing, White River Junction, Vermont, 2013

Seamon, D. and A. Zajonc (eds), *Goethe's Way of Science: A Phenomenology of Nature*, State University of New York Press, Albany, New York, 1998

Seddon, G., *Landprints: Reflections on Place and Landscape*, Cambridge University Press, Melbourne, 1997

Seddon, G. and M. Davis (eds), *Man and Landscape in Australia: Towards an Ecological Vision*, AGPS, Canberra, 1976

Sheldrick, J., *Nature's Line: George Goyder – Surveyor, Environmentalist, Visionary*, Wakefield Press, Adelaide, 2013

Shiva, V., *Making Peace with the Earth: Beyond Resource, Land and Food Wars*, Spinifex Press, Melbourne, 2012

—— *Monocultures of the Mind: Perspectives on Biodiversity and Biotechnology*, Zed Books, London, 1993

—— *Soil Not Oil: Climate Change, Peak Oil, and Food Insecurity*, Spinifex Press, Melbourne, 2009

—— *Staying Alive: Women, Ecology and Development*, Zed Books, London, 1988

—— *Stolen Harvest: The Hijacking of the Global Food Supply*, South End Press, Cambridge, Massachusetts, 2000

—— *The Violence of the Green Revolution: Ecological Degredation and Political Conflict*, Zed Books, London, 1991

Smith, A., *The Wealth of Nations*, A. Skinner (ed.), Penguin, New York, 1986

Smith, J. R., *Tree Crops: A Permanent Agriculture*, Devin-Adair, Old Greenwich, Connecticut, 1977

Smuts, J., *Holism and Evolution*, Greenwood Press, Westport, Connecticut, 1973

Soule, J. D. and J. K. Piper, *Farming in Nature's Image: An Ecological Approach to Agriculture*, Island Press, Washington DC, 1992

Speth, G., *The Bridge at the Edge of the World: Capitalism, the Environment, and Crossing from Crisis to Sustainability*, Yale University Press, New Haven, 2008

Stanner, W. E. H., *After the Dreaming: The 1968 Boyer Lectures*, ABC, Sydney, 1969

Steinbeck, J., *The Grapes of Wrath*, Penguin, New York, 1992

Steiner, R., *Spiritual Foundations for the Renewal of Agriculture*, M. Gardner (ed.), C. E. Creeger (trans.), Bio-Dynamic Farming & Gardening Association, Inc., Junction City, Oregon, 1993

Swain, T. and D. Rose (eds), *Aboriginal Australians and Christian Missions: Ethnographic and Historical Studies*, The Australian Association of the Study of Religions, Adelaide, 1988

Taylor, B., *Encyclopaedia of Religion and Nature*, Continuum, London, 2005

Thomas, L., *The Lives of a Cell: Notes of a Biology Watcher*, The Viking Press, New York, 1974

Thomas Jnr, W. L. (ed.), *Man's Role in Changing the Face of the Earth*, Vol. 1, University of Chicago Press, Chicago, 1956

Tongway, D. J. and J. A. Ludwig, *Restoring Disturbed Landscapes: Putting Principles into Practice*, Island Press, Washington DC, 2011

Bibliography

USDA, *Soils and Men: Yearbook of Agriculture*, US Government Printing Office, Washington DC, 1938

Vickers-Rich, P. and T. Rich, *Wildlife of Gondwana: Dinosaurs and Other Vertebrates from the Ancient Supercontinent*, Reed Books, Sydney, 1993

Vitek, W. and W. Jackson (eds), *Rooted in the Land: Essays on Community and Place*, Yale University Press, New Haven, 1996

Voisin, A., *Better Grassland Sward: Ecology, Botany and Management*, Crosby Lockwood & Son, London, 1960

—— *Grass Productivity*, Crosby Lockwood & Son, London, 1961

Walker, B. and D. Salt, *Resilience Practice: Building Capacity to Absorb Disturbance and Maintain Function*, Island Press, Washington DC, 2012

—— *Resilience Thinking: Sustaining Ecosystems and People in a Changing World*, Island Press, Washington DC, 2006

Welton, M. R. (ed.), *In Defense of the Lifeworld: Critical Perspective on Adult Learning*, State University of New York Press, New York, 1995

Wenger, E., *Communities of Practice: Learning, Meaning, and Identity*, Cambridge University Press, Cambridge, 1998

West-Eberhard, M. J., *Developmental Plasticity and Evolution*, Oxford University Press, New York, 2003

Wheeler, P. A. and R. B. Ward, *The Non-Toxic Farming Handbook*, Acres USA, Metairie, Louisiana, 1998

White, C., *Revolution on the Range: The Rise of a New Ranch in the American West*, Island Press/Shearwater Books, Washington, 2008

White, H. F. and C. S. Hicks, *Life from the Soil*, Longmans Green & Co., Melbourne, 1953

White, M., E, *After the Greening: The Browning of Australia*, Kangaroo Press, Sydney, 1994

—— *Earth Alive!: From Microbes to a Living Planet*, Rosenberg Publishing Ltd., 2003

Whitehead, A. N., *The Concept of Nature*, Cambridge University Press, Cambridge, 1929

—— *Process and Reality*, MacMillan, New York, 1929

Whitney, M., *Soil and Civilization: A Modern Concept of the Soil and the Historical Development of Agriculture*, Van Nostrand, Berkeley, California, 1925

Wilber, K., *Sex, Ecology and Spirituality: The Spirit of Evolution*, Shambhala, Boston, 1995

Williams, N. S. G. et al. (eds), *Land of Sweeping Plains: Managing and Restoring the Native Grasslands of South-Eastern Australia*, CSIRO Publishing, Clayton, South Australia, 2015

Willis, H., *Foundations of Natural Farming: Understanding Core Concepts of Ecological Agriculture*, Acres USA, Austin, Texas, 2008

Wilson, E. O., *Biophilia: The Human Bond with Other Species*, Harvard University Press, Cambridge, Massachusetts, 1984

Wulf, A., *The Invention of Nature: The Adventures of Alexander von Humboldt, the Lost Hero of Science*, John Murray, London, 2015

Yeomans, P. A., *The Challenge of Landscape: The Development and Practice of Keyline*, Keyline Press, Sydney, 1958

—— *The City Forest: The Keyline Plan for the Human Environment Revolution*, Keyline Publishing, Sydney, 1971

—— *The Keyline Plan*, P. A. Yeomans, Sydney, 1954

—— *Water for Every Farm: Using the Keyline Plan*, Murray Book Distributors, Sydney, 1978

Zietsman, J., *Man, Cattle and Veld*, Beef Power LLC, Tampa, Florida, 2014

Articles, Essays, Papers and Poems

'A Brief History of Bio-Dynamics – An Australian Perspective', *Biodynamics Magazine*, No. 1, December 2003

Adamson, D., 'A Southern Hemisphere View of Nature', *Plants in Action*, no. 1, 2010, http://plantsinaction.science.uq.edu.au/edition1

Allen, P. L. and C. E. Sachs, 'The Poverty of Sustainability: An Analysis of Current Positions', *Agriculture & Human Values*, 9, no. 4, September 1992

Altieri, M. A., 'Agroecology: A New Research and Development Paradigm for World Agriculture', *Agriculture, Ecosystems & Environment*, 27, November 1989

Ampt, P. and S. Doornbos, 'Communities in Landscapes Project: Benchmark Study of Innovators', University of Sydney, Gulgong, Central West Catchment, New South Wales, December 2010, http://www.pasturecropping.com/images/PDF/Sydney%20 University%20Winona%20Data%20Final.2011.pdf

Amrine, F., 'Discovering a Genius: Rudolf Steiner at 150', Anthroposophical Society in America, 2011, http://www.anthroposophy.org/fileadmin/vision-in-action/being -human-2011-01-Amrine-Discovering.pdf

Angus, J. F. and A. J. Good, 'Dryland Cropping in Australia', *Challenges & Strategies for Dryland Agriculture*, CSSA Special Publication, Madison, Wisconsin, 2004

Arabena, K., 'Regeneration: Healthy People in Healthy Landscapes – Addressing Wicked Problems of Our Time', Public Health Association of Australia 42nd Annual Conference, Keynote Address, Melbourne, 2013

Australian Landcare Magazine, 'Interview with Jon Taylor: *Australian Farm Journal*', March 2004

Bahn, M., et al. 'Does Photosynthesis Affect Grassland Soil-Respired CO_2 and Its Carbon Isotope Composition on a Diurnal Timescale?', *New Phytologist*, 182, no. 2: 451–460, February 2009

Baker, G., 'Wingless Grasshopper', Agfact AE1, NSW Agriculture, 2005

Baker, M., 'The Tree Whisperer', *Sydney Morning Herald Weekend Magazine*, 26 April 2014, http://www.smh.com.au/environment/the-tree-whisperer-20140420-36ze4.html

Balfour, E., 'Towards a Sustainable Agriculture: The Living Soil', Address to the International Federation of Organic Agriculture Movements Conference, Switzerland, Journey to Forever (website), 1977, http://journeytoforever.org/farm_library/balfour _sustag.html

Bateson, P., 'The Rise and Rise of Epigenetics. Guest Essay', in F. Capra and P. L. Luisi, *The Systems View of Life: A Unifying Vision*, Cambridge University Press, Cambridge, 2014

Beeson, C. F. C., 'The Moon and Plant Growth', *Nature*, 158, October 1946

Belsky, A. J., 'Does Herbivory Benefit Plants? A Review of the Evidence', *The American Naturalist*, 127, no. 6, June 1986

Bergsmann, O., 'The Vienna Report: Influence of Pathogenic Locations on People', Research Project, Institute of Consumer Research and Product Marketing, Vienna, 1989

Bibliography

Berry, W., 'It All Turns on Affection', Jefferson Lecture, National Endowment for the
  Humanities, Washington DC, 2012

—— 'Standing by Words', *Essays by Wendell Berry*, North Point Press, New York, 1983,
  reprinted in *The Trumpeter*, 2, no. 2, Spring 1985

Bhanoo, S. N., 'Farming Had an Earlier Start, a Study Says', *New York Times*, 27 July 2015, http://
  www.nytimes.com/2015/07/28/science/farming-had-an-earlier-start-a-study-says.html

Birch, T., '"We Are Still Here": Remorse, the National Psyche and Country', The Wheeler
  Centre, 18 June 2014, http://www.wheelercentre.com/notes/785becc41b5d

Bordenstein, S. R. and K. R. Theis, 'Host Biology in Light of the Microbiome: Ten
  Principles of Holobionts and Hologenomes', *PLOS Biol*, 13, no. 8, August 2015

Bouchenak-Khelladi, Y. et al., 'The Origins and Diversification of C_4 Grasses and
  Savanna-Adapted Ungulates', *Global Change Biology*, 15, no. 10, January 2009

Bowman, D. M. J. S. et al., 'Fire in the Earth System', *Science*, 324, no. 5926, April 2009

Bradbury, K. 'Gondwana Link: A Landscape Scale Restoration Project in South-West
  WA.' *Global Restoration Network Report. 'Top 25' Ecological Restoration Projects in
  Australasia*. 2010, http://www.gondwanalink.org/Glink_GRNreport.pdf

Bredenkap, G. J. et al., 'On the Origin of Northern and Southern Hemisphere Grasslands',
  Plant Ecology, 163, no. 2, December 2002

Brevik, E. C. and A. E. Hartemink, 'Early Soil Knowledge and the Birth and Development
  of Soil Science', *Catena*, 83, no. 1, October 2010

Brown, K. S. et al., 'Fire as an Engineering Tool of Early Modern Humans', *Science*, 325,
  no. 5942, August 2009

Brown, T. A. et al., 'The Complex Origins of Domesticated Crops in the Fertile Crescent',
  Trends in Ecology & Evolution, 24, no. 2, February 2008

Bruce, S. E. et al., 'Pasture-Cropping: Effect on Biomass, Total Cover, Soil Water and
  Nitrogen', Winona: Merino and Kelpie Studs, 2016, http://www.pasturecropping.com
  /14-articles

Butler, C. D., 'Globalization, Population, Ecology and Conflict', *Health Promotion Journal
  of Australia*, 18, no. 2, 2007

Carey, N., 'Beyond DNA: Epigenetics', Natural History, 2012, http://www.naturalhistory
  magazine.com/features/142195/beyond-dna-epigenetics

Carson, R., 'Help Your Child to Wonder', *Woman's Home Companion*, July 1956

Cawood, M., 'Glomalin Key to Earth's Health', *The Land*, 8 May 2008

—— 'Grazing Key to Ecology', *The Land*, 3 December 2009

—— 'Low-Risk Perennial Grains', *The Land*, 17 March 2011

—— 'A Wool and CO_2 Winner', Farm Weekly, 22 March 2016, http://www.farmweekly.
  com.au/news/agriculture/general/news/a-wool-and-co2-winner/2751985.aspx

—— 'Pasture Crop Trials Fuelling Corn Progress', *The Land*, 17 June 2010

—— 'Production and Regeneration at Wellington', *The Land*, 23 September 2010

—— 'Ocean Cycles Drive Carbon Cycles', *The Land*, 31 April 2016

—— 'Simplifying Towards Complexity: The Good Land Project', 4 December 2015,
  http://goodlandproject.com.au/2015/06david-marsh-simplicity-for-complexity

—— 'SLM's Holistic Hook Nabs Investors', Farmonline, 3 June 2014, http://www
  .farmonline.com.au/story/3574412/slms-holistic-hook-nabs-investors

Chadwick, M. A. et al., 'Feeding Pregnant Ewes a High-Salt Diet or Saltbush Suppresses Their Offspring's Postnatal Renin Activity', *Animal*, 3, no. 7, July 2009

Chazan, M., 'The Human Use of Fire', Live Science, 2012

Chivers, I., 'Splendour in the Grass: New Approaches to Cereal Production', ABC Radio, *The Conversation*, 2012, cited in B. Pascoe, *Dark Emu: Black Seeds – Agriculture or Accident?*, Magabala Books, Broome, 2014

Choi, C. Q., 'Humans Used Fire One Million Years Ago', Live Science, 2 April 2012, http://www.livescience.com/19425-earliest-human-fire.html

Cock, S., 'A Comparison of Soil and Plant Root Characteristics in Irrigated Summer Pasture from Two Different Farming Systems', undergraduate thesis, LaTrobe University, School of Agriculture, 1991

Comis, D., 'Glomalin: What Is It ... And What Does It Do?', *Agricultural Research*, July 2008

Commonwealth of Australia, 'Australia: State of the Environment Report', Department of the Environment and Energy, Canberra, 1996

Compassion in World Farming, 'Nutritional Benefits of Higher Welfare Animal Products', Surrey, United Kingdom, July 2012, https://www.ciwf.org.uk/media/5234769/Nutritional-benefits-of-higher-welfare-animal-products-June-2012.pdf

Cooke, M. L., 'The Future of the Great Plains: Report of the Great Plains Committee', US Government Printing Office, Washington DC, 1936

Croft, M., 'The Accidental Food Sovereignty Activist', in N. Rose (ed.), *Fair Food: Stories from a Movement Changing the World*, UQP, Brisbane, 2015

Davis, D. R., 'Declining Fruit and Vegetable Nutrient Composition: What is the Evidence?', *HortScience*, 44, no. 1, February 2009

—— 'Trade-Offs in Agriculture and Nutrition', *Food Technology*, 59, no. 3, March 2005

Davis, D. R. et al., 'Changes in USDA Food Composition Data for 43 Garden Crops, 1950 to 1999', *Journal of the American College of Nutrition*, 23, no. 6, 2004

Demeter, 'A Brief History of Biodynamics: An Australian Perspective', *Biodynamic Growing*, no. 1, December 2003, http://www.biodynamic.com.au/content/contentfiles/pdf/bd_history.pdf

Descartes, R., 'Discourse on Method', in R. Descartes, *The Philosophical Works of Descartes*, E. S. Haldane and C. R. T. Ross (eds), 2 vols, Dover, New York, 1955

Diaz, S. et al., 'The Plant Traits that Drive Ecosystems: Evidence from Three Continents', *Journal of Vegetation Science*, 15, no. 3, February 2004

Digby, S. N. et al., 'Offspring Born to Ewes Fed High Salt During Pregnancy Have Altered Responses to Oral Salt Loads', *Animal*, 4, no. 1, January 2009

Dodd, G., 'Gut Feeling', *Sydney Alumni Magazine*, Issue 5, Semester One 2017

Doran-Browne, N. A. et al., 'Carbon-Neutral Wool Farming in South-Eastern Australia', *Animal Production Science*, 56, no. 3, February 2016

Doughty, C. E. et al., 'Global Nutrient Transport in a World of Giants', *Proceedings of the National Academy of Sciences of the United States of America*, 113, no. 4, January 2015, http://www.pnas.org/cgi/doi/10.1073/pnas.1502549112

Dregne, H. E., 'Desertification of Arid Lands', in F. El-Baz and M. H. A. Hassan (eds), *The Physics of Desertification*, Martinus Nijhoff Publishers, Dordrecht, Netherlands, 2012

Bibliography

Earl, J. M. and C. E. Jones, 'The Need for a New Approach to Grazing Management – Is Cell Grazing the Answer?', *The Rangeland Journal*, 18, no. 2, January 1996

Edwards, E. J. et al., 'The Origin of C_4 Grasslands: Integrating Evolutionary and Ecosystem Science', *Science*, 328, no. 5978, April 2010

Eshuis, J. and M. Stuiver, 'Learning in Context through Conflict and Alignment: Farmers and Scientists in Search of Sustainable Agriculture', *Agriculture and Human Values*, 22, no. 2, June 2005

Fan, M-S. et al., 'Evidence of Decreasing Mineral Density in Wheat Grain over the Last 160 Years', *Journal of Trace Elements in Medicine and Biology*, 22, no. 4, 2008

FAO, 'Combating Micronutrient Deficiencies: Food-Based Approaches', Brian Thompson and Leslie Amoroso (eds), Rome, 2011

Feller, C. L. et al., '"The Principles of Rational Agriculture" by Albrecht Daniel Thaer (1752–1828): An Approach to the Sustainability of Cropping Systems at the Beginning of the 19th Century', *Journal of Plant Nutrition and Soil Science*, 166, no. 6, December 2003

Finlayson, G., 'Beyond Sustainability for the Australian Pastoral Industry: Improving Profitability for the Family Farmers in the Rangelands', Nuffield Australia Farming Scholars, Griffith, New South Wales, 2008, http://nuffieldinternational.org/live/Report/AU/2008/graham-finlayson

Fischer, J. et al., 'Reversing a Tree Regeneration Crisis in an Endangered Ecoregion', *Proceedings of the National Academy of Sciences of the United States of America*, 106, no. 25, June 2009

Fischer, R. A. et al., 'Chapter 11: Resource Use Efficiency, Sustainability and Environment', in T. Fischer, D. Byerlee and G. Edmeades, *Crop Yields and Global Food Security: Will Yield Increase Continue to Feed the World?*, Australian Centre for International Agricultural Research, Canberra, 2014, http://aciar.gov.au/publication/mn158

Fließbach, A. et al., 'Soil Organic Matter and Biological Soil Quality Indicators After 21 Years of Organic and Conventional Farming', *Agriculture, Ecosystems and Environment*, 118, nos 1–4, January 2007

Foley, J. A. et al., 'Solutions for a Cultivated Planet', *Nature*, 478, October 2011

'The Foodies Chat with Jeremy of Eden Valley Flour', We Love Perth, 11 April 2012, http://weloveperth.net.au/the-foodies-chat-with-jeremy-of-eden-valley-flour

Foster, J. B., 'Marx's Theory of Metabolic Rift: Classical Foundations for Environmental Sociology', *American Journal of Sociology*, 105, no. 2, September 1999

Fraser, C. E. et al., 'Farming and Mental Health Problems and Mental Illness', *International Journal of Social Psychiatry*, 51, no. 4, December 2005

Frawley, K., 'Evolving Visions: Environmental Management and Nature Conservation in Australia', in S. Dovers (ed.), *Australian Environmental History: Essays and Cases*, Oxford University Press, Melbourne, 1994

Gammage, B., 'The Biggest Estate on Earth', Paper for Writing the Australian Landscape Conference, National Library of Australia, 2013, http://www.nla.gov.au/content/the-biggest-estate-on-earth

Gatschet, J., 'Natural Systems Agriculture at The Land Institute', The Land Institute, September 1997

Gattinger, A. et al., 'Enhanced Top Soil Carbon Stocks under Organic Farming',
Proceedings of the National Academy of Sciences, 109, no. 44, October 2012

George, S., 'How to Win the War of Ideas: Lessons from the Gramscian Right', *Dissent*, 44,
no. 3, Summer 1997

Gepts, P., 'Crop Domestication as a Long-Term Selection Experiment', *Plant Breeding
Reviews*, Vol. 24, Pt 2, John Wiley & Sons, 2004

Gersick, C. J. G., 'Revolutionary Change Theories: A Multilevel Exploration of the Punctuated Equilibrium Paradigm', *The Academy of Management Review*, 16, no. 1, January 1991

Gill, A. M., 'Fire Pulses in the Heart of Australia: Fire Regimes and Fire Management in
Central Australia', Report to Environment Australia (per CSIRO Plant Industry),
August 2000

—— 'Underpinnings of Fire Management for Biodiversity Conservation in Reserves', Fire
and Adaptive Management, Report No. 73, Victorian Government Department of
Sustainability and Environment, Melbourne, 2008

Glikson, A., 'Fire and Human Evolution: The Deep-Time Blueprints of the
Anthropocene', *Anthropocene*, 3, November 2013

Global Footprint Network, 'World Footprint: Do We Fit on the Planet?', 2009, http://
www.footprintnetwork.org/en/index.php/GFN/page/world_footprint

Goldstein, W. et al., 'Comparisons of Conventional, Organic, and Biodynamic Methods',
Demeter, http://www.demeter-usa.org/downloads/Demeter-Science-Comparisons.pdf

Goodman, D., 'Organic and Conventional Agriculture: Materialising Discourse and
Agro-Ecological Managerialism', *Agriculture and Human Values*, 17, September 2000

Granstedt, A. and L. Kjellenberg, 'Organic and Biodynamic Cultivation: A Possible Way
of Increasing Humus Capital, Improving Soil Fertility and Providing a Significant
Carbon Sink in Nordic Conditions', 16th IFOAM World Congress, Modena, Italy,
16–20 June 2008, Organic Eprints, http://orgprints.org/12625

Grayson, R., 'A Short and Incomplete History of Permaculture', Pacific Edge, 26 July 2007,
http://pacific-edge.info/2007/07/a-short-and-incomplete-history-of-permaculture

Guilliatt, R., 'How a Row over Land Clearing Left Compliance Officer Glen Turner
Dead', *Weekend Australian Magazine*, 13 September 2014

Gupta, A. K., 'Origin of Agriculture and Domestication of Plants and Animals Linked to
Early Holocene Climate Amelioration', *Current Science*, 87, no. 1, July 2004

Haikai, T., 'Terraquaculture: Farming Living Waters Flowing Through the Landscape',
Terraquaculture website, 2008, http://www.terraquaculture.com

Haken, H., 'Synergistics: Is Self-Organization Governed by Universal Principles?', in E.
Jantsch (ed.), *Evolutionary Vision: Toward a Unifying Paradigm of Physical, Biological, and
Sociocultural Evolution*, Westview Press, Boulder, Colorado, 1981

Haynes, G., 'Elephant Landscapes: Human Foragers in the World of Mammoths, Mast-
odonts, and Elephants', The World of Elephants, International Congress, Rome, 2001

Heathcote, R. L., 'The Visions of Australia, 1770–1970', in A. Rapoport (ed.), *Australia as
Human Setting*, Angus & Robertson, Sydney, 1994

Heckman, J., 'A History of Organic Farming: Transitions from Sir Albert Howard's War
in the Soil to USDA National Organic Program', *Renewable Agriculture and Food Systems*,
21, no. 3, September 2006

Hicks, C. S., 'Keyline Farming and the Australian Future', *Keyline*, 1, no. 1, Keyline Publications, Sydney, 1955

Hicks, S., 'The Relation of the Health of the Soil to the Health of Man', *Transactions and Proceedings of the Royal Society of New Zealand 1868–1961*, 77, 1948

Hill, S., 'Yeomans' Keyline Design for Sustainable Soil, Water, Agrosystems and Biodiversity Conservation: A Personal Social Ecology Analysis', The 2002 Australian Academy of Science Fenner Conference on the Environment: Agriculture for the Australian Environment, Canberra, 2003

Hodges, J., 'Foundations, Fallacies and Assumptions of Science for Livestock Development', FAO International Symposium on Sustainable Improvement of Animal Production and Health, Vienna, 2010

Hoffman, M. T. and R. M. Cowling, '"Enough To Be Considered Useful": John Acocks' Contribution to South African Botany', *South African Journal of Botany*, 69, no. 1, March 2003

Holmgren, D., 'Migrant Plants and Animals: Ecological Imperialism or Co-Evolution', unpublished manuscript, researched 1990s

Hope, A. D., 'Australia', in H. Heseltine (ed.), *The Penguin Book of Australian Verse*, Penguin, Melbourne, 1972

IFOAM (International Foundation for Organic Agriculture), 'Alex Podolinsky', 2015, http://www.ifoam.bio/en/alex-podolinsky

Ikerd, J., 'Understanding and Managing the Multi-Dimensions of Sustainable Agriculture', paper presented at the Southern Region Sustainable Agriculture Professional Development Programme Workshop, Gainesville, Florida, 1997

Isherwood, J., cited in USDA, Natural Resources Conservation Service, 'Soils – Soil Quotations', https://www.nrcs.usda.gov/wps/portal/nrcs/detail/soils/edu/?cid=nrcs 142p2_054312

Ive, J. and R., 'Achieving Production and Environmental Benefits in a Challenging Landscape', Proceedings of the 22nd Annual Conference of the Grassland Society of New South Wales, 2007

Jackson, W., 'Analogy: A Driver and Natural Systems Agriculture', The Land Institute, 4 October 1999, https://landinstitute.org/media-coverage/analogy-driver-natural -systems-agriculture

—— 'The Ecosystem as a Conceptual Tool for Agriculture and Culture', The Land Institute, 2009

—— 'Natural Systems Agriculture: A Truly Radical Alternative', *Agriculture, Ecosystems and Environment*, 88, no. 2, February 2002, The Land Institute, https://landinstitute .org/wp-content/uploads/2002/01/NSA-Jackson-2002.pdf

James, S. R. et al., 'Hominid Use of Fire in the Lower and Middle Pleistocene: A Review of the Evidence', *Current Anthropology*, 30, no. 1, February 1989

Jefferson, R., 'The Hologenome. Agriculture, environment and the developing world: a future of PCR', Cold Spring Harbor, New York, 1994

Jehne, W., 'How Relationships between Healthy Soils, Healthy Food and Healthy People Govern Our Health and Future', talk at Rahamim Ecology Centre, 23 May 2012

—— 'Regenerating the Earth's Soil Carbon Sponge', draft science paper for *Healthy Soils Australia* proposal, the Virgin Earth Challenge, September 2014

—— 'Restoring Water Cycles to Naturally Cool Climate and Reverse Global Warming', paper for *Soils for Life*, unpublished, 2015

Joffre, R., S. Rambal and J. P. Ratte, 'The Dehesa System of Southern Spain and Portugal as a Natural Ecosystem Mimic', *Agroforestry Systems*, 45, 1999

Johal, G. S. and D. M. Huber, 'Glyphosate Effects on Diseases of Plants', *European Journal of Agronomy*, 31, no. 3, October 2009

Johnson, C. N., 'Ecological Consequences of Late Quaternary Extinctions of Megafauna', *Proceedings: Biological Sciences*, 276, no. 1667, March 2009

—— 'The Remaking of Australia's Ecology', *Science*, 309, no. 5732, July 2006

Johnson, J. 'Ecological Restoration of Cleared Agricultural Land in Gondwana Link: Lifting the Bar at "Peniup"', *Ecological Management and Restoration,* 11, no. 1: 16-26, March 2010

Jones, C.G., et al. 'Organisms as Ecosystem Engineers', *Oikos,* 69, no. 3: 373-386, April 1994

Jones, C. E., 'Carbon, Air and Water: Is That All We Need?', Managing the Carbon Cycle, Katanning Workshop, 21–22 March 2007

—— 'Carbon and Catchments: Inspiring REAL CHANGE in Natural Resource Management', Managing the Carbon Cycle, National Forum, 22–23 November 2006

—— 'Carbon that Counts', New England and North-West Landcare Adventure, Guyra, New South Wales, 2011

—— 'Liquid Carbon Pathway', extract from *Australian Farm Journal*, 338, 3 July 2008

Jones, R., 'Fire-Stick Farming', *Australian Natural History*, 16, no. 7, September 1969

Joost, S. and R. Negrini, 'Early Stirrings of Landscape Genomics: Awaiting Next-Next Generation Sequencing Platforms Before Take-Off', FAO International Symposium on Sustainable Improvement of Animal Production and Health, Vienna, 2010

Jurskis, V. and J. Turner, 'Eucalypt Dieback in Eastern Australia: A Simple Model', *Australian Forestry*, 65, no. 2, 2002

Kellogg, C., 'Soil and Society', in USDA, *Soils and Men: Yearbook of Agriculture*, US Government Printing Office, Washington DC, 1938

Kingsley, J. Y. et al., 'Exploring Aboriginal People's Connections to Strengthen Human-Nature Theoretical Perspectives', *in Advances in Medical Sociology*, vol. 15 of *Ecological Health: Society, Ecology & Health*, 2013

Kingsnorth, P., 'The Axis and the Sycamore', *Orion Magazine*, 36, no. 1, Jan/Feb 2017

Kinsella, J., 'Why They Stripped the Last Trees from the Banks of the Creek', in J. Kinsella, *The Silo: A Pastoral Symphony – Poems*, Fremantle Arts Centre Press, Fremantle, 1995

Kirschenmann, F., 'Spirituality in Agriculture', paper prepared for the Concord School of Philosophy, Concord, Massachusetts, 8 October 2005

Kiviat, E., 'Ecosystem Services of *Phragmites* in North America with Emphasis on Habitat Functions', *AoB Plants, no. 5*, 2013: plt008

Korten, D., 'The Great Turning: From Empire to Earth Community', *Yes! A Journal of Positive Futures*, Summer 2006

Kremer, R., 'New Peer Review Study Finds Commonly Used Pesticides (Glyphosate) Induce Disease', Real Food Forager, 7 May 2013, http://realfoodforager.com/new-peer-review-study-finds-commonly-used-pesticides-glyphosate-induce-disease

Kummer, C., 'The Nestlé Health Offensive', *MIT Technology Review Business Report*, 26 May 2015

Laibow, R. E., 'Nutrition and Health', paper for Natural Solutions Foundation, 2005, http://www.healthfreedomusa.org

Lambert, C., 'The Way We Eat Now', *Harvard Magazine*, May–June 2004, http://harvard magazine.com/2004/05/the-way-we-eat-now.html

Land & Water Australia, 'Land, Water and Wool Northern Tablelands Project 2006, LWAust Case Studies, "Wool Production and Diodiversity Working Together for Jon and Vicki Taylor"', Government Printing Office, Canberra, 2006

Larrson, P., 'Plant Pollination Synchronised with Full Moon', Stockholm University, 2 April 2015, http://www.su.se/english/about/news-and-events/plant-pollination -synchronised-with-full-moon-1.231179

Law, A. et al., 'Using Ecosystem Engineers as Tools in Habitat Restoration and Rewilding: Beaver and Wetlands', *Science of the Total Environment*, 605–606: 1021 – 1030, December 2017

Lawrence, J. E. et al., 'Hyporheic Zone in Urban Streams: A Review and Opportunities for Enhancing Water Quality and Improving Aquatic Habitat by Active Management', *Environmental Engineering Science*, 30, no. 8: 480-501, August 2013

Lefroy, E. C., 'Fragility, Health and Design: Conceptual Challenges for Australian Agriculture', Fenner Conference on the Environment, 2002

Lefroy, E. C. et al., 'What Can Agriculture Learn from Natural Ecosystems?', *Agroforestry Systems*, 45, no. 1–3, 1999

Leonard, W. R., 'Food for Thought: Into the Fire', *Scientific American*, 287, no. 6, December 2002

Leopold, A., 'Engineering and Conservation', 1938, in Flader and Callicott (eds), *The River of the Mother of God and Other Essays by Aldo Leopold*, University of Wisconsin Press, Madison, Wisconsin, 1991

—— 'The Land-Health Concept and Conservation', in J. B. Callicott and E. T. Freyfogle (eds), *Aldo Leopold: For the Health of the Land – Previously Unpublished Essays and Other Writings*, Island Press–Shearwater Books, Washington, 1999

Lincoln, A., 'Address to the Wisconsin State Agricultural Society, Milwaukee, Wisconsin', 30 September 1859, in A. Delbanco (ed.), *The Portable Abraham Lincoln*, Penguin, New York, 1993

Lloyd, D. and E., 'Grain Growing: Using Knowledge from the Past to Shape the Future', unpublished paper

Locke, J., 'The First Treatise' and 'The Second Treatise', in P. Laslett (ed.), *Two Treatises of Government*, 2nd edition, Cambridge University Press, Cambridge, 1967

Lowdermilk, W. C., 'Conquest of the Land Through Seven Thousand Years', *Agricultural Information Bulletin*, 99, no. 1, 1953

Lytton-Hitchins, J. A. et al., 'The Soil Condition of Adjacent Bio-Dynamic and Conventionally Managed Dairy Pastures in Victoria, Australia', *Soil Use and Management*, 10, no. 2, June 1994

MacGregor, A. J. et al., 'Natural and Post-European Settlement Variability in Water Quality of the Lower Snowy River Floodplain, Eastern Victoria, Australia', *River Research and Applications*, 21, nos. 2–3, 2005

Marsh, D., 'Farming from First Principles', in E. C. Lefroy et al., *Biodiversity: Integrating Conservation and Production – Case Studies from Australian Farms, Forests and Fisheries*, CSIRO, Melbourne, 2008
—— Rural Conservation Feature Article, *National Parks Association of NSW Journal*, December 2004
Marsh, D., cited in M. Cawood, 'Simplifying Towards Complexity: The Good Land Project', 4 December 2015, http://goodlandproject.com.au/2015/06david-marsh-simplicity-for-complexity
Martin, G., 'The Role of Small Ground-Foraging Mammals in Topsoil Health and Biodiversity: Implications to Management and Restoration', *Ecological Management and Restoration*, 4, no. 2, September 2003
Massy, C., 'Transforming the Earth: A Study in the Change of Agricultural Mindscapes', PhD thesis, Australian National University, 2013
McCosker, T., 'Cell Grazing: The First 10 Years in Australia', *Tropical Grasslands*, 34, no. 3, September 2000
McDonald, A., 'Relationship of Nutrition to Prevention of Diseases', Northwestern Medicine, Northwestern University, http://www.feinberg.northwestern.edu/nutrition/tools-resources/sbm.html
McHarg, I. L., 'Man, Planetary Disease', Address to North American Wildlife & Natural Resources conference, Portland, Oregon, *Vital Speeches of the Day*, October 1971
Meadows, D., 'Leverage Points: Places to Intervene in a System', *Whole Earth*, Winter 1997
Meghani, Z., 'Values, Technologies, and Epistemology', *Agriculture and Human Values*, 25, no. 1, January 2008
Meuret, M. and F. Provenza, 'When Art and Science Meet: Integrating Knowledge of French Herders with Science of Foraging Behaviour', *Rangeland Ecology and Management*, 68, no. 1, 2015
Mezirow, J., 'Transformation Theory of Adult Learning', in M. R. Welton (ed.), *In Defense of the Lifeworld: Critical Perspectives on Adult Learning*, State University of New York Press, New York, 1995
Miller, G. H. et al., 'Ecosystem Collapse in Pleistocene Australia and a Human Role in Megafaunal Extinction', *Science*, 309, no. 5732, July 2005
Morris, G. D., 'Sustaining National Water Supplies by Understanding the Dynamic Capacity that Humus Has to Increase Soil Water-Holding Capacity', thesis submitted for Master of Sustainable Agriculture, University of Sydney, July 2004
Moorsel, S.J. van, et al., 'Community Evolution Increases Plant Productivity at Low Diversity', *Ecology Letters*, 21, no. 1: 128-137, November 2017
Mulcaster, G., 'Making a Mess of His Life', *The Age*, 12 February 2004, http://www.theage.com.au/articles/2004/02/11/1076388417401.html
Mulvaney, R. L., S. A. Khan and T. R. Ellsworth, 'Need for a Soil-Based Approach in Managing Nitrogen Fertilizers for Profitable Corn Production', *Soil Science Society of America Journal*, 70, no. 1, 2006
Murphy, K. M., P. G. Reeves and S. S. Jones, 'Relationship Between Yield and Mineral Nutrient Concentrations in Historical and Modern Spring Wheat Cultivars', *Euphytica*, 163, no. 3, Octover 2008

Bibliography

Murray, L., 'Noonday Axeman', in *The Ilex Tree*, University of Sydney Press, Sydney, 1965
—— 'Rainwater Tank', in *Collected Poems*, Duffy & Snellgrove, Sydney, 2002
Namirrkki, I., 'Our Spirits Lie in the Water', in H. Perkins (ed.), *Crossing Country: The Alchemy of Western Arnhem Land Art*, Art Gallery of New South Wales, Sydney, 2004
New South Wales Department of Industry and Investment, 'Prime Facts: Wingless Grasshoppers', April 2010, based on G. Baker, 'Wingless Grasshopper', Agfact AE1, NSW Agriculture, 2005
Nixon, R., 'Slow Violence and the Environmentalism of the Poor', *Journal of Commonwealth and Postcolonial Studies*, Vols 13.2–14.1, 2006–07
Norton, R., 'Nitrogen Loss Pathways', International Plant Nutrition Institute, 11 October 2012, http://anz.ipni.net/article/ANZ-3002
Olea, L. and A. San Miguel-Ayanz, 'The Spanish Dehesa: A Traditional Mediterranean Silvopastoral System Linking Production and Nature Conservation', Paper for 21st General Meeting of the European Grassland Federation, Badajoz (Spain), April 2006
Oosthoek, J., 'Worlds Apart? The Scottish Forestry Tradition and the Development of Forestry in India', *Journal of Irish and Scottish Studies*, 3, no. 1, 2010
Orians, G. H. and A. V. Milewski, 'Ecology of Australia: The Effects of Nutrient-Poor Soils and Intense Fires', *Biology Review of the Cambridge Philosphical Society*, 82, no. 3, August 2007
Orr, D., 'Four Challenges of Sustainability', spring seminar to School of Natural Resources, University of Vermont, 2003
Osborne, C. P., 'Atmosphere, Ecology and Evolution: What Drove the Miocene Expansion of C_4 Grasslands?', *Journal of Ecology*, 96, no. 1, January 2008
Owen, D. F. and R. G. Wiegert, 'Mutualism Between Grasses and Grazers: An Evolutionary Hypothesis', *Oikos*, 36, no. 3, March 1981
Owen-Smith, N., 'Pleistocene Extinctions: The Pivotal Role of Megaherbivores', *Paleobiology*, 13, no. 3, June 1987
Pahl-Wostl, C. and M. Hare, 'Processes of Social Learning in Integrated Resources Management', *Journal of Community and Applied Social Psychology*, 14, no. 3, May 2004
Parker, C. B., 'The Phosphorus Balance of a Conventional and a Biodynamic Dairy Farm', undergraduate thesis, La Trobe University, School of Agriculture, 1992
Paull, J., 'The Lost History of Organic Farming in Australia', *Journal of Organic Systems*, 3, no. 2, 2008
—— 'Organic Farming: The Arrival and Uptake of the Dissident Agriculture Meme in Australia', *Journal of Organics*, 2, no. 1, 2015
—— 'The Soil Association and Australia: From *Mother Earth* to Eve Balfour', *Mother Earth*, 4, Spring, 2011, http://orgprints.org/20947/1/Paull2011MotherEarth.pdf
Petty, A. M., V. de Koninck and B. Orlove, 'Cleaning, Protecting, or Abating? Making Indigenous Fire Management "Work" in Northern Australia', *Journal of Ethnobiology*, 35, no. 1, 2015
Philpott, T., 'New Research: Synthetic Nitrogen Destroys Soil Carbon, Undermines Soil Health', Grist, 24 February 2010, http://grist.org/article/2010-02-23-new-research -synthetic-nitrogen-destroys-soil-carbon-undermines
Planet Ark, 'Climbing Trees: Getting Aussie Kids Back Outdoors', 2011, http://treeday .planetark.org/documents/doc-535-climbing-trees-media-summary-2011-06-17-final.pdf

—— 'Missing Trees: The Inside Story of an Outdoor Nation', 2013, http://treeday.planet ark.org/documents/doc-1049-missing-trees-key-findings.pdf

—— 'Planting Trees: Just What the Doctor Ordered', 2012, http://treeday.planetark.org /documents/doc-812-planting-trees-report-2012-06-25-final.pdf

Ploeg, R. R. van der, W. Böhm and M. B. Kirkham, 'History of Soil Science: On the Origin of the Theory of Mineral Nutrition of Plants and the Law of the Minimum', *Soil Science Society of America Journal*, 63, no. 5, 1999

Porter, F., 'Alex Podolinsky: Biodynamic Pioneer and Man of the Soil', *Biodynamic Growing*, 11, December 2008

Pretty, J. N., 'Agricultural Sustainability: Concepts, Principles and Evidence', *Philosophical Transactions of the Royal Society B*, 363, no. 1491, February 2008

'Prospect Pastoral Co.', Soils for Life Case Study, Soils for Life, September 2012, http:// soilsforlife.org.au/cs-prospect-pastoral-company

Provenza, F. et al., 'Complex Creative Systems: Principles, Processes, and Practices of Transformation', *Rangelands*, 35, no. 5, October 2013

Provenza, F. et al., 'The Wisdom Body: Nutrition, Health, and Nature's Pharmacopeia', paper for invited synthesis for *Evolutionary Biology*, undated, as personal communication, 30 March 2015

Provenza, F., M. Meuret and P. Gregorini, 'Our Landscapes, Our Livestock, Ourselves: Restoring Broken Linkages among Plants, Herbivores, and Humans with Diets that Nourish and Satiate', *Appetite*, 95, December 2015

Raupp, J., 'The Long-Term Trial in Darmstadt: Mineral Fertilizer, Composted Manure and Composted Manure Plus All Biodynamic Preparations', in J. Raupp (ed.), *Main Effects of Various Organic and Mineral Fertilization on Soil Organic Matter Turnover and Plant Growth: EC-Concerted Action 'Fertilization Systems in Organic Farming'*, Publications of the Institute for Biodynamic Research, Darmstadt, Germany, Vol. 5, 1995

Reganold, J. P., 'Soil Quality and Profitability of Biodynamic and Conventional Farming Systems: A Review', *American Journal of Alternative Agriculture*, 10, no. 1, March 1995

Retallack, G. J., 'Global Cooling by Grassland Soils of the Geological Past and Near Future', *Annual Review of Earth Planet Sciences,* 41: 69-86, May 2013

Revillini, D., et al., 'Ecosystems, Evolution and Plant Soil Feedbacks: The Role of Locally Adapted Mycorrhizas and Rhizobacteria in Plant-Soil Feedback Systems', *Functional Ecology*, April, 2016

Rinaudo, T., 'The Development of Farmer Managed Natural Regeneration', The Permaculture Research Institute News, 24 September 2008, http://permaculturenews .org/2008/09/24/the-development-of-farmer-managed-natural-regeneration

Ripl, W., 'Water: The Bloodstream of the Biosphere', *Philosophical Transactions of the Royal Society B*, 358, no. 1440, December 2003

Roberts, R. G. et al., 'New Ages for the Last Australian Megafauna: Continent-Wide Extinction about 46,000 Years Ago', *Science*, 292, no. 5523, June 2001

Rockström, J. et al., 'Planetary Boundaries: Exploring the Safe Operating Space for Humanity', *Ecology and Society*, 14, no. 2, 2009, http://www.ecologyandsociety.org /vol14/iss2/art32

Rodale Institute, 'Regenerative Organic Agriculture and Climate Change: A Down-to-Earth Solution to Global Warming', 2014, http://rodaleinstitute.org/assets/RegenOrgAgricultureAndClimateChange_20140418.pdf

Rolls, E., 'The End, or New Beginning?', in S. Dovers, *Environmental History and Policy: Still Settling Australia*, Oxford University Press, Melbourne, 2000

—— 'More a New Planet than a New Continent', in S. Dovers (ed.), *Australian Environmental History: Essays and Cases*, Oxford University Press, Melbourne, 1994

Rosenberg, E. et al., 'The Role of Microorganisms in Coral Health, Disease and Evolution', *Nature Reviews Microbiology*, 5, March 2007

Rúa, M. A. et al., 'Home-Field Advantage? Evidence of Local Adaptation Among Plants, Soil, and Arbuscular Mycorrhizal Fungi: Through Meta-Analysis', *BMC Evolutionary Biology* 16: 122, doi 10.1186/s12862-016-0698-9, 2016

Rule, S. et al., 'The Aftermath of Megafaunal Extinction: Ecosystem Transformation in Pleistocene Australia', *Science*, 335, no. 6075, March 2012

Russell, D. B. and R. L. Ison, 'The Research-Development Relationship in Rural Communities: An Opportunity for Contextual Science', in D. B. Russell and R. L. Ison, *Agricultural Extension and Rural Development: Breaking out of Knowledge Transfer Traditions: A Second-Order Systems Perspective*, Cambridge University Press, Cambridge, 2000

Russell-Smith, J. and A. C. Edwards, 'Seasonality and Fire Severity in Savanna Landscapes of Monsoonal Northern Australia', *International Journal of Wildland Fire*, 15, no. 4, December 2006

Russell-Smith, J. et al., 'Managing Fire Regimes in North Australia: Applying Aboriginal Approaches to Contemporary Global Problems', *Frontiers in Ecology and the Environment*, 11, 2013

Ryan, P., 'Overview of Non-Vascular Plants, Lichens, Fungi and Algae in the Goulburn Broken Catchment: Their Status, Threats and Management', a background paper for the Goulburn Broken Catchment Management Authority Regional Catchment Strategy Review Process, Canberra, 2002

Rydin, C. and K. Bolinder, 'Moonlight Pollination in the Gymnosperm *Ephedra* (Gnetales)', *Biology Letters*, 1 April 2015, http://rsbl.royalsocietypublishing.org/content/11/4/20140993

Sage, R. F., 'The Evolution of C_4 Photosynthesis', *New Phytologist*, 161, no. 2, December 2004

Samsel, A. and S. Seneff, 'Glyphosate's Suppression of Cytochrome P450 Enzymes and Amino Acid Biosynthesis by the Gut Microbiome: Pathways to Modern Diseases', *Entropy*, 15, no. 4, 2013

Sauer, C. O., 'The Agency of Man on the Earth', in W. L. Thomas (ed.), *Man's Role in Changing the Face of the Earth*, Vol. 1, University of Chicago Press, Chicago, 1956

Savory, A., 'For the Love of Fire', unpublished paper, personal communication, 18 August 2015

—— 'Holistic Resource Management: A Conceptual Framework for Ecologically Sound Economic Modelling', *Ecological Economics*, 3, no. 3, September 1991

Schirmer, J. et al., 'Healthier Land, Healthier Farmers: Considering the Potential of Natural Resource Management as a Place-Focused Farmer Health Intervention', *Health and Place*, 24, November 2013

Schwartz, M. K. et al., 'Landscape Genomics: A Brief Perspective', in S. A. Cushman and F. Huettmann (eds), *Spatial Complexity, Informatics, and Wildlife Conservation*, Springer, Japan, 2009

Schwenke, G., 'Nitrogen Volatilisation: Factors Affecting How Much N Is Lost and How Much Is Left Over Time', 25 July 2014, Grains Research and Development Corporation, https://grdc.com.au/research-and-development/grdc-update-papers/2014/07/factors-affecting-how-much-n-is-lost-and-how-much-is-left-over-time

Scolaro, N., 'Archie Roach Sings from the Soul', *Dumbo Feather*, no. 50, 2017

Séralini, G-E. et al., 'Long Term Toxicity of a Roundup Herbicide and a Roundup-Tolerant Genetically Modified Maize', *Food and Chemical Toxicology*, 50, no. 11, 2012

SLM, 'Environmental and Social Impact', http://slmpartners.com/about/environmental-impact

—— 'SLM Australia Livestock Fund', http://slmpartners.com/activities/slm-australia-livestock-fund

Smith, T. et al., 'Implications of the Synergies Between Systems Theory and Permaculture for Learning About and Acting Towards Sustainability', 2007 ANZSEE Conference, Re-Inventing Sustainability: A Climate for Change', 3–6 July 2007

Spencer, L., 'Cultural Keyline: The Life and Work of Dr Neville Yeomans', PhD thesis, James Cook University, Queensland, 2005

Speybroeck, L. V., 'From Epigenesis to Epigenetics: The Case of C. H. Waddington', *Annals of the New York Academy of Sciences*, 981, December 2002

Sprengel, cited in R. R. van der Ploeg, W. Böhm and M. B. Kirkham, 'History of Soil Science: On the Origin of the Theory of Mineral Nutrition of Plants and the Law of the Minimum', *Soil Science Society of America Journal*, 63, no. 5, 1999

Spriggs, E. L. et al., 'C_4 Photosynthesis Promoted Species Diversification during the Miocene Grassland Expansion', *Plos One*, 9, no. 5, 16 May 2014, http://journals.plos.org/plosone/article?id=10.1371/journal.pone.0097722

Stebbins, G. L., 'Coevolution of Grasses and Herbivores', *Annals of the Missouri Botanical Garden*, 68, no. 1, 1981

Steffen, W. et al., 'The Anthropocene: From Global Change to Planetary Stewardship', *AMBIO: A Journal of the Human Environment*, 40, no. 7, 2011

Steiner, F., 'Healing the Earth: The Relevance of Ian McHarg's Work for the Future', *Philosophy and Geography*, 7, no. 1, February 2004

SupremeFulvic.com, 'Fulvic Acid: A Substance Critical to Human Health', http://www.supremefulvic.com/documents/pdf/8b.fulvic.acid.report.addinfo.pdf

Sustainable Pulse, 'German Beer Industry in Shock over Glyphosate Contamination', 25 February 2016, http://sustainablepulse.com/2016/02/25/german-beer-industry-in-shock-over-probable-carcinogen-glyphosate-contamination

Thistleton, J., 'Beetle-Mania Gives Bush Flies the Brush', *Canberra Times*, 13 February 2015

Thomas, D., 'The Mineral Depletion of Foods Available to US as a Nation (1940–2002) – A Review of the 6th Edition of McCance and Widdowson', *Nutrition and Health*, 19, nos. 1–2, July 2007

Thompson, B. and L. Amoroso, 'Combating Micronutrient Deficiencies: Food-Based Approaches', Food and Agriculture Organization and CAB International, Rome, 2011

Bibliography

Thompson, J. L., 'Really Useful Knowledge: Linking Theory and Practice', in B. Connolly et al. (eds), *Radical Learning for Liberation*, Maynooth Adult and Community Education, National University of Ireland, Maynooth, Kildare, 1996

Tongway, D. and N. Hindley, 'Landscape Function Analysis: A System for Monitoring Rangeland Function', *African Journal of Range and Forage Science*, 21, no. 2, 2004

Tothill, J. C., 'A Review of Fire in the Management of Native Pastures with Particular Reference to North-Eastern Australia', *Tropical Grasslands*, 5, no. 1, 1971

Turinek, M. et al., 'Biodynamic Agriculture Research Progress and Priorities', *Renewable Agriculture and Food Systems*, 24, no. 2, June 2009

Turner, D., 'The Incarnation of Nambirrirrma', in T. Swain and D. Rose (eds), *Aboriginal Australians and Christian Missions: Ethnographic and Historical Studies*, The Australian Association of the Study of Religions, Adelaide, 1988

Ullrich, H., 'Rudolf Steiner (1861–1925)', *Prospects: The Quarterly Review of Comparative Education*, UNESCO, International Bureau of Education, Paris, vol. xxiv (3/4), 1994

Wallace, A. and C. Dowling, 'Carbon Farming and Nitrogen Fertilizer, Opportunity or Threat?', Grain Industry Association of Western Australia, 2015

Watson-Gegeo, K. A., 'Mind, Language, and Epistemology: Toward a Language Socialization Paradigm for SLA', *The Modern Language Journal*, 88, no. 3, Autumn 2004

Weatherly, R., 'Ecosystem Decline in Isolated Habitats', Australian Society of Animal Production, Annual Conference, Geelong, October 2009

Weatherstone, J., 'Lyndfield Park: Looking Back – Moving Forward', Greening Australia/ Land & Water Australia, pamphlet, 2003

Webb, J. A. and M. Domanski, 'Fire and Stone', *Science*, 325, no. 5942, September 2009

Weber, K. T. and S. Horst, 'Desertification and Livestock Grazing: The Roles of Sedentarization, Mobility and Rest', *Pastoralism: Research, Policy and Practice*, 1, no. 19, December 2011

Weisser, W. W. et al., 'Biodiversity Effects of Ecosystem Functioning in a 15-Year Grassland Experiment: Patterns, Mechanisms, and Open Questions', *Basic and Applied Ecology* 23: 1-73, September 2017

Welch, R. M., 'The Impact of Mineral Nutrients in Food Crops on Global Human Health', *Developments in Plant and Soil Sciences*, Vol. 98, 2002

Wendon, E., 'A Report Documenting the Impact of Different Agricultural Management Actions: No Till Versus Conventional Till, on Soil Biological Populations and Diversity', for University of Sydney Soil Resource Management, Assignment Three: Soil Research Issue, 2007

Wentworth Group of Concerned Scientists, 'Optimising Carbon in the Australian Landscape: How to Guide the Terrestrial Carbon Market to Deliver Multiple Economic and Environmental Benefits', 22 October 2009, http://wentworthgroup.org/2009/10 /optimising-carbon-in-the-australian-landscape

Western Australian Department of Agriculture and Food, 'Wingless Grasshoppers and Their Control', 6 May 2016, http://www.agric.wa.gov.au/spring/wingless-grasshoppers -and-their-control

Westoby, M. et al., 'Opportunistic Management for Rangelands Not at Equilibrium', *Journal of Range Management*, 42, no. 4, July 1989

Whalley, W., 'Grassland Regeneration and Reconstruction: The Role of Grazing Animals', *Ecological Management and Restoration*, 6, no. 1, 2006

'What is Geopathic Stress?', Geomantica, http://www.geomantica.com/articles/what-is -geopathic-stress

White Jnr, L., 'The Historical Roots of our Ecologic Crisis', *Science*, 155, no. 3767, March 1967

Williams, B., J. Walker and H. Tane, 'Drier Landscapes and Rising Watertables: An Eco Hydrological Paradox', *Natural Resource Management*, 4, 2001

Williams, J., 'Farming in the Future: Some Ways Forward', paper, Wentworth Group, March 2006

Wit, M. M. de and A. Iles, 'Toward Thick Legitimacy: Creating a Web of Legitimacy for Agroecology', *Elementa: Science of the Anthropocene*, 4, 2016

World Health Organization, 'Diet, Nutrition and the Prevention of Chronic Diseases', Report of a WHO/FAO Expert Consultation, Geneva, 2003

Yassoglou, N. J. and C. Kosmas, 'Desertification in the Mediterranean Europe: A Case in Greece', Rala Report No. 200, 2000

Yeomans, A. J., 'The Late Percival Alfred ("P. A.") Yeomans: A Man Before His Time', frontispiece to P. A. Yeomans, *The Challenge of Landscape*, online edition, Soiland Health.org, 1993, http://soilandhealth.org/wp-content/uploads/01aglibrary /010126yeomansII/010126homage.html

Zhang, W. et al., 'Ecosystem Services and Dis-Services to Agriculture', *Ecological Economics*, 64, no. 2, December 2007

Television and Video

Dean, B. and M. Butler, *First Footprints*, ABC TV, Screen Australia, Contact Films, Screen NSW et al., 2013

Escher, J., 'Epigenetics and the Multigenerational Effects of Nutrition, Chemicals, and Drugs', talk at Ancestral Health Symposium, California, 11 August 2014, YouTube, http://www.youtube.com/watch?v=k4LezkjNwnY

Randall, B., 'The Land Owns Us', video, Creative Spirits, https://www.creativespirits .info/aboriginalculture/land/meaning-of-land-to-aboriginal-people

Smith, R., *Australia: The Time Traveller's Guide*, ABC TV, Essential Media et al., 2012

Williams, J., interview, 'Peter Andrews', *Australian Story*, ABC, 6 June 2005

'Wool Industry Spreading Tree Message', *Landline*, ABC TV, 23 September 2007

NOTES

Note: Sources that appear in the Bibliography are not fully cited here.

FRONTISPIECE

1 J. L. Thompson, p. 21.
2 Jackson, *Becoming Native to this Place*, p. 26.

INTRODUCTION: FROM THE GROUND UP

1 Jackson, *Becoming Native to this Place*, p. 22.
2 Hawken, *Blessed Unrest*, p. 26.
3 Holocene: that unique period of post ice-age climatic conditions over the last 10,000 years or so during which agriculture evolved and humanity began its march towards modern civilisation.
4 Rockström et al., p. 32.
5 T. Berry, *The Great Work*, p. 104.
6 This new macro approach to Earth's life-support systems is timely because it enables the issue of global sustainability to be seen in the context of disruptions to the equilibrium of the self-organising system and self-regulating capacity of the planet.
7 McNeill, p. 3.
8 Rockström et al., p. 2; Foley et al., 'Solutions for a Cultivated Planet', *Nature*, p. 338.
9 Martin Luther King, 'Beyond Vietnam', Riverside Church, New York City, 4 April 1967.

Part I: Into the Anthropocene

CHAPTER 1: A GONDWANAN ARK

1 Margulis and Sagan, p. 243.
2 Flannery, *The Future Eaters*.
3 R. Smith, *Australia: The Time Traveller's Guide* (TV series).
4 M. E. White, *After the Greening*, p. 13.
5 P. Clarke, p. 53, citing Cook, 1770.
6 Gammage, in Dean and Butler, *First Footprints* (TV series).
7 B. Maynard, personal communication with author, 20 May 2016.
8 P. Clarke, p. 16.
9 Flood, p. 138.

10 Ibid., pp. 136–7, 159.

11 Gammage (article), p. 2; Gammage (book), pp. 124–5.

12 L. Kelly, pp. xii, 14.

13 Ibid., p. 14.

14 B. Maynard, personal communication with author, 20 May 2016.

15 Gammage, in Dean and Butler, *First Footprints* (TV series).

16 D. Rose, p. 67.

17 Pascoe; Pascoe, personal communication with author, 10 May 2018.

18 D. Rose, p. 7.

19 Cited in ibid., p. 9.

20 Kingsley et al., p. 682; Gammage (article), p. 1.

21 Randall, 'The Land Owns Us'.

22 Gammage (book), p. 131, note 28, and pp. 132–3.

23 Ibid., p. 139.

24 Stanner, p. 34.

25 Hancock, p. 33.

26 Stanner, p. 42.

27 Diamond, *Guns, Germs and Steel*.

28 Griffiths and Robin, p. 4.

29 C. Muir, p. 184.

30 Murray, 'Noonday Axeman'.

CHAPTER 2: EMERGENCE OF THE MECHANICAL MIND

1 Jackson, *New Roots for Agriculture*, p.2.

2 Guilliatt, p. 13.

3 Merchant, *The Death of Nature*, p. 268.

4 Ibid., pp. 20, 249.

5 Ibid., p. xvi.

6 Frankfort, 1949.

7 Plumwood, p. 192.

8 Ibid., p. 143.

9 Saul, p. 13.

10 See, for example, Brock; and van der Ploeg et al., 'History of Soil Science', *Soil Science Society of America Journal*.

11 Sprengel, cited in van der Ploeg et al.

12 Kellogg, p. 880.

13 Brock, p. ix.

14 Foster.

15 P. A. Yeomans, *The City Forest*, p. 88.

16 De Wit and Iles, 2016.

17 Shiva, *Making Peace with the Earth*, p. 3.

18 Jackson, *Consulting the Genius of the Place*, pp. 43–4.

19 Darwin, *The Formation of Vegetable Mould*.

20 For a comprehensive treatment of this history, see, for example, Conford.

21 Following Leopold's landmark book *A Sand Country Almanac* (1949), Wendell Berry published *The Unsettling of America* in 1977. In this, Berry gave us some of the best modern critiques of industrial agriculture. In time, he was joined by the likes of Jim Hightower, Susan George, Ingolf Vogeler, Frances Moore Lappé, J. Baird Callicott and dozens of others. In 1970, Charles Walters founded his influential journal and publishing house *Acres USA*, which, through its invaluable regular platform of ideas, articles and conferences, provided a stage for the emerging contesting eco-agriculture genre and movement.

22 For the comprehensive story of early Australian organic farming, see Paull.

23 Rodale Institute, 'Regenerative Organic Agriculture and Climate Change', p. 7.

24 Ibid., pp. 7–8.

Part II: Regenerating the Five Landscape Functions

CHAPTER 3: AN INDIVISIBLE, DYNAMIC WHOLE

1 Leopold, 'The Land-Health Concept and Conservation', p. 225.

2 Bride (ed.), pp. 154, 156.

3 Ibid., p. 164.

4 Ibid., p. 159.

5 Ibid., p. 160.

6 Ibid., pp. 167–8. 'Silk grass' was probably an undesirable annual *Vulpia* species, sometimes known as 'sixty-day grass' or 'rat's-tail fescue'.

7 Ibid., pp. 168–9.

8 Hope, p. 190.

9 The C_3 and C_4 biogeochemical pathways (where 'C' stands for carbon, and '3' and '4' denote different combinations of carbon atoms) relate to different photosynthetic and therefore different chemical and physiological pathways in plants. Thus, when the C_4 grasslands eventually rose to dominance in the period three to eight million years ago, the plant world suddenly had two different major plant types whose function covered the whole year, providing year-round green feed. This constituted a huge evolutionary step in life on Earth, as it allowed grassland communities to spread and occupy novel suitable niches. Today, while C_4 grasslands dominate around thirty per cent of the Earth's land surface, they however provide a high percentage of human food – such as cereal crops (aka annual C_4 weeds) for agriculture. That is, our entire history has been closely tied to both grasslands and their derived domesticated plants such as cereals. (Sage, 'The Evolution of C_4 Photosynthesis', *New Phytologist*; Spriggs et al., 'C_4 Photosynthesis Promoted Species Diversification', *Plos One*; Bouchenak-Khelladi et al., 'The Origins and Diversification of C_4 Grasses and Savanna-Adapted Ungulates', *Global Change Biology*; Osborne, 'Atmosphere, Ecology and Evolution', *Journal of Ecology*; Edwards et al., 'The Origin of C_4 Grasslands', *Science*; Bredenkap et al., 'On the Origin of Northern and Southern Hemisphere Grasslands', *Plant Ecology*.)

10 Orr, *Ecological Literacy*, p. 199; Lopez, p. 65.

11 Orr, *Ecological Literacy*, p. 93.

12 Ibid; Capra and Luisi, p. 291.

Regenerating the Solar-Energy Function

CHAPTER 4: AN UPSIDE-DOWN WORLD

1 N. S. G. Williams et al. (eds), p. 3.
2 'Dry sheep equivalent', or 'dse', is a standard measurement to compare different animal types, ages and weights. A dse is based on one fifty-kilogram Merino wether. Thus a pregnant Merino ewe is one and a half dses; a 'dry' (i.e. not pregnant or lactating) cow is ten dses.

CHAPTER 5: OUT OF AFRICA

1 Foley et al., 'Solutions for a Cultivated Planet', *Nature*; Zhang et al., 'Ecosystem Services and Dis-Services to Agriculture', *Ecological Economics*.
2 N. S. G. Williams et al. (eds), p. 3.
3 Weber and Horst.
4 Sauer, pp. 56, 61.
5 Plato, *Critias*, 111 a–d, W. R. M. Lamb (trans.), Laudator Temporis Acti website, http://laudatortemporisacti.blogspot.com.au; Montgomery.
6 A professional practitioner of ergonomics (the science of optimising equipment and systems for human use).
7 Voisin, *Grass Productivity*; Anderson, *Corn Laws*; Anderson, *Agriculture and Rural Affairs*.
8 'Brittleness scale' relates to the idea that regardless of total rainfall, all environments fall on a scale between brittle and non-brittle. A highly brittle environment – such as a tropical savannah with a short, distinct wet season and a long dry season – typically has 'unreliable precipitation, regardless of volume'; poor annual distribution of precipitation; high rates 'of chemical (oxidation) and physical (weathering) decay in old plant and animal material'; 'very slow successional development from bare and smooth soil surfaces'; and, 'with a lack of adequate physical disturbance for years, successional communities become simpler, less diversified, and less stable.' A highly non-brittle environment exhibits converse traits (Savory, *Holistic Resource Management*, pp. 509–10).
9 A. Savory, personal communication with author, 4 May 2014.
10 A. Savory, personal communication with author, 24 September 2016.
11 Acocks, Introduction.
12 Hoffman and Cowling, pp. 1–5.
13 A. Savory, personal communication with author, 24 September 2016.
14 Zietsman; A. Savory, personal communication with author, 24 September, 2016.
15 A. Savory, personal communication with author, 4 May, 2014.
16 Terry McCosker, personal communication with author, 29 January 2018; Hazel Parsons, personal communication with author, 30 January 2018.
17 Terry McCosker, personal communication with author, 30 October, 2018.

CHAPTER 6: MAKE MISTAKES BUT DON'T DO NOTHING

1 W. Berry, *The Unsettling of America*, p. 47.

2 J. Armstrong, personal communication with the author, 18 April 2017.

3 'Beetaloo, Mungabroom for sale', Beef Central, 29 August 2016, http: www.beef central.com/property/beetaloo-mungabroom-for-sale/.

4 M. N. G. Boyce, 'Nomination of Mr John Dunnicliff for Recognition under the Australian Honours System', p. 1.

5 Ibid.

6 Nossal, in Ibid., p. 23.

7 Cranston, M., 'Dunnicliff's Beetaloo Cattle Station Could Fetch $200m with Blundy Support', *Financial Review*, 29 August 2016, http://www.afr.com/real-estate/ dunnicliffs-beetaloo-cattle-station-could-fetch-200m-with-Blundy-support-20160826-gr280w.

8 Cawood, 'SLM's Holistic Hook Nabs Investors'; 'SLM Australia Livestock Fund'.

9 Stanley, C., 'SLM Partners seeks $100m for sustainable farming in Chile', *PFI Agri Investor*, 24 November 2015, http://slmpartners.com/slm-partners-seeks-100m-for -sustainable-farming-in-chile.html.

10 Ovis 21, http://en.ovis21.com.

11 Finlayson, pp. 7, 44.

12 J. L. Thompson, p. 21.

Regenerating the Water Cycle

CHAPTER 7: WATER, WATER EVERYWHERE

1 Leopold, 'The Land Ethic', *A Sand County Almanac*, pp. 201–226.

2 Margulis and Sagan, pp. 32, 34.

3 Davie.

4 A. J. Yeomans.

5 Mulligan and Hill, p. 195.

6 Spencer.

7 Mulligan and Hill, p. 193.

8 Bennett, p. 2; A. J. Yeomans.

9 Hill, p. 40.

10 A. J. Yeomans.

11 P. A. Yeomans, *The Challenge of Landscape*, p. 166.

12 Ibid., p. 25.

13 Ibid., p. 27.

14 C. S. Hicks, pp. 2–7.

15 Hill, in Mulligan and Hill, pp. 201–2.

16 Rowan Wright, personal communication with author, 25 January 2017.

17 Rix Wright notes, per Bronwyn Rix, 28 April 2016.

18 Carol and Mervyn Hardie, personal communication with author, February 2017.

19 'Subtle energies': a range of esoteric, hard to measure energies emanating from lunar and planetary influences, and also from within our own Earth. (See, for example, http://christinagrant.com/what-are-subtle-energies/).

20 Morris, p. 11.

CHAPTER 8: CALL OF THE REED WARBLER

1 Archie Roach, cited in N. Scolaro, pp. 77–9.
2 Cathcart.
3 The Landcare movement began in 1986 through a farmers' group in Victoria, Australia. It subsequently developed into a national movement with over 4000 community groups, and later internationally. The movement (part-funded by the federal government) initially comprised local groups attempting to address land degradation in local, catchment-specific sites. Landcare later broadened to include rural-farming, lifestyle and community-development issues.
4 Ratcliffe.
5 Andrews, p. 15.
6 Ibid., p. 46.
7 Ibid., p. 47.
8 Ibid., p. 38.
9 Ibid., pp. 63, 69.
10 Ibid., p. 175.
11 Ibid., p. 214.
12 Ibid., pp. 13, 74.
13 Limnology: the study of freshwater lakes and ponds.
14 J. Williams, *Australian Story*.
15 Charles Massy interview with Peter Marshall, 3 February, 2018.
16 Peter Marshall, personal communication with author, 7 February, 2018.
17 Peter Marshall, personal communication with author, 5 February, 2018.
18 Lawrence et al.; Edwards; M. E. White, *Earth Alive!*.
19 Peter Marshall, personal communication with author, 12 January, 2018.
20 Law et al.; Jones et al.
21 Kiviat.
22 Wikipedia, *Ciénegas*, accessed 7 February, 2018.

Regenerating the Soil-Mineral Cycle
CHAPTER 9: FROM STARDUST TO STARDUST

1 W. Berry, *The Unsettling of America*, p. 2.
2 Kellogg, p. 863.
3 Andrews, pp. 20, 24.
4 Lowenfels and Lewis, p. 31.
5 SupremeFulvic.com, 'Fulvic Acid'; W. R. Jackson, *Humic, Fulvic and Microbial Balance*; Christman and Gjessing.
6 'Glomalin': the sticky aggregation of soil biota excretions that stimulate healthy soil development and which helps recycle the scarce element of phosphorus in Australian soils.
7 Jehne lecture, ANU Sustainable Rural Systems, 14 July 2015.
8 Jehne lecture notes, 14 July 2015, p. 5.
9 W. Jehne, personal communication with author, 10 March 2017.

10 Jehne lecture notes, 14 July 2015, pp. 5, 6, 9.

11 Ibid., p. 9.

12 W. Jehne, personal communication with author, 9 March 2017.

13 Lowenfels and Lewis, p. 16.

14 The comments about bacteria and their role also apply to the recently discovered and related ancient order of Archea – which are bacteria lookalikes but have their own basic order on the tree of life.

15 M. Stapper, 'From Green Revolution to Agroecology', *Arena* 122: 33-36, 2013.

16 Earl and Jones, pp. 327–50.

17 C. E. Jones, 'Carbon and Catchments', Managing the Carbon Cycle, National Forum, p. 1.

18 Ibid., p. 1; and see C. E. Jones, 'Liquid Carbon Pathway', extract from *Australian Farm Journal*.

19 Interview with K. Nichols in Comis, 'Glomalin', *Australian Grain*, pp. 12–13; Cawood, 'Glomalin Key to Earth's Health', *The Land*, p. 26.

20 Interview with K. Nichols in Comis, 'Glomalin', *Australian Grain*, pp. 12–13.

21 Rua et al., 'Home-field advantage?', 122; Revillini et al., 'Ecosystems, evolution and plant feedbacks'.

CHAPTER 10: FARMING WITHOUT FARMING

1 Howard, *An Agricultural Testament*, p. 4.

2 Hyams; J. Isherwood, cited in United States Department of Agriculture, Natural Resources Conservation Service, 'Soil Quotations'.

3 J. R. Smith.

4 See, for example, Hillel, and Lowdermilk.

5 Faulkner, p. 22.

6 Gliessman, p. 3.

7 Bhanoo; Pascoe; Gerritsen.

8 Chivers.

9 Cawood, 'Pasture Crop Trials Fuelling Corn Progress', *The Land*, p. 28.

10 W. Berry, Foreword to W. Jackson, *New Roots for Agriculture*, p. ix.

11 W. Jackson, *New Roots for Agriculture*, p. 34; Gatschet, 'Natural Systems Agriculture at The Land Institute', The Land Institute; W. Jackson, 'Analogy: A Driver and Natural Systems Agriculture', The Land Institute.

12 W. Jackson, *Becoming Native to this Place*, p. 116; W. Jackson, 'The Ecosystem as a Conceptual Tool for Agriculture and Culture', The Land Institute.

13 Jackson in Barber, pp. 41–2.

14 Cluff.

15 'Root-pruning' means that whenever a crop or grass is grazed or cut, roughly the equivalent of roots under the ground to green foliage above dies off. Providing the cropping system is ecologically based, such root die-off puts soil organic matter, and thus carbon, into the soil and feeds the soil biota.

16 Wendon, 'A Report Documenting the Impact of Different Agricultural Management Actions', University of Sydney Soil Resource Management; Ampt and Doornbos,

'Communities in Landscapes Project', University of Sydney, Gulgong; C. E. Jones, 'Carbon that Counts', New England and North-West Landcare Adventure; Bruce et al., 'Pasture Cropping', Winona: Merino and Kelpie Studs.

17 D. Freudenberger, personal communication with author, 17 November 2016.

CHAPTER 11: DANCING UNDER THE MOON

1 Berman, p. 23.
2 Animism: 'A trope for beliefs that the natural world is inspirited' (i.e. inhabited by nature spirits: 'that a sacred reality exists' and is different from everyday profane realities) 'and is manifested at special times and places, usually through natural entities and place.' (B. Taylor, pp. xiii–xiv).
3 Goethe, 1791, in Wulf, p. 28.
4 Seamon and Zajonc (eds), pp. 4, 10.
5 Schilthuis, p. 41.
6 Ullrich, pp. 555–72.
7 R. Steiner, *Spiritual Foundations*.
8 Lovel, p. 5.
9 Pfeiffer, p. 7.
10 Podolinsky, p. 32.
11 Pfeiffer, p. 52.
12 Podolinsky.
13 R. Steiner, *Spiritual Foundations*, p. 46.
14 Ibid., p. 91.
15 See, for example: Endres; Larrson, 'Plant Pollination Synchronised with Full Moon', Stockholm University; Rydin and Bolinder, 'Moonlight Pollination in the Gymnosperm *Ephedra* (Gnetales)', *Biology Letters*; Beeson, 'The Moon and Plant Growth', *Nature*.
16 Pfeiffer, pp. 53–4.
17 See, for example, Kolisko.
18 See, for example: Wallace and Dowling, 'Carbon Farming and Nitrogen Fertilizer, Opportunity or Threat?', Grain Industry Association of Western Australia; Schwenke, 'Nitrogen Volatilisation', Grains Research and Development Corporation.
19 Pfeiffer, p. 49.
20 R. Steiner, *Spiritual Foundations*, pp. 11, 46.
21 D. Outhred, personal communication with author, 2 February, 2018.
22 IFOAM, 'Alex Podolinsky'; 'A Brief History of Bio-Dynamics', *Biodynamics Magazine*; Paull, 'The Soil Association and Australia'.
23 See, for example: Turinek et al., 'Biodynamic Agriculture Research Progress and Priorities', *Renewable Agriculture and Food Systems*; Lytton-Hitchins et al., 'The Soil Condition of Adjacent Bio-Dynamic and Conventionally Managed Dairy Pastures in Victoria, Australia', *Soil Use and Management*; Raupp, 'The Long-Term Trial in Darmstadt'; Fließbach et al., 'Soil Organic Matter and Biological Soil Quality Indicators After 21 Years of Organic and Conventional Farming', *Agriculture, Ecosystems and Environment*; Reganold, 'Soil Quality and Profitability of Biodynamic

Notes

and Conventional Farming', *American Journal of Alternative Agriculture*; Granstedt and Kjellenberg, 'Organic and Biodynamic Cultivation', 16th IFOAM World Congress.

Regenerating Dynamic Ecosystems

CHAPTER 12: KEEP A GREEN BOUGH IN YOUR HEART

1 'Mutualism' is where two organisms of different species exist in a relationship where each individual benefits from the activity of the other (e.g. ruminant sheep and cattle and their gut flora). 'Symbiosis' involves 'the living together and sometimes merging of different species of organisms'. Such a relationship can be mutualistic, parasitic and/or commensal (the latter being a relationship between organisms where one benefits from the other without affecting it).

2 Margulis and Sagan, pp. 15–16, 121.

3 Ibid., p. 18.

4 Ibid., pp. 32, 35.

5 Savory and Butterfield, p. 102.

6 Ibid., pp. 101–2.

7 Wulf, pp. 87–9.

8 Weisser et al., pp. 2, 4.

9 See, for example, van Moorsel, S et al. 2018. 'Community evolution increases plant productivity at low diversity.' In: *Ecology Letters* 21: 128-137.

10 T. Mitchell, vol. 2, p. 271.

11 Niel Black journal, cited in Kiddle, p. 49.

12 Psyllids, also known as 'plant lice', are tiny sap-sucking insects. They have very host-specific feeding preferences, especially for native plants.

13 A. Leopold, 'Engineering and Conservation', p. 254.

14 Weatherly, p. i.

CHAPTER 13: BLESSED ARE THE MEEK

1 Capra, p. 11.

2 Leach, p. 54.

3 Lindenmeyer; Commonwealth of Australia.

4 Oosthoek, pp. 69–80.

5 While elements of this story will be told briefly in Chapter 19, due to the now national recognition and development of agroforestry, I will not deal with agroforestry in this book, except to briefly sketch its context for this story.

6 Nuberg et al., p. 2.

7 Reid and Stephen, p. 6.

8 Jurskis and Turner; *Land & Water Australia*; John Taylor in Norton and Reid.

9 Jurskis and Turner, 'Eucalypt Dieback in Eastern Australia', *Australian Forestry*, pp. 87–98.

10 J. Taylor in *Land & Water Australia*, pp. 4–5; J. Taylor in Norton and Reid, p. 209; J. Taylor, interview with author, 2010.

11 *Land & Water Australia*, p. 5.

12 'Wool Industry Spreading Tree Message', *Landline*; *Land & Water Australia*.

13 *Australian Landcare Magazine*, p. 37.

14 Norton and Reid, p. 211.

15 *Dehesa*: A clever and sustainable use of polycultural agroforestry that occurs on the famous Iberian cultural farming oak forest landscapes. Described as complex 'agro-sylvo-pastoral systems', the *Dehesa* of southern and central Spain and the *Montado* of southern Portugal are both private and communal lands of closely managed, over-storey oak forests and other species (including beech and pine trees, plus savannah grasslands, and sometimes scrub with understorey). These cultural landscapes cover around 20,000 square kilometres of the Iberian Peninsula (two million hectares), some dating back 800 years to the Middle Ages. Under this mixed forest, sheep, cattle and pigs graze in what is a rich mixed grazing and cropping landscape. These landscapes provide a variety of foods such as meat from cattle, sheep and goats, grains (cereals), wild game, mushrooms and honey, and of course the famous *Jamón Ibérico*: widely regarded as the finest ham in the world and made from black *Ibérico* pigs who graze on the acorns. Other products are wool, various timbers, cork, charcoal, tannins and firewood. The *Dehesa/Montado* also provides a crucial habitat for both common and also rare wildlife, while hunting is another enterprise of these cultural landscapes. (Wikipedia: https://en.wikipedia.org/wiki /Dehesa, accessed 20 December 2015; Ecoagriculture Partners, Washington DC. EcoAgriculture Snapshots.org: 'The Dehesa and the Montado. Eco Agriculture land management systems in Spain and Portugal'; Joffre et al., 'The Dehesa System of Southern Spain and Portugal as a Natural Ecosystem Mimic', cited in *Agroforestry Systems*, pp. 57–79; Olea and San Miguel-Ayanz, 'The Spanish Dehesa', Paper for 21st General Meeting of the European Grassland Federation.)

CHAPTER 14: LISTEN TO THE LAND

1 J. Muir, *My First Summer in the Sierra*, p. 110.

2 Petty, de Koninck and Orlove, 'Cleaning, Protecting, or Abating?', *Journal of Ethnobiology*, pp. 140–62; Brown et al, 'Fire as an Engineering Tool of Early Modern Humans', *Science*, pp. 859–62; Choi, 'Humans Used Fire One Million Years Ago', Live Science; Leonard, 'Food for Thought', *Scientific American*; Lambert, 'The Way We Eat Now', *Harvard Magazine*.

3 Savory, *Holistic Resource Management*, pp. 125, 130.

4 Ibid., pp. 127–8.

5 Latz, *The Flaming Desert*; and see Savory, 'For the Love of Fire', unpublished paper.

6 Orians and Milewski.

7 Ibid., p. 395.

8 Notwithstanding the fact that E. M. Curr in 1883 had previously used this term (*Recollections of Squatting in Victoria*).

9 Adamson.

10 Bahn et al., 'Does Photosynthesis Affect Grassland Soil-Respired CO_2 and Its Carbon Isotope Composition on a Diurnal Timescale?', *New Phytologist*; Peter Marshall, personal communication with author, February 3, 2018.

11 Whalley, 'Grassland Regeneration and Reconstruction', *Ecological Management and Restoration*, pp. 3–4; Martin, 'The Role of Small Ground-Foraging Mammals in Topsoil Health and Biodiversity', *Ecological Management & Restoration*, pp. 114–19.

12 Johnson, *Australia's Mammal Extinctions*, pp. 75, 90; Miller et al., 'Ecosystem Collapse in Pleistocene Australia and a Human Role in Megafaunal Extinction', *Science*, pp. 287–90.

13 Flannery, *The Future Eaters*; Flood; Mulvaney and Kamminga.

14 Flannery, *Mammals of New Guinea*.

15 G. Retallack, personal communication with author, 9 January 2018; Retallack 2013.

16 Retallack, 2013, pp. 69–86.

17 Terry McCosker, personal communication with author, 29 January 2018.

18 Ibid.

19 Gill, 'Fire Pulses in the Heart of Australia', Report to Environment Australia, p. 211.

20 L. White, 'The Historical Roots of Our Ecologic Crisis', *Science*, p. 1203.

21 Weatherstone, 'Lyndfield Park', Greening Australia/Land & Water Australia, p. 4; J. Weatherstone, interview with author, 2010.

Regenerating the Landscape: Role of the Human–Social

CHAPTER 15: A MYSTERIOUS DIALOGUE

1 G. P. Marsh, p. 91.

2 For background to this issue, see, for example: Ratcliffe; Sheldrick; C. Muir.

3 Regarding 'dialogue', from the Greek, 'dia' means 'through' and not 'to' or 'at'. For physicist and philosopher David Bohm, true dialogue doesn't involve someone trying to win but is rather a 'common participation' where everybody wins (Bohm, p. 7). He sees dialogue as a stream of meaning, which comprises an empathic process involving real listening and, when coupled with humility, a capacity to think holistically and challenge one's own assumptions and world view: to be able to look in places previously ignored. Such dialogue becomes a caring relationship.

4 Elements of this story are sketched in my book *The Australian Merino: The Story of a Nation*, 2007.

5 See my story of these lessons in my book *Breaking the Sheep's Back*, 2011.

6 The concept of 'additive gene functions' comes from a mechanistic, reductionist view of how genetics works via the school of 'quantitative genetics'. This paradigm holds, for example, that animal production traits are composed of many scores of genes all contributing a small amount to the trait, and that simple measurements will identify the best animals. By contrast, non-additive gene effects involve more holistic, complex genetic actions and interactions, and incorporate elements such as epigenetics and major gene effects.

7 Joost and Negrini, 'Early Stirrings of Landscape Genomics', FAO International Symposium on Sustainable Improvement of Animal Production and Health, pp. 137–8; Schwartz et al., 'Landscape Genomics', in Cushman and Huettmann (eds), *Spatial Complexity, Informatics, and Wildlife Conservation*, pp. 165–74; Hodges,

'Foundations, Fallacies and Assumptions of Science for Livestock Development', FAO International Symposium on Sustainable Improvement of Animal Production and Health, pp. 135–6.

8 Shiva, *Making Peace with the Earth*, p. 12.
9 Chadwick, Vercoe, Williams and Revell, 'Feeding Pregnant Ewes a High-Salt Diet or Saltbush Suppresses their Offspring's Postnatal Rennin Activity', *Animal*.
10 Andrews, p. 24.
11 Hughes, personal communication with author, 14 March 2017.
12 Ibid., 17 March 2017.
13 Leopold, 'The Land-Health Concept and Conservation', in Callicott and Freyfogle (eds), p. 22; Callicott and Freyfogle (eds), p. 219.
14 D. and E. Lloyd, 'Grain Growing', unpublished paper.
15 Lloyds, interview with author, 2010; 'The Foodies Chat with Jeremy of Eden Valley Flour', We Love Perth.
16 Carson, *Silent Spring*, p. 2.

CHAPTER 16: DESIGN WITH NATURE

1 Brown and Harris, p. 70.
2 Orr, *The Nature of Design*, p. 7.
3 F. Steiner, 'Healing the Earth', *Philosophy & Geography*.
4 McHarg, *Design with Nature*, pp. 2–5.
5 Ibid., p. 5.
6 McHarg, 'Man, Planetary Disease', Address to North American Wildlife & Natural Resources conference, pp. 634–40.
7 *The New York Times*, 'Obituary to Ian McHarg', 12 March 2001.
8 L. Mumford, in McHarg, *Design with Nature*, p. viii.
9 T. Haikai; T. Hakai, personal communication with author, 4 April 2018.
10 P. A. Yeomans, *The City Forest*, p. 55.
11 Ibid., p. 109.
12 Ibid., pp. 109–10.
13 Ibid., pp. 94–5.
14 Ibid., p. 93.
15 Wootton, interview with author, 2010.
16 Campbell, Jack and Mathews, 'Report to Ian Potter Foundation on Potter Farmland Plan Project: 1984–1988', January 1989, unpublished.
17 Mulligan and Hill, p. 204.
18 Dawborn and Smith (eds), p. 21.
19 Mulcaster, 'Making a Mess of His Life', *The Age*.
20 Ibid.
21 Dawborn and Smith (eds), p. 22.
22 Mulligan and Hill, p. 203.
23 Dawborn and Smith (eds), p. 23.
24 Mulligan and Hill, pp. 203–4.
25 D. Holmgren, personal communication with author, 20 September 2016.

Notes

26 Grayson, 'A Short and Incomplete History of Permaculture', Pacific Edge.

27 H. T. and E. C. Odum.

28 Holmgren, *Essence of Permaculture*.

29 Holmgren, personal communication with author, May 2015.

30 Holmgren, cited in Capra and Luisi, p. 449.

31 Ibid., p. 442.

32 Holmgren, *Permaculture: Pathways and Principles*; see also Smith et al., 'Implications of the Synergies Between Systems Theory and Permaculture for Learning About and Acting Towards Sustainability', 2007 ANZSEE Conference.

CHAPTER 17: AGRI-CULTURE - SOURCE OF A HEALTHY CULTURE, SOCIETY AND MOTHER EARTH

1 T. Berry, *The Great Work*, p. iii.

2 'Prospect Pastoral Co.', Soils for Life Case Study, http://soilsforlife.org.au/cs-prospect-pastoral-company; interview with D. Haggerty, 2010, and with I. and D. Haggerty, January 2016.

3 Barber, p. 377.

Part III: Transforming Ourselves – Transforming Earth

CHAPTER 18: THE BIG PICTURE - CO-CREATING WITH LANDSCAPES

1 T. Berry, *The Great Work*, p. 26.

2 Kloof: a deep ravine or cleft (Afrikaans).

3 Humboldt, cited in Wulf, p. 7.

4 Darwin, *Origin of Species*, cited in J. J. Clarke, p. 27.

5 Smuts, pp. 317, 319.

6 Popper, *The Self and its Brain*, cited in J. J. Clarke.

7 J. J. Clarke, p. 13.

8 Capra and Luisi, p. 66.

9 M. Mitchell, pp. 13, 137.

10 L. Thomas, Foreword, Margulis and Sagan, p. 9.

11 Capra and Luisi, p. 305.

12 Whitehead, *The Concept of Nature*, p. 178.

13 Goodwin, pp. 168–9.

14 Lovelock.

15 Capra and Luisi, p. 161.

16 Ibid., p. 309.

17 Gunderson and Holling (eds); Walker and Salt.

18 Rosen, p. 94.

19 Provenza et al., 'Complex Creative Systems', *Rangelands*, p. 8.

20 Stebbins, 'Co-Evolution of Grasses and Herbivores', *Annals of the Missouri Botanical Garden*, pp. 75–86.

21 Provenza, Meuret and Gregorini, 'Our Landscapes, Our Livestock, Ourselves', *Appetite*, p. 502.

22 Provenza et al., 'The Wisdom Body: Nutrition, Health, and Nature's Pharmacopeia', paper for invited synthesis for Evolutionary Biology, undated personal communication.

23 Meuret and Provenza (eds), *The Art and Science of Shepherding*, pp. 174–7.

24 Meuret and Provenza, 'When Art and Science Meet', *Rangeland Ecology & Management*, pp. 1–17.

25 Barber, p. 32.

26 P. Bateson, in Capra and Luisi, p. 200; for background, see also West-Eberhard.

27 Steinbeck, p. 54.

28 Provenza et al., 'Complex Creative Systems', *Rangelands*, p. 10.

29 Maynard and Revell, Self Herding and Self Shepherding (website), http://www.selfherding.org/rangelands-self-herding.html; Stress Free Stockmanship (website), http://www.stressfreestockmanship.com.au; EMU Project (website), http://www.emuproject.org.au.

30 Provenza et al., 'Complex Creative Systems', *Rangelands*, p. 12.

CHAPTER 19: TRANSFORMING OURSELVES

1 Speth, p. 65.

2 Jackson, *Becoming Native to This Place*, p. 26.

3 McKernan, pp. 2–4.

4 Shiva, *Monocultures of the Mind*, p. 7.

5 Mezirow, 'Transformation Theory of Adult Learning', in Welton (ed.), *In Defense of the Lifeworld*, pp. 39–70.

6 Russell and Ison, 'The Research-Development Relationship in Rural Communities', in *Agricultural Extension and Rural Development*, p. 21; Watson-Gegeo, 'Mind, Language, and Epistemology', *The Modern Language Journal*, p. 338.

7 Haken, 'Synergistics', in Jantsch; Gersick, 'Revolutionary Change Theories', *The Academy of Management Review*, p. 30.

8 Boyd; Wilber.

9 Lave and Wenger; Wenger; Eshuis and Stuiver, 'Learning in Context through Conflict and Alignment', *Agriculture and Human Values*, p. 143; Russell and Ison.

10 Bandura; Pahl-Wostl and Hare, 'Processes of Social Learning in Integrated Resources Management', *Community and Applied Social Psychology*, pp. 193–206.

11 Wenger, pp. 13, 15; Nelson; Meghani, paper, p. 29.

12 Pretty, 'Agricultural Sustainability', *Philosophical Transactions of the Royal Society B*, p. 451.

13 Lincoln, 'Address to the Wisconsin State Agricultural Society', in Delbanco (ed.), p. 192.

14 Hawken, *The Ecology of Commerce*; Marsh, Rural Conservation Feature Article, *National Parks Association of NSW Journal*.

15 Marsh, cited in Cawood, 'Simplifying Towards Complexity', The Good Land Project; Marsh, 'Farming from First Principles', in Lefroy et al., pp. 55–62.

16 R. W. Emerson, cited in J. Porte, p. 77.

17 Speth, p. 204.

18 D. Pratt, cited in J. Lindsay, Testimonial to McCosker, RCS 25th Anniversary Conference, July 2010.

CHAPTER 20: HEALING EARTH

1 Hawken, *Blessed Unrest*, pp. 4–5.
2 MacGregor et al., 'Natural and Post-European Settlement Variability in Water Quality of the Lower Snowy River Flood Plain, Eastern Victoria, Australia', *River Research & Applications*, pp. 201–13.
3 W. Berry, 'Standing by Words', *The Trumpeter*, pp. 11–16.
4 Jackson, *Becoming Native to This Place*, p. 15.
5 Butler, 'Globalization, Population, Ecology and Conflict', *Health Promotion Journal of Australia*, pp. 87–91.
6 Saul, *Voltaire's Bastards*, p. 582.
7 Steffen et al., 'The Anthropocene', *AMBIO*, pp. 739, 761.
8 Martin Luther King, 4 April 1967.
9 Cawood, 'Rain Falls, Mulga Rises, Carbon Sinks', *The Land*, p. 14.
10 Interview by the author with microbiologist/soil scientist W. Jehne, February 2015.
11 Wentworth Group of Concerned Scientists, 'Optimising Carbon in the Australian Landscape'.
12 Retallack 2013, p. 69.
13 Ibid., pp. 79-80.
14 Ibid., p. 69.
15 Ibid., p. 70.
16 Ibid., p. 70.
17 G. J. Retallack, 'Cenozoic Expansion of Grasslands and Global Cooling', *The Journal of Geology* 190, 2001: pp. 407–426.
18 Jehne, 'Restoring Water Cycles to Naturally Cool Climate and Reverse Global Warming', paper for *Soils for Life*, p. 7.
19 Jehne, 'Regenerating the Earth's Soil Carbon Sponge', draft science paper for *Healthy Soils Australia* proposal, the Virgin Earth Challenge.
20 Williams, Walker and Tane, 'Drier Landscapes and Rising Watertables: An Eco Hydrological Paradox', *Natural Resource Management*, pp. 10–18.
21 Ripl, 'Water: The Bloodstream of the Biosphere', *Philosophical Transactions of the Royal Society B*, p. 1922.
22 Ibid., p. 1928.
23 Ibid., p. 1927.
24 Ibid., pp. 1932–3, 1929.
25 Philpott, 'New Research: Synthetic Nitrogen Destroys Soil Carbon, Undermines Soil Health', Grist.
26 Cawood, 'More Carbon Taken in by Wool Operation Than Given Out', Farm Weekly; Doran-Browne et al., 'Carbon-Neutral Farming in South-Eastern Australia', *Animal Production Science*, pp. 417–22; J. and R. Ive, 'Achieving Production and Environmental Benefits in a Challenging Landscape', Proceedings of the 22nd Annual Conference of the Grassland Society of NSW.

CHAPTER 21: HEALING OURSELVES

1 Lieberman, p. 366.

2 Ibid., pp. xi–xii

3 Ibid., p. 176.

4 Ibid., p. 168.

5 Ibid., p. 252.

6 Robinson, p. 7.

7 See for example: Davis, 'Declining Fruit and Vegetable Nutrient Composition';
   Thomas, 'The mineral depletion of foods available to us as a nation (1940–2002)';
   Davis, 'Trade-offs in agriculture and nutrition'; Davis et al., 'Changes in USDA food
   composition data for 43 garden crops, 1950 to 1999'.

8 See, for example: Thompson and Amoroso, 'Combating Micronutrient Deficiencies',
   Food and Agriculture Organization and CAB International; World Health Organization, 'Diet, Nutrition and the Prevention of Chronic Diseases', p. 12.

9 McDonald, 'Relationship of Nutrition to Prevention of Diseases', Northwestern
   Medicine.

10 Holmes, in G. Dodd, p. 6.

11 W. Berry, 'It All Turns on Affection', Jefferson Lecture.

12 Laibow, 'Nutrition and Health', paper for Natural Solutions Foundation, pp. 7–8.

13 Ibid.; Compassion in World Farming, 'Nutritional Benefits of Higher Welfare Animal
   Products'.

14 Pollan, p. 84.

15 Hawken, *Blessed Unrest*, p. 104.

16 T. Berry, *The Great Work*, p. 67.

17 Kummer, 'The Nestlé Health Offensive', *MIT Technology Review Business Report*, pp. 5–6.

18 Shiva, *Making Peace with the Earth*, pp. 26, 130.

19 I. Zilber-Rosenberg and E. Rosenberg, 'Role of microorganisms in the evolution of
   animals and plants: the hologenome theory of evolution', *FEMS Microbiol Rev.*, August
   2008, 32 (5), pp. 723–5; see also Speybroeck, pp. 61–81; Margulis, *Symbiosis in Cell
   Evolution*; Rosenberg E. et al., pp. 355–362; and Bordenstein and Theis.

20 Jefferson.

21 Escher, 'Epigenetics and the Multigenerational Effects of Nutrition, Chemicals, and
   Drugs', talk at Ancestral Health Symposium, YouTube.

22 Carey, 'Beyond DNA', *Natural History*.

23 K. M. Murphy et al.; see also Welch, pp. 381–90.

24 Nixon, 'Slow Violence and the Environmentalism of the Poor', *Journal of Commonwealth and Postcolonial Studies*, pp. 14–37.

25 Carey, 'Beyond DNA', *Natural History*.

26 Interview with the author, August 2010.

27 Jackson, *Consulting the Genius of the Place*, p. 129.

28 C. M. Benbrook, 'Trends in Glyphosate Herbicide Use in the United States and
   Globally', *Environmental Sciences Europe*, 28, no. 1: 3, 2016.

29 Sustainable Pulse, 'German Beer Industry in Shock over Glyphosate Contamination'.

30 Kremer, 'New Peer Review Study Finds Commonly Used Pesticides (Glyphosate)
   Induce Disease', Real Food Forager.

31 Samsel and Seneff, 'Glyphosate's Suppression of Cytochrome Enzymes and Amino Acid Biosynthesis by the Gut Microbiome', *Entropy*, p. 1417.

32 Ibid., p. 1445.

33 Balfour, *The Living Soil*, p. 174.

34 Fraser et al., 'Farming and Mental Health Problems and Mental Illness', *International Journal of Social Psychiatry*, pp. 340–9; Schirmer et al., 'Healthier Land, Healthier Farmers', *Health and Place*, p. 97.

35 Wilson.

36 Planet Ark, 'Climbing Trees'.

37 Planet Ark, 'Missing Trees'; and see also Planet Ark, 'Planting Trees'.

38 Louv, *Last Child in the Woods*, pp. 26, 33, 36.

39 Ibid., p. 5.

40 Ibid., pp. 46–7, 103.

41 Lieberman, pp. 294, 316.

42 Ibid., pp. 295, 297, 309.

43 Carson, 'Help Your Child to Wonder', *Woman's Home Companion*, p. 46.

44 Louv, *Last Child in the Woods*, pp. 224, 226.

45 George, 'How to Win the War of Ideas', *Dissent*.

46 Allen and Sachs, 'The Poverty of Sustainability', *Agriculture & Human Values*, p. 31.

47 J. Muir, *Travels in Alaska*, p. 5.

48 A. J. Heschel, quoted in Louv, *Last Child in the Woods*, pp. 291–2.

49 W. Berry, *The Unsettling of America*, pp. 104–5.

CHAPTER 22: TOWARDS AN EMERGENT FUTURE

1 Arabena, *Becoming Indigenous to the Universe*, pp. xiv, 53–4.

2 Arabena, 'Regeneration', unpublished paper, p. 5.

3 Maturana and Varela.

4 Margulis and Sagan, p. 29.

5 McHarg, *Design with Nature*, p. 196.

6 T. Berry, *The Great Work*, p. 60.

7 D. Marsh, personal communication with author, 27 November 2016.

8 Orr, 'Four Challenges of Sustainability', University of Vermont.

9 Kirschenmann, 'Spirituality in Agriculture', paper prepared for the Concord School of Philosophy, p. 7.

10 See Birch, '"We Are Still Here"', The Wheeler Centre.

11 Korten, 'The Great Turning', *Yes! A Journal of Positive Futures*, p. 6.

12 Friedman, Introduction.

13 P. Kingsnorth. The Dark Mountain Project is a network of writers, artists and thinkers rebelling against the contemporary ecological, economic and social crises. Having stopped believing the stories modern civilisation tells itself, they are creating new stories and art so that our cultural responses can reflect this multi-crisis reality of the Mechanical mind.

14 Meadows, 'Leverage Points', *Whole Earth*.

15 Kingsnorth; 'echo-location' from mythologist Martin Shaw, http://drmartinshaw.com/essays.

16 T. Berry, *The Dream of the Earth*, p. xi.

17 T. Berry, *The Great Work*, p. 59.

18 Murray, 'Rainwater Tank'.

INDEX

Note: Page numbers preceded by *ci* refer to images in the color insert.

Index

Index

Index

Index

Index

Index

Index

ABOUT THE AUTHOR

Theo Schoo

Charles Massy gained a Bachelor of Science at the Australian National University (ANU) in 1976 before going farming for thirty-five years and developing the prominent Merino sheep stud Severn Park. While continuing to farm sheep and cattle, concern at ongoing land degradation and humanity's sustainability challenge led him to return to ANU in 2009 to undertake a PhD in human ecology. Charles was awarded an Order of Australia Medal for his service as chair and director of a number of research organizations and statutory wool boards. He has also served on national and international review panels in sheep and wool research and development and genomics. Charles has authored several books on the Australian sheep industry, the most recent being the widely acclaimed *Breaking the Sheep's Back*, which was short-listed for the Prime Minister's Australian Literary Awards in Australian History in 2012.

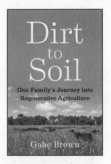

the politics and practice of sustainable living

CHELSEA GREEN PUBLISHING

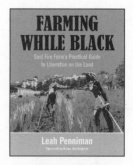

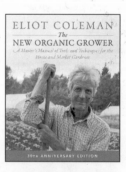